Scientific Basis for Nuclear Waste Management XXXVI

MATERIALS RESEARCH SOCIETY
SYMPOSIUM PROCEEDINGS VOLUME 1518

Scientific Basis for Nuclear Waste Management XXXVI

Symposium held November 25–30, 2012, Boston, Massachusetts, U.S.A.

EDITORS

Neil Hyatt
The University of Sheffield
Sheffield, U.K.

Kevin M. Fox
Savannah River National Laboratory
Aiken, South Carolina, U.S.A.

Kazuya Idemitsu
Kyushu University
Fukuoka, Japan

Christophe Poinssot
French Nuclear and Alternative Energies Commission
Bagnols sur Ceze, France

Karl R. Whittle
The University of Sheffield
Sheffield, U.K.

Materials Research Society
Warrendale, Pennsylvania

CAMBRIDGE
UNIVERSITY PRESS

Shaftesbury Road, Cambridge CB2 8EA, United Kingdom

One Liberty Plaza, 20th Floor, New York, NY 10006, USA

477 Williamstown Road, Port Melbourne, VIC 3207, Australia

314–321, 3rd Floor, Plot 3, Splendor Forum, Jasola District Centre, New Delhi – 110025, India

103 Penang Road, #05–06/07, Visioncrest Commercial, Singapore 238467

Cambridge University Press is part of Cambridge University Press & Assessment, a department of the University of Cambridge.

We share the University's mission to contribute to society through the pursuit of education, learning and research at the highest international levels of excellence.

www.cambridge.org
Information on this title: www.cambridge.org/9781605114958

Materials Research Society
506 Keystone Drive, Warrendale, PA 15086
http://www.mrs.org

First published 2013

Sponsor: AFOSR/AOARD (Air Force Office of Scientific Research/Asian Office of Aerospace Research and Development) "AFOSR/AOARD support is not intended to express or imply endorsement by the U.S. Federal Government"

CODEN: MRSPDH

A catalogue record for this publication is available from the British Library

ISBN   978-1-605-11495-8   Hardback

CONTENTS

CERAMIC WASTEFORMS - BETA DECAY

TECHNETIUM SOLUTIONS

SPENT NUCLEAR FUEL

WASTE REPOSITORIES

*Invited Paper

PREFACE

Symposium LL, Scientific Basis for Nuclear Waste Management XXXVI, was held November 25–30 at the 2012 MRS Fall Meeting in Boston, Massachusetts.

This Symposium continues to set the research agenda in the field of radioactive waste management, charting the development of waste processing, conditioning, packaging and disposal. Symposium XXXVI featured 77 presentations, delivered over four days during the 2012 MRS Fall Meeting, from participants in Australia, Austria, Finland, France, Japan, Russia, Spain, Sweden, Switzerland, the United Kingdom, and United States of America.

A highlight of Symposium XXXVI was the excellent tutorial on "Analysis of Radioactive Nuclear Materials", led by Claude Degueldre of the Paul Scherrer Institute and University of Geneva, Switzerland. Symposium XXXVI featured a special session devoted to Fukushima Dai-Ici, following the Tohoku Earthquake on March 11, 2011, reporting on efforts to stabilize the facility and decontaminate affected surroundings, drawing over a hundred attendees. The Symposium also included sessions reporting on advances in glass and ceramic wasteforms, conditioning of technetium, management of spent nuclear fuel, and geological disposal, plus a special joint session with Symposium HH, on radiation effects in nuclear materials. Each paper in this volume provides a snapshot of the exciting recent developments in each of these areas and the international progress toward achieving the safe, timely and cost-effective management and disposal of radioactive wastes.

Symposium XXXVII will be held in Barcelona in September 2013, chaired by Prof. Joan de Pablo, and we look forward to meeting participants at this event and future symposia.

Neil Hyatt
Kevin Fox
Kazuya Idemitsu
Christophe Poinssot
Karl Whittle

August 2013

ACKNOWLEDGMENTS

We are grateful to all the presenters and participants of the Symposium XXXVI for their contribution to the continuing success of this series of meetings. In particular, we are grateful to the invited speakers at the Symposium, for their high quality and engaging presentations: John Vienna, Bernard Boullis, Stephan Bourg, Akira Tokuhiro, Kaname Miyahara, Toshiki Sasaki, Chris Stanek, Rod Ewing, Lou Vance, Patrick Landais, and Joan de Pablo. We also extend our thanks to the Symposium assistant, Paul Heath, for supporting the smooth delivery of the meeting, and the Session Chairs for their skillful management of discussion: Kevin Fox, Kazuya Idemitsu, John Vienna, Rod Ewing, Neil Hyatt, Christophe Poinssot, Karl Whittle, Lou Vance, Claire Corkhill, Daniel Gregg, Michael Ojovan, Claude Degueldre, Marc Robinson, Martin Stennett, and Joan de Pablo. We acknowledge the hard work and support of the numerous reviewers of papers published in this Proceedings volume, who provided timely and careful reviews of the draft manuscripts. Finally, we thank the staff of the Materials Research Society for their dedication and assistance in organizing the Symposium and preparation of this volume.

ACKNOWLEDGMENTS

MATERIALS RESEARCH SOCIETY SYMPOSIUM PROCEEDINGS

Volume 1477 — Low-Dimensional Bismuth-based Materials, 2012, S. Muhl, R. Serna, A. Zeinert, S. Hirsekor, ISBN 978-1-60511-454-5

Volume 1478 — Nanostructured Carbon Materials for MEMS/NEMS and Nanoelectronics, 2012, A.V. Sumant, A.A. Balandin, S.A. Getty, F. Piazza, ISBN 978-1-60511-455-2

Volume 1479 — Nanostructured Materials and Nanotechnology—2012, 2012, C. Gutiérrez-Wing, J.L. Rodríguez-López, O. Graeve, M. Munoz-Navia, ISBN 978-1-60511-456-9

Volume 1480 — Novel Characterization Methods for Biological Systems, 2012, P.S. Bermudez, J. Majewski, N. Alcantar, A.J. Hurd, ISBN 978-1-60511-457-6

Volume 1481 — Structural and Chemical Characterization of Metals, Alloys and Compounds—2012, 2012, A. Contreras Cuevas, R. Pérez Campos, R. Esparza Muñoz, ISBN 978-1-60511-458-3

Volume 1482 — Photocatalytic and Photoelectrochemical Nanomaterials for Sustainable Energy, 2012, L. Guo, S.S. Mao, G. Lu, ISBN 978-1-60511-459-0

Volume 1483 — New Trends in Polymer Chemistry and Characterization, 2012, L. Fomina, M.P. Carreón Castro, G. Cedillo Valverde, J. Godínez Sánchez, ISBN 978-1-60511-460-6

Volume 1484 — Advances in Computational Materials Science, 2012, E. Martínez Guerra, J.U. Reveles, A. Aguayo González, ISBN 978-1-60511-461-3

Volume 1485 — Advanced Structural Materials—2012, 2012, H. Calderon, H.A. Balmori, A. Salinas, ISBN 978-1-60511-462-0

Volume 1486E — Nanotechnology-enhanced Biomaterials and Biomedical Devices, 2012, L. Yang, M. Su, D. Cortes, Y. Li, ISBN 978-1-60511-463-7

Volume 1487E — Biomaterials for Medical Applications—2012, 2012, S. Rodil, A. Almaguer, K. Anselme, J. Castro, ISBN 978-1-60511-464-4

Volume 1488E — Concrete with Smart Additives and Supplementary Cementitious Materials, 2012, L.E. Rendon Diaz Miron, B. Martinez Sanchez, K. Kovler, N. De Belie, ISBN 978-1-60511-465-1

Volume 1489E — Compliant Energy Sources, 2013, D. Mitlin, ISBN 978-1-60511-466-8

Volume 1490 — Thermoelectric Materials Research and Device Development for Power Conversion and Refrigeration, 2013, G.S. Nolas, Y. Grin, A. Thompson, D. Johnson, ISBN 978-1-60511-467-5

Volume 1491E — Electrocatalysis and Interfacial Electrochemistry for Energy Conversion and Storage, 2013, T.J. Schmidt, V. Stamenkovic, M. Arenz, S. Mitsushima, ISBN 978-1-60511-468-2

Volume 1492 — Materials for Sustainable Development—Challenges and Opportunities, 2013, M-I. Baraton, S. Duclos, L. Espinal, A. King, S.S. Mao, J. Poate, M.M. Poulton, E. Traversa, ISBN 978-1-60511-469-9

Volume 1493 — Photovoltaic Technologies, Devices and Systems Based on Inorganic Materials, Small Organic Molecules and Hybrids, 2013, K.A. Sablon, J. Heier, S.R. Tatavarti, L. Fu, F.A. Nüesch, C.J. Brabec, B. Kippelen, Z. Wang, D.C. Olson, ISBN 978-1-60511-470-5

Volume 1494 — Oxide Semiconductors and Thin Films, 2013, A. Schleife, M. Allen, S.M. Durbin, T. Veal, C.W. Schneider, C.B. Arnold, N. Pryds, ISBN 978-1-60511-471-2

Volume 1495E — Functional Materials for Solid Oxide Fuel Cells, 2013, J.A. Kilner, J. Janek, B. Yildiz, T. Ishihara, ISBN 978-1-60511-472-9

Volume 1496E — Materials Aspects of Advanced Lithium Batteries, 2013, V. Thangadurai, ISBN 978-1-60511-473-6

Volume 1497E — Hierarchically Structured Materials for Energy Conversion and Storage, 2013, P.V. Braun, ISBN 978-1-60511-474-3

Volume 1498 — Biomimetic, Bio-inspired and Self-Assembled Materials for Engineered Surfaces and Applications, 2013, M.L. Oyen, S.R. Peyton, G.E. Stein, ISBN 978-1-60511-475-0

Volume 1499E — Precision Polymer Materials—Fabricating Functional Assemblies, Surfaces, Interfaces and Devices, 2013, C. Hire, ISBN 978-1-60511-476-7

Volume 1500E — Next-Generation Polymer-Based Organic Photovoltaics, 2013, M.D. Barnes, ISBN 978-1-60511-477-4

Volume 1501E — Single-Crystalline Organic and Polymer Semiconductors—Fundamentals and Devices, 2013, S.R. Parkin, ISBN 978-1-60511-478-1

Volume 1502E — Membrane Material Platforms and Concepts for Energy, Environment and Medical Applications, 2013, B. Hinds, F. Fornasiero, P. Miele, M. Kozlov, ISBN 978-1-60511-479-8

Volume 1503E — Colloidal Crystals, Quasicrystals, Assemblies, Jammings and Packings, 2013, S.C. Glotzer, F. Stellacci, A. Tkachenko, ISBN 978-1-60511-480-4

Volume 1504E — Geometry and Topology of Biomolecular and Functional Nanomaterials, 2013, A. Saxena, S. Gupta, R. Lipowsky, S.T. Hyde, ISBN 978-1-60511-481-1

MATERIALS RESEARCH SOCIETY SYMPOSIUM PROCEEDINGS

MATERIALS RESEARCH SOCIETY SYMPOSIUM PROCEEDINGS

Volume 1534E — Low-Dimensional Semiconductor Structures, 2012, T. Torchyn, Y. Vorobie, Z. Horvath, ISBN 978-1-60511-511-5

Prior Materials Research Society Symposium Proceedings available by contacting Materials Research Society

Glass Wasteforms I

Mater. Res. Soc. Symp. Proc. Vol. 1518 © 2013 Materials Research Society
DOI: 10.1557/opl.2013.129

The Use of High Durability Alumino-Borosilicate Glass for the Encapsulation of High Temperature Reactor (HTR) Fuel

Paul G. Heath, Martin C. Stennett, Owen J. McGann, Russell J. Hand and Neil C. Hyatt
Department of Materials Science and Engineering, University of Sheffield, Sheffield, South Yorkshire, S1 3JD, United Kingdom.

ABSTRACT

The development of suitable waste forms for waste produced by generation IV reactors is of critical concern for future operations. To date no accepted disposal route for Tri-Structural Isotropic (TRISO) High Temperature Reactor (HTR) fuel exists. Alumino-silicate glass has been studied for its ability to encapsulate TRISO particle fuels. This glass was selected for its high aqueous durability. Encapsulation was achieved by cold pressing and sintering of glass powders mixed with HTR fuel. Sintering profiles capable of eliminating interconnected porosity in the composites were developed. The chemical compatibility and wetting of the glass matrix with the fuel were analysed along with the aqueous durability of the sintered glass matrix. Composites sintered under a controlled atmosphere produced unfractured monoliths with minimal chemical interaction between the glass and the TRISO particles. The Product Consistency Test (PCT) durability assessment indicated the sintered alumino-borosilicate glass was approximately an order of magnitude more durable than an equivalent R7T7 borosilicate glass. These results suggest sintered alumino-borosilicate glass-TRISO particle composites may provide a potential disposal route for spent TRISO particle fuel.

INTRODUCTION

Generation IV nuclear HTRs, such as the 'pebble bed' reactor and the U.S. next generation nuclear plant, plan to utilize Tri-Structural Isotropic (TRISO) fuel particles [1]. TRISO particles are multi-layered sub mm spheres consisting of a uranium dioxide (UO_2) core which is coated in porous graphite (BC), pyrolytic carbon (IPyC), silicon carbide (SiC) and a pyrolytic carbon (OPyC) as illustrated in figure 1. No reprocessing route or disposal pathway has been accepted for this waste stream. It has been suggested that direct disposal of TRISO fuel compacts may be possible [2], however due to the additional volume provided by the graphite compact this may be unfeasible for commercial operations [3]. Previous work has studied the encapsulation of TRISO particles in sintered borosilicate and Soda-Lime Silicate (SLS) glass matrices after separation from their fuel arrays [4, 5]. For the encapsulation of radioactive wastes the durability of the matrix should be considered a priority. This study has qualified the potential for higher durability alumino-borosilicate glasses to be used as encapsulants for TRISO fuel particles. The ABS-1 glass was chosen as the matrix material due to its low Coefficient of Thermal Expansion (CTE) and high aqueous durability [6]. Density measurements were used to optimise sintering parameters. Scanning Electron Microscopy (SEM) was employed to study ABS-1–TRISO particle interactions. The Product Consistency Test (PCT) was used to access aqueous durability of the sintered ABS-1 matrix [7].

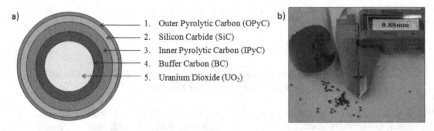

a)
1. Outer Pyrolytic Carbon (OPyC)
2. Silicon Carbide (SiC)
3. Inner Pyrolytic Carbon (IPyC)
4. Buffer Carbon (BC)
5. Uranium Dioxide (UO₂)

b)
0.88mm

Figure 1. a) Schematic illustration of the isostructural layers in a TRISO particle and b) image of simulant particles used in this study.

EXPERIMENT

Non-active simulant TRISO particles, fabricated with a stabilised zirconium dioxide core in place of uranium dioxide, were provided by the Pebble Bed Modular Reactor Company (South Africa). ABS-1 was produced using chemical grade reagents in an alumina crucible by heating to 1650 ° C with a one hour dwell. The glass was annealed at 570 °C for one hour and cooled at 1 °C min^{-1}. Glass transition temperature was measured using Differential Thermal Analysis (DTA). The CTE was measured using dilatometry. The glass was powdered to a sub 75 μm size fraction (40 ± 20 μm measured using laser particle size analysis) for green body formation. The composition was measured using X-ray Fluorescence (XRF) analysis. Theoretical density was calculated using helium pyconometry. Characteristics are displayed in Table 1.

Green bodies were formed by uniaxially pressing glass powder in a 10 mm die. Glass pellets were sintered at 700 °C, 750 °C and 800 °C for 15 minutes, 30 minutes and 60 minutes under a 5% H₂/N₂ atmosphere. Densification on sintering was measured using the Archimedes principle. Composite green bodies were formed by mixing simulant TRISO particles (10 wt %) with powdered ABS-1 and pressing as above. Composites were sintered at 750 °C for 30 minutes under a 5% H₂/N₂ atmosphere. Composites were sectioned, then ground and polished to a 1 μm finish, for SEM analysis using diamond abrasives.

A 14 day PCT test was conducted using sintered glass powders with a surface area to volume ratio of 2270 m^{-1}. Sampling was performed in triplicate at 3, 7 and 14 days. Leachate samples were analysed using Inductively Coupled Plasma-Atomic Emission Spectroscopy and normalised as per the PCT ASTM standard [7]. PhreeqC geochemical modeling was performed to evaluate potential saturation effects. The results were compared with equivalent data for a soda-lime silicate (SLS) and a boro-silicate glass (R7T7) from an earlier study of sintered SLS and borosilicate glass compositions [5].

Table 1. Composition and physical properties of ABS-1 glass.

Glass Composition (wt % ± 0.05%)	SiO₂	Al₂O₃	B₂O₃	Na₂O	CaO	MgO
	72.20	8.60	8.20	7.10	2.60	0.10
Glass transition temperature (° C)	610 ± 5					
CTE (x 10^{-6} K^{-1})	7.2 ± 0.1					
Density (g cm^{-3})	2.360 ± 0.001					

DISCUSSION

The retention of the TRISO OPyC layer is desirable during high temperature encapsulation. In the event of water ingress, the OPyC acts as a barrier to fission product release from the SiC layer. It also provides a barrier to SiC oxidation during processing as such reducing the potential for release of radioactive volatiles. To minimise the risk of OPyC oxidation during processing, the minimum time and temperature required to achieve a density in excess of 95% of the theoretical density was determined. As seen in Table 2 optimal conditions were obtained by sintering at 750 °C for 30 minutes.

Table 2. Percentage of theoretical density attained sintering ABS-1 green bodies.

Sintering Regime	700 °C	750 °C	800 °C
15 minutes	81 ± 1 %	85 ± 1 %	90 ± 1 %
30 minutes	95 ± 1 %	97 ± 1 %	97 ± 1 %
60 minutes	94 ± 1 %	97 ± 1 %	97 ± 1 %

Composite ABS-1-TRISO green bodies showed a volume reduction during sintering of ~30%. Circumferential cracking and composite lamination evidenced in earlier work using SLS glasses was not apparent upon visual inspection of ABS-1 sintered composites. Sectioned composites were examined using back-scattered SEM. Complete retention of the OPyC layer was observed as illustrated in Figure 2. The use of alumino-borosilicate glasses has eliminated the radial cracking previously seen in SLS-TRISO particle composites [5]. The lower CTE of ABS-1 provides a reduction in the CTE mismatch between the glass and the TRISO particles. This lowers tensile stresses created in the glass matrix on cooling and reduces the driving force for radial cracking.

Despite the sintered glass matrix density exceeding 95% of the theoretical density, Figure 2 a) shows that a degree of residual porosity remains. This does not appear to be interconnected and so should not prove excessively detrimental to waste form performance. Figure 2 b) illustrates the effective wetting produced between the ABS-1 and the OPyC layer. This coating is essential to the integrity of an encapsulated TRISO particle waste stream.

Figure 2. Back-scattered SEM images of a) TRISO particle embedded in ABS-1 b) OPyC-ABS-1 interface of the composite.

The PCT test was performed on powdered sintered glass samples. The normalised mass losses from the 14 day sampling of ABS-1 are displayed in Table 3. Sodium showed the highest normalised dissolution rate indicative of an ion exchange mechanism. The solution reaches steady state dissolution with respect to silicon after 3 days. Similarities were displayed in the normalised dissolution rate of network formers silicon and boron with a slower rate observed for aluminium.

The low aluminium concentration in solution may be explained by the precipitation of an alteration phase. Evidence suggesting the formation of an alteration phase is also provided by the low calcium and magnesium dissolution rates. The decrease in calcium concentration between 3 and 7 days (see Figure 3) may be due to the precipitation of a calcium bearing phase on the glass surface. This hypothesis is supported by PhreeqC geochemical modeling of the leachate solutions [8]. The saturation indices for the zeolite phases phillisite-Ca ($CaAl_2Si_5O_{14}.5H_2O$) and chabazite ($CaAl_2Si_4O_{12}.6H_2O$), were positive for the 7 and 14 day samples. These phases are known to form as a result of glass dissolution [9]. If these zeolite phases are present they may act as a passivation barrier to both ion exchange and network hydration of the glass, further lowering dissolution rates.

Table 3. Normalised mass losses and normalised dissolution rates for ABS-1 after PCT testing to 14 days (precision ± 5 %).

Element	Normalised Elemental Mass Loss ± 5% $(g\ m^{-2})$	Normalised Elemental Dissolution Rate ± 5% $(g\ m^{-2}\ day^{-1})$
Si	3.14×10^{-2}	2.25×10^{-3}
Al	1.87×10^{-2}	1.34×10^{-3}
B	3.04×10^{-2}	2.17×10^{-3}
Na	5.49×10^{-2}	3.92×10^{-3}
Ca	1.07×10^{-2}	7.64×10^{-4}
Mg	3.90×10^{-2}	2.79×10^{-3}

The durability of ABS-1 was compared with glasses previously used in the formation of glass-TRISO particle composites. A borosilicate, R7T7 (wt% SiO_2:60.5/B_2O_3:16.1/Na_2O: 7/CaO:5.8/Li_2O:5.3/Al_2O_3:2.6/ZnO:2.4/ZrO_2:0.3) and an SLS, (wt% SiO_2:69/Na_2O: 15/CaO:15/B_2O_3:1) were used for this comparison. PCT analysis was performed using sintered glass powders under the same conditions as in the current study.

Figure 3 displays the vast improvement in durability provided by ABS-1 when compared to R7T7 and SLS. It was shown that across common elements, ABS-1 has a normalised mass loss approximately an order of magnitude lower than the other compositions. As the matrix in all glasses tested consists of a silicate network the dissolution rate of silica can be considered the limiting factor is the degradation of the matrix. Table 4 shows the calculated normalised dissolution rate for silicon and boron after 14 days. The rate of silicon dissolution is approximately an order of magnitude lower for ABS-1 than for SLS and R7T7.

6

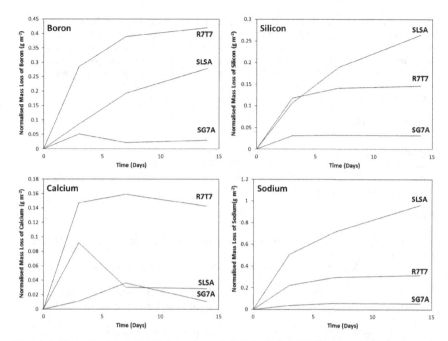

Figure 3. Normalised mass losses from sintered ABS-1, SLS and R7T7 glasses subjected to a 14 day PCT test at 90 °C (sampling times of 3, 7 and 14 days, precision ± 5 %).

Table 4. Comparison of normalised mass losses and dissolution rates for silicon and boron from sintered ABS-1, R7T7 and SLS subjected to a 14 day PCT test at 90 °C (precision ± 5 %).

	ABS-1	R7T7	SLS
Normalised Mass Loss - Silicon $(g\ m^{-2})$	3.14×10^{-2}	2.64×10^{-1}	1.46×10^{-1}
Normalised Dissolution Rate Silicon $(g\ m^{-2}\ day^{-1})$	2.25×10^{-3}	1.88×10^{-2}	1.05×10^{-2}
Normalised Mass Loss - Boron $(g\ m^{-2})$	3.04×10^{-2}	4.21×10^{-1}	2.79×10^{-1}
Normalised Dissolution Rate Boron $(g\ m^{-2}\ day^{-1})$	2.17×10^{-3}	3.01×10^{-2}	1.99×10^{-2}

CONCLUSIONS

Alumino-borosilicate glass has been shown to have superior properties when compared to glass compositions previously investigated for TRISO immobilisation. Improvements include the elimination of composite circumferential cracking, composite lamination and radial cracking in SLS composites. Wetting of the OPyC layer by the glass is also improved in comparison to SLS glasses. The normalised dissolution rate of silicon and boron from the ABS-1 glass is over an order of magnitude lower than that of glass compositions previously tested for this purpose.

A small amount of non interconnected porosity is retained in the composites after sintering, which may be detrimental to the quality of the waste form. However it should be possible to completely remove this porosity using alternative processing methodologies.

ACKNOWLEDGMENTS

The authors would like to thank Paul Lythgoe for performing the ICP-AES analysis, the Nuclear First DTC and EPRSC for funding and the PBMR Company for the kind provision of simulant TRISO fuel particles. Neil Hyatt is grateful to the Royal Academy of Engineering and the Nuclear Decommissioning Authority for funding.

REFERENCES

1. E. Cartlidge, Nuclear's New Generation, in: Physics World, Institute of Physics, October (2010).
2. J. Fachinger, M. den Exter, B. Grambow, S. Holgersson, C. Landesman, M. Titov, T. Podruhzina, Nucl. Eng. Des. **236**, 543-554 (2006).
3. International Atomic Energy Agency, Report; Technical, institutional and economic factors important for developing a multinational radioactive waste repository, June (1998).
4. A. Abdelouas, S. Noirault, B. Grambow, J. Nucl. Mater. **358**, 1-9 (2006).
5. P.G. Heath, C. L. Corkhill, M.C. Stennett, R. J. Hand, W. C. Meyer, N. C. Hyatt, *Submitted:* J. Nucl. Mater. Septemeber (2012).
6. S. Gahlert, G. Ondrackek, in Radioactive Waste forms for the Future, edited by R. C. Ewing and W. Lutze (North-Holland, Amsterdam, 1988) p. 162-191
7. ASTM Standard C 1285, *Test Methods for Determining Chemical Durability of Nuclear, Hazardous and Mixed Waste Glasses: The Product Consistency Test (PCT),* (2002).
8. D.L. Parkhurst, U.S. Geological Survey Water-Resources Investigations Report 95-4227 (1995)
9. R. A. Howie, J. Zussman, W.S. Wise, *Framework Silicates: Silica Minerals, Feldspathoids and Zeolites,* 2nd Ed. (Geological Society, London, 2004) p. 600

Mater. Res. Soc. Symp. Proc. Vol. 1518 © 2013 Materials Research Society
DOI: 10.1557/opl.2013.94

Long-term aqueous alteration kinetics of an alpha-doped SON68 borosilicate glass

Magaly Tribet[1], Séverine Rolland[1], Sylvain Peuget[1], Magali Magnin[1], Véronique Broudic[1], Arne Janssen[2], Thierry Wiss[2], Christophe Jégou[1] and Pierre Toulhoat[3,4]

[1]CEA/DEN/DTCD/SECM, BP 17171, 30207 Bagnols sur Cèze cedex, France
[2]European Commission, Joint Research Centre, Institute for Transuranium Elements, Karlsruhe, Germany
[3]CNRS/ISA, Institut des Sciences Analytiques, Université de Lyon, 5 rue de la Doua, 69100 Villeurbanne cedex, France
[4]INERIS, Parc Technologique Alata, BP 2, 60550 Verneuil en Halatte, France,

ABSTRACT

The long-term behavior of nuclear glass subjected to alpha radiation by minor actinides must be investigated with a view to geological disposal. This study focuses on the effect of alpha radiation on the chemical reactivity of R7T7 glass with pure water, mainly on the residual alteration rate regime. A glass specimen doped with 0.85 wt% ^{239}PuO$_2$ (α emitter) is leached under static conditions in argon atmosphere at 90°C and at a high surface-area-to-volume ratio (S/V = 20 cm^{-1}). The alteration rate is monitored by the release of glass alteration tracer elements (B, Na and Li). Radiation effects on the leached glass and its gel network are characterized by SEM and TEM analyses. Plutonium release is also measured by radiometry and its chemical oxidation state is assessed by measuring the pH and redox potential of the leachates. The results do not highlight any significant effect of alpha radiation on the residual alteration of this doped glass. This observation is consistent with SEM and TEM characterizations, which show that a protective layer can be formed under alpha radiation. Very low concentrations of soluble plutonium are measured in the leachate. These Pu releases are three orders of magnitude lower than the boron release, indicating strong plutonium retention.

INTRODUCTION

In France, fission products and minor actinides are currently immobilized by vitrification in a borosilicate glass known as "R7T7". One of the major functions of the glass matrix is to retain the radioactive elements in the event of water intrusion in the repository, expected after several thousand years. The alteration of SON68 glass (nonradioactive surrogate of R7T7 glass) has been extensively studied during the last thirty years and the key mechanisms and kinetics of borosilicate glass alteration have been determined [1]: first interdiffusion, then hydrolysis of the glass network, followed by a rate drop and finally a residual rate regime in which glass dissolution results from the formation of secondary aluminosilicate crystalline phases such as phyllosilicates and the interdiffusion of water and solvated ions through the "passivating reactive interphase" (PRI). This long-term alteration rate would be the dominant alteration phenomenon under geological disposal conditions.

However, under disposal conditions, the glass packages and the surrounding environment, including the solution, will be subjected to alpha radiation. This could modify the glass leachability through structural damage to the matrix [2] and/or changes in solution chemistry via radiolysis [3]. Moreover, the behavior of the actinides such as Pu and Np, which will account for most of the radioactivity of the glass package after the first 1000 years, must be known to assess the long-term safety of a repository site. Although the model currently used to assess the long-term safety of a deep geological repository assumes that radionuclides are released into solution at the same rate as the glass matrix, experimental data suggest that actinide release rates are orders of magnitude lower than the glass corrosion rate [4-6].

Despite the numerous studies carried out to assess the impact of radiation on nuclear glass leaching behavior [4-12], even on alpha-doped glasses [7, 10-12], no data are available specifically concerning the residual rate stage. In this study, we thus focus on the effects of alpha radiation on the residual rate and actinide behavior.

EXPERIMENTAL DETAILS

Doped glass powder

The ^{239}Pu doped glass was fabricated in 1985 in a glove box in the CEA's Vulcain facility at Marcoule [11]. It was elaborated at 1160°C for 3 h 20 min, then annealed for one hour at 525°C and cooled at a rate of less than 50°C·h^{-1}. The glass, whose chemical composition is detailed in previous papers [11-12], contains 0.85 wt% of ^{239}PuO$_2$. At the beginning of the leach test, its alpha activity (5 × 10^7 Bq·g^{-1}) was representative of R7T7 glass after around 1000 years [12-13]. The glass dose rate was calculated from its alpha activity assuming an average alpha radiation energy of 5.25 MeV, and was about 150 Gy·h^{-1}.

Glass samples were ground and sieved to recover the 63−125 µm size fraction. The specific surface area of the glass powder sample was estimated as 645 ± 50 cm^2·g^{-1} based on tests with ground and sieved inactive SON68 glass powder.

Leaching conditions

The leach test was performed in the CEA's Atalante facility at Marcoule. Prior to the leaching experiment, 9.44 g of glass powder was introduced in a glass vessel filled with 30 mL of ultrapure water in order to easily transfer it into the stainless steel leaching reactor (volume of this reactor: 350 mL). Then, the solution volume was adjusted to 305 mL by adding ultrapure water. The leach test was thus carried out under static conditions at an initial glass-surface-area-to-solution-volume (S/V) ratio of 20 cm^{-1}.

The reactor watertightness was ensured by a Viton® seal. After reactor closure, the solution was degassed by argon bubbling for one hour. Then an argon overpressure of 4 bars was applied to prevent any ingress of air into the reactor and to allow leachate sampling at regular intervals. Finally, the temperature was adjusted to 90°C by using a heating belt. The total duration of the leach test was 1174 days.

A reference experiment was also performed on a non radioactive glass for 1550 days under the same conditions (initial S/V = 26 cm^{-1}, 90°C), but outside the shielded cell.

Solution and solid analyses

The leachate pH and redox potential were measured on an aliquot at room temperature by means of a combination gel-filled pH electrode (Checker) and a combination Ag/AgCl redox electrode (3 mol·L^{-1} KCl) with a platinum probe.

Most of the leachate samples were analyzed directly. Three samples were filtered (0.45 µm) and ultrafiltered (10 000 Dalton) to check for colloids, as previously described [13]. Cation analyses were performed with a Jobin Yvon JY66 ICP-AES system. The analytical uncertainty was ±3% for concentrations above 10 mg·L⁻¹, ±5% between 1 and 10 mg·L⁻¹ and ±20% for concentrations below 1 mg·L⁻¹. The releases in solution are expressed in terms of normalized mass losses (NL in g·m⁻²) [5].

Plutonium releases were measured by α spectrometry with a Canberra Alpha Analyst spectrometer. Only (^{239}Pu + ^{240}Pu) isotopes, measured simultaneously, were considered because they represent more than 99% of the glass activity.

Altered glass powder samples taken at the end of the leach test were dried for 1 day at 50°C. Scanning electron microscopy (SEM) analyses were carried out on raw grains with a Philips CP XL 40 (ITU, Karlsruhe, Germany). Transmission electron microscopy was also performed on ground altered glass powder with a FEI Tecnai G² at ITU, modified during its fabrication to enable the examination of radioactive samples.

RESULTS AND DISCUSSION

Glass alteration

Glass alteration was traced by soluble elements released from the glass into solution, especially boron, sodium and lithium. Figure 1 shows the variation of normalized mass losses calculated from these releases for the ^{239}Pu-doped glass and the reference experiment.

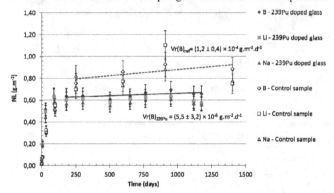

Figure 1: Normalized mass losses (NL) of soluble elements versus time for the ^{239}Pu-doped glass leach test and the reference experiment. Residual rate calculated from a linear regression on the boron NL, expressed with an uncertainty corresponding to a 68% confidence interval.

The data show that congruence is conserved between boron, sodium and lithium, indicating that the mobility of alkali elements is not affected by irradiation. The data also indicate that overall tracer element releases under irradiation vary in the same way as the reference experiment, versus time. The NL variation versus time shows a major drop in the glass dissolution rate after 50 days for the ^{239}Pu-doped glass, and after 200 days for the reference experiment. This difference could be due to the presence of fines in the ^{239}Pu-doped glass powder, resulting in a higher S/V ratio and reaching the residual rate regime quicker than expected. For the reference experiment the residual rate is established after 250 days of leaching compared with only 100 days for the doped glass. For both leach tests, the residual rate values were calculated from a linear regression on the normalized boron mass losses. These values and

11

their uncertainties are detailed in Figure 1. The overall similarity suggests that the residual rate of the [239]Pu-doped glass is not affected by irradiation.

Figure 2 shows TEM and SEM images of the altered layer formed during leaching of the [239]Pu-doped glass. The different phases of this altered layer can be seen in the TEM images. The same phases were observed on the nonradioactive altered glass powder. From the pristine glass to the solution, a dense inner layer corresponding to the hydrated glass is observed, then a very porous zone formed by recondensation of hydrolyzed species (the gel), and finally secondary phases, called phyllosilicates, formed by precipitation of elements from solution [14]. These secondary phases cover the whole altered glass surface, as seen by SEM. Alteration layer thickness estimated from TEM analysis (270 - 350 nm) is the same order of magnitude than calculated from solution releases (250 nm).

Furthermore, data concerning the evolution of the silicon concentration in solution over time indicate that the silicon release reached a steady state corresponding to the value of saturation of R7T7 glass with respect to orthosilicic acid (H_4SiO_4) measured for the inactive SON68 glass under the same leaching conditions ($pH_{90°C}$ ~8) by S. Gin [15].

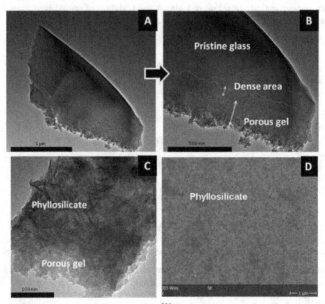

Figure 2: A-C) Bright-field TEM images of the [239]Pu-doped glass altered layer (HV=200 kV). D) SEM image of the leached [239]Pu-doped glass.

The results obtained from solution and solid analyses indicate that the altered layer formed under irradiation has the same chemical and protective properties as its inactive surrogate, leading to the same residual alteration rate.

Plutonium behavior

According to the literature [5], under the [239]Pu-doped glass leaching conditions (Eh/SHE = 380 mV and $pH_{90°C}$ ~8.6), the predominant form of plutonium in the leachate is Pu(IV). Figure 3-A shows the evolution of the plutonium concentration in the leachate during the leaching experiment. The data indicate the presence of colloids in solution. The released plutonium could be controlled by a solid phase with a solubility limit around 5×10^{-9} mol·L^{-1}, but this value is higher than the solubility limit of the Pu(IV) hydroxide phase known to be about 10^{-11} mol·L^{-1} [16]. However, similar concentrations (about 10^{-8} mol·L^{-1}) have already been measured in plutonium-doped glass leachates [11, 17-18]. Various hypothesis can be proposed to explain these higher concentrations. Small polynuclear species such as oligomers and colloids could be formed with diameters < 1 nm [16], or complex(es) with dissolved glass species could predominate and induce higher plutonium concentrations [19]. However these data show that plutonium releases remain three orders of magnitude lower than boron releases (figure 3-B), suggesting strong plutonium retention in the alteration layer and/or sorption on the walls of the leaching reactor.

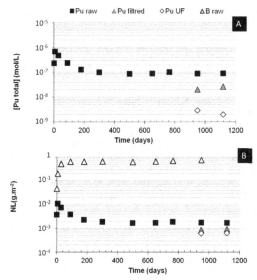

Figure 3: Total, filtered and ultrafiltered (UF) Pu concentrations versus time (A- mol·L^{-1} and B-g·m^{-2}).

CONCLUSIONS

The residual alteration regime of a [239]Pu-doped SON 68 glass was investigated. Data obtained from leachate and the solid analyses show that the residual rate was not affected by irradiation. Moreover, the morphology and protective properties of the altered layer formed under these conditions are similar to those of its inactive surrogate. Concerning plutonium, the presence of colloids was evidenced and the soluble Pu concentration shows that Pu release was not controlled by a simple hydroxide phase solubility. However, the Pu release remained very low, about three orders of magnitude lower than the boron release, indicating strong plutonium retention.

In terms of the long term behavior of the glass in a geological disposal, these results show that the alpha activity expected after around one thousand years (i.e. at the moment of water ingress) does not have a significant impact on the glass alteration rate in pure water. In prospect, evaluating the potential adsorbed plutonium on the reactor walls will allow us to calculate the Pu retention factor. Then, further experiments would be necessary to assess the relative impact of groundwater and environmental materials surrounding a geological repository.

ACKNOWLEDGMENTS

The authors would like to thank the ACTINET-I3 project and AREVA for their financial support.

REFERENCES

1. P. Frugier, S. Gin, Y. Minet, T. Chave, B. Bonin, N. Godon, J.E. Lartigue, P. Jollivet, A. Ayral, L. De Windt, G. Santarini, *J. Nucl. Mater.* **380**, 8 (2008).
2. S. Peuget, J-N Cachia, C. Jégou, X. Deschanels, D. Roudil, V. Broudic, JM Delaye and J-M. Bart, *J Nucl Mater.* **354**, 1-13 (2006).
3. C. Ferradini, J.P. Jay-Gerin, *Can. J. Chem.* **77**, 1542–1575 (1999).
4. O. Ménard, T. Advocat; J.P. Ambrosi, A. Michard, *Applied Geochemistry.***13**, 105-126 (1998).
5. P. Jollivet and G. Parisot, *J Nucl Mater.* **345**, 46–64 (2005).
6. E. Valcke, M. Gysemans, H. Moors, P. Van Iseghem, N. Godon, P. Jollivet, *Mat. Res. Soc. Symp. Proc* **932**, 999-1006 (2006).
7. T. Advocat, P. Jollivet, J.L. Crovisier, *J Nucl Mater.* **298**, 55-62 (2001).
8. A. Abdelouas, K. Ferrand, B. Grambow, *Mat. Res. Soc. Symp. Proc* **807**, 175-179 (2004).
9. C. M. Jantzen, I. Kaplan, N. E. Bibler, D. K. Peeler, M. J. Plodinec, *J Nucl Mater.* **378**, 244 (2008).
10. D. M. Wellman, J.P. Icenhower, W. J. Weber, *J Nucl Mater.* **340**, 149-162 (2005).
11. S. Fillet, JL. Nogues, E. Vernaz and N. Jacquet-Francillon, *Mat. Res. Soc. Symp. Proc* **50**, 211 (1985)
12. S. Peuget, V. Broudic, C. Jégou, P. Frugier, D. Roudil, X. Deschanels, H. Rabiller, P.Y. Noel, *J. Nucl. Mater.* **362**, 474 (2007)
13. S. Rolland, M. Tribet, P. Jollivet, C. Jégou, V. Broudic, C. Marques, H. Ooms and P. Toulhoat, *J Nucl Mater, in press.*
14. N. Valle, A. Verney-Carron, J. Sterpenich, G. Libourel, E. Deloule, P. Jollivet, *Geochimica and Cosmochimica Acta* **74**, 3412–3431(2010)
15. S. Gin, J.P. Mestre, *J. Nucl. Mater.* **295**, 1,83-96 (2001)
16. R. Knopp, *Radiochimica Acta* **86**, 101-108(1999)
17. V. Pirlet, *J Nucl. Mater.* **298**, 47-54 (2001)
18. Y. Kohara, T. Ashida, M. Yui, *J. Nucl. Sci. Technol.* **34**, 1107-1109 (1997)
19. A.B. Yusov, A.M. Fedosseev, C.H. Delegard, *Radiochimica Acta* **92**, 869-881 (2004)

Mater. Res. Soc. Symp. Proc. Vol. 1518 © 2013 Materials Research Society
DOI: 10.1557/opl.2013.71

Corrosion and Alteration of Lead Borate Glass in Bentonite Equilibrated Water

Atsushi MUKUNOKI[1], Tamotsu CHIBA[1], Takahiro KIKUCHI[1], Tomofumi SAKURAGI[2], Hitoshi OWADA[2], Toshihiro KOGURE[3]

1 JGC Corporation, Minato-Mirai 2-3-1, Nishi-ku, Yokohama, Japan
2 Radioactive Waste Management Funding and Research Center, Tsukishima 1-15-7, Chuo-ku, Tokyo, Japan
3 Department of Earth and Planetary Science, Graduate School of Science, The University of Tokyo, 7-3-1, Hongo, Bunkyo-ku, Tokyo, 113-0033, Japan

ABSTRACT

The development of an iodine immobilization technique that can fix radioactive iodine in waste form for a long period and constrain its leaching into pore water is necessary in order to secure the long-term safety of geological disposal of transuranic (TRU) waste. Lead borate glass vitrified at a low temperature is regarded as a promising material for immobilizing the Iodine-129 that is recovered from spent AgI filters generated by reprocessing plants in Japan and which may have a significant effect on the long-term safety of geological disposal.

Batch leaching tests were conducted to understand glass dissolution behavior in various solutions that account for geological disposal conditions. Boron dissolved at the highest rate in all types of solutions to be used as an index element for measuring the glass dissolution rate. On the other hand, lead dissolved in these solutions at a much lower rate. These results are consistent with an electron micro-probe analysis (EPMA) of the altered glass surfaces that indicated the depletion of boron and enrichment of lead near the surfaces.

The altered glass surfaces were further examined by scanning and transmission electron microscopy (SEM/TEM) and X-ray diffraction (XRD). SEM/TEM observation showed formation of a porous altered layer consisting of fine crystallites on the pristine glass and euhedral crystals on the altered layer. XRD analysis indicated that the fine crystallites and euhedral crystals are hydrocerussite, $Pb_3(CO_3)_2(OH)_2$, which was predicted by geochemical calculation as the precipitate for the experimental system.

INTRODUCTION

A recent review of TRU waste states that Iodine-129 is a key radionuclide in the safety assessment for the geological disposal of TRU waste.[1] It is anticipated that the development of an iodine immobilization technique that can fix iodine in waste form for a long period and constrain its leaching into pore water will offer an alternative that would reduce the release rates of Iodine-129 from primary containment under typical geological conditions.

Low temperature vitrification with $BiPbO_2I$ (BPI) is one immobilization candidate.[2] Initially, H_2 is used as a reductant to release iodine from a spent iodine filter. Then, the iodide ion is fixed into the BPI by reaction with an inorganic anion exchanger, $BiPbO_2NO_3$ (BPN). [3-4] The ion exchange reaction is as follows:

$$BiPbO_2NO_3 + I^- \rightarrow BiPbO_2I + NO_3^- \qquad (1)$$

In the vitrification process, a low melting temperature of 540°C is used to avoid iodine volatilization during the vitrification process.

Previous studies have suggested that the dissolution mechanism of iodine from BPI glass depends on the composition of the leaching solution.[5-6] For dissolution in simulated bentonite pore water and seawater, iodine dissolution is congruent with that of boron, a soluble element among BPI glass constituents, but is incongruent in $Ca(OH)_2$ solution.

In this study, leaching tests of BPI glass were carried out using simulated bentonite pore water, and geochemical simulation was done in order to understand the equilibrium conditions of lead minerals. In addition, the altered glass surface was further examined by EPMA, SEM/TEM and XRD.

EXPERIMENTS

To prepare BPI vitrified waste forms, we mixed BPI powder and lead borate glass frit and melted them at 540 °C by using an electric muffle furnace. The composition of the glass frit is PbO: B_2O_3: ZnO at ratios of 65:30:5 [mol%] and the target composition of the BPI vitrified waste form is shown in Table 1.

Table 1. Composition of BPI vitrified waste form

Elements	I	B	Pb	Zn	Bi	O
mol%	1.0	19.8	22.1	1.6	1.0	54.5
wt%	2.0	3.5	75.2	1.8	3.3	14.3

The leaching tests were carried out in an inert glove box in which the concentrations of oxygen and carbon dioxide were maintained at less than 1 ppm. The leachant was prepared by adding reductant (electrolytic iron powder) to ensure reducing conditions. For the preparation of simulated bentonite pore water, deionized water was equilibrated with bentonite (Kunigel-V1) for 30 days followed by filtration through a 0.45 μm filter. Table 2 shows the composition, pH and Eh of simulated bentonite pore water as measured by ICP-AES and ion chromatography.

Table 2. Composition, pH and Eh of simulated bentonite pore water

Element	Na	K	Ca	Si	Al	Mg	IC	Cl⁻	SO_4^{2-}	pH	Eh[mv]
Conc. (mg/L)	95.0	1.14	1.82	32.6	13.1	2.85	27.9	1.34	59.9	10.1	-650

RESULTS AND DISCUSSION

Leaching Behavior of Boron, Iodine and Lead

Based on the results of the leaching tests, we calculated normalized elemental mass loss rates (NLR) from the solution concentration by using the following equation:

$$NLR_i = \frac{\Delta C_i}{f_i \Delta t} \cdot \frac{V}{S} \qquad (2)$$

where ΔC_i denotes the difference in concentration of element i in the solution, f_i the original fraction of element i in the glass, Δt the difference in time, V the solution volume, and S the surface area of the glass.

The normalized mass loss rate of iodine (NLR_I) is compared with those of boron (NLR_B) and lead (NLR_{Pb}) in Figure 1. These data are the averages of data from duplicate tests and show that in simulated bentonite pore water, iodine dissolution was congruent with boron dissolution, but that lead dissolution was incongruent with these other two elements.

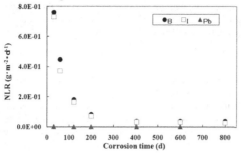

Figure 1. NLR values of boron, iodine and lead in simulated bentonite pore water

Observation by EPMA

Figure 2 shows the SEM and EPMA images and composition of main elements from a cross-section of the block immersed for 800 days in simulated bentonite pore water. The concentration of boron and iodine in the altered glass surface is lower than their concentration in the pristine glass. On the other hand, the concentration of lead in the altered glass surface is almost as high as its concentration in the pristine glass. This observation is consistent with the leaching behavior of boron, iodine and lead, as described above. When bonds of boron or iodine, which are both soluble elements, are cut by contact with water molecules, these substances are released into liquid phase. On the other hand, when bonds of lead are cut, part of the lead is released into liquid phase but part of the lead remains in solid phase through recombination of the bonds. The characteristics of the alteration layer, such as the lead enrichment and depletion of boron and iodine, can be explained by these mechanisms.

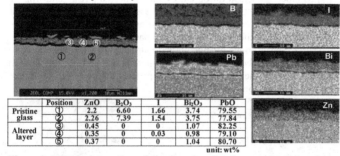

	Position	ZnO	B_2O_3	I	Bi_2O_3	PbO
Pristine glass	①	2.2	6.60	1.66	3.74	79.55
	②	2.26	7.39	1.54	3.75	77.84
Altered layer	③	0.45	0	0	1.07	82.25
	④	0.35	0	0.03	0.98	79.10
	⑤	0.37	0	0	1.04	80.70

unit: wt%

Figure 2. EPMA observation of the altered surface layer and pristine glass

Geochemical simulation

The incongruent dissolution of lead can be explained by precipitation of lead mineral at the glass surface. In order to understand equilibrium conditions of lead minerals in bentonite

17

pore water, we performed a geochemical simulation by using PHREEQC. For thermodynamic data, JAEA-TDB version 100331c1 provided by the Japan Atomic Energy Association (JAEA) was chosen. Table 3 shows the input data for the simulation. All of these values are consistent with those observed in the leaching tests. The equilibrium condition of minerals can be thought of as a saturation index (SI) defined as follows:

$$SI_i = \log \frac{IAP_j}{K_j} \qquad (3),$$

where SI_j denotes the saturation index of mineral j, IAP_j denotes the product of the relevant aqueous species composing mineral j, and K_j denotes the equilibrium constant of mineral j.

Table 4 shows calculated saturation indices of possible lead minerals. It was found that cerussite ($PbCO_3$) has an SI of 0 and stays in equilibrium. Also, hydrocerussite ($Pb_3(CO_3)_2(OH)_2$) can be regarded as the second possible mineral in equilibrium, though it is slightly oversaturated.

Table 3. Input data for geochemical simulation

pH	pe	Temperature (K)						
9.8	-10.9	298						
Element	Na	K	Ca	Si	Al	Mg	C(+4)	Cl	S(+6)	Pb
Conc. (mg/l)	95	1.14	1.82	32.6	13.1	2.85	27.9	1.34	59.9	1

Table 4. Conditions of possible lead minerals in bentonite pore water

Mineral	Pb(OH)$_2$ (am)	Hydro-cerussite	PbCl$_2$ (s)	Cerussite	PbOHCl (cr)	PbO (litharge)	PbO (massicot)	Anglesite
SI	-2.63	1.61	-13.28	0.01	-4.45	-2.26	-2.54	-4.71

SEM/TEM and XRD analysis

Observation by SEM/TEM

The BPI glass block immersed for 60 days visually showed a rusty and whitish surface. Imaging by SEM (Hitachi S-4500) indicated that the surface was covered with euhedral crystals (Figure 3a). Most of the crystals had facets of regular triangles and these facets were generally parallel to the glass surface. On the other hand, SEM images from a cross-section of the block showed an altered surface layer on the pristine glass (Figure 3b). TEM examination (JEOL JEM 2010 operated at 200 kV) revealed that the layer was porous and consisted of fine grains that showed a crystalline feature in the electron diffraction pattern (Figure 3c). X-ray microanalysis using an energy dispersive spectrometer (EDS) detected Pb, C and O in the altered layer.

Figure 3. (a) SEM image of the surface of the BPI glass block immersed for 60 days, showing trigonal crystals covering the surface. (b) Cross-sectional SEM image of the same glass block. (c) Cross-sectional TEM image of the altered layer, indicating a porous structure with fine crystallites. The inset at the top-right is the electron diffraction pattern from the altered layer.

The XRD pattern was measured from the surface of the BPI glass subjected to the leaching test for 60 days. The measurement was performed using a conventional Bragg-Brentano type diffractometer (Rigaku RINT Ultima[+]) with monochromated CuKα radiation and having values of 0.5°, 0.5° and 0.3 mm for the divergence, anti-scatter, and receiving slits, respectively. The results in Figure 4 show a number of sharp peaks in the pattern. These peaks are well explained by those calculated using the crystallographic parameters of hydrocerussite, Pb_3 $(CO_3)_2(OH)_2$, as reported by Martinetto et al.,[7] which are also indicated in the figure. However, the intensity distributions in the experimental and calculated patterns do not correspond well. In particular, the peaks in the experimental pattern indexed at 00l were extremely intense compared to the calculated peaks. This is owing to the preferred orientation of the crystals on the surface. In the SEM image (Figure 3a), the crystals were oriented with the triangle facet parallel to the surface. Given that hydrocerussite belongs to the trigonal system, the trigonal c-axis of hydrocerussite should be normal to the surface, which results in the extremely intense 00l peaks in the XRD pattern measured by a Bragg-Brentano type diffractometer. Moreover, the peak widths for the 00l reflections are much smaller than those for the other reflections (Figure 4). This phenomenon can be explained by assuming that the 00l peaks come mainly from large crystals, namely the surface euhedral crystals with their notable orientation (Figure 3a), whereas the other peaks come from small crystals in the altered layer (Figures 3b and 3c). In other words, the crystals on the surface and in the altered layer are both considered to be hydrocerussite.

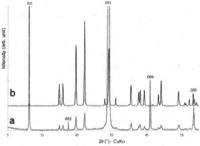

Figure 4. (a) Experimental XRD pattern from the surface of BPI glass immersed in the solution for 60 days. (b) Calculated pattern for hydrocerussite using crystallographic parameters reported by Martinetto et al.[7]

CONCLUSIONS

In a batch corrosion test of BPI glass, it was observed that for dissolution in simulated bentonite water, boron dissolution was congruent with iodine dissolution, but lead dissolution was incongruent with these other two elements. These results are consistent with EPMA analysis of the altered glass surfaces, which indicates the depletion of boron and enrichment of lead near the surfaces. According to the geochemical simulation using PHREEQC, cerussite ($PbCO_3$) has an SI of 0 and stays in equilibrium and hydrocerussite (Pb_3 $(CO_3)_2(OH)_2$) can be regarded as the second possible mineral in equilibrium, although it is slightly oversaturated.

In order to confirm the above results, we examined the altered glass surfaces by use of SEM/TEM and XRD. SEM/TEM observation showed formation of a porous altered layer consisting of fine crystallites on the pristine glass and euhedral crystals on the altered layer. XRD analysis indicated that the fine crystallites and euhedral crystals are hydrocerussite, which was predicted by geochemical calculation as the precipitate for the experimental system.

ACKNOWLEDGMENTS

The authors are grateful to Prof. Takahashi of Hiroshima University for EPMA analysis. This research is a part of "Research and development of processing and disposal technique for TRU waste contain I-129 and C-14 (FY2011)" under a grant from the Japanese Ministry of Economy, Trade and Industry (METI).

REFERENCES

1. Federation of Electric Power Companies (FEPC) and Japan Atomic Energy Agency (JAEA), "Second Progress Report on Research and Development for TRU Waste Disposal in Japan," *Repository Design, Safety Assessment Means of Implementation in the Generic Phase* (2007).
2. H. Tanabe et al., "Development of New Waste Forms to Immobilize Iodine-129 Released from a Spent Fuel Reprocessing Plant," *Advances in Science and Technology*, Vol. 73, 158–170 (2010).
3. H. Kodama, A. Dyer, M. J. Hudson, P. A. Williams, *Progress in Ion Exchange*, The Royal Society of Chemistry, Cambridge, UK, p. 39 (1997).
4. H. Kodama, N. Kabay, "Reactivity of Inorganic Anion Exchanger $BiPbO_2NO_3$ with Fluoride Ions in Solution," *Solid State Ionics*, 141–142 (2001).
5. A. Mukunoki, T. Chiba, "Development of an Iodine Immobilization Technique by Low Temperature Vitrification with $BiPbO_2I$," *The 11th International Conference on Environmental Remediation and Radioactive Waste Management* (ICEM07-7142) (2007).
6. A. Mukunoki, T. Chiba, "Further development of Iodine Immobilization Technique by Low Temperature Vitrification with $BiPbO_2I$," *The 12th International Conference on Environmental Remediation and Radioactive Waste Management* (ICEM2009-16268) (2009).
7. P. Martinetto, M. Anne, E. Dooryhée, P. Walter and G. Tsoucaris, *Acta Cryst.*, **C58**, i82–i84 (2002).

Mater. Res. Soc. Symp. Proc. Vol. 1518 © 2013 Materials Research Society
DOI: 10.1557/opl.2013.276

Vitrification of high molybdenum feeds in the presence of reprocessing waste liquor

Rick Short[1], Barbara Dunnett[1], Nick Gribble[1], Hannah Steel[2], Carl James Steele[2]

[1] National Nuclear Laboratory, Sellafield, Seascale, Cumbria, UK, CA20 1PG
[2] Sellafield Ltd, Sellafield, Seascale, Cumbria, UK, CA20 1PG

ABSTRACT

At Sellafield, the Post Operational Clean Out (POCO) of solids from the base of the highly active waste storage tanks, in preparation for decommissioning, will result in a high molybdenum stream which will be vitrified using the current Waste Vitrification Plant (WVP). In order to minimise the number of containers required for POCO, the high molybdenum feed could be co-vitrified by addition to reprocessing waste, using the borosilicate glass formulation currently utilised on WVP. Co-vitrification of high molybdenum feeds has been carried out using non-active simulants, both in the laboratory and on the Vitrification Test Rig (VTR) which is a full scale working replica of a WVP processing line.

In addition, a new borosilicate glass formulation containing calcium has been developed by NNL which allows a higher incorporation of molybdenum through the formation of a durable $CaMoO_4$ phase, after the solubility limit of molybdenum in the glass has been reached. Vitrification of the high molybdenum feed in the presence of varying quantities of reprocessing waste liquor using the new glass formulation has been carried out in the laboratory. Up to ~10 wt% MoO_3 could be incorporated without any detrimental phase separation in the product glass, but increasing the fraction of reprocessing waste was found to decrease the MoO_3 incorporation. Soxhlet and static powder leach tests have been performed to assess the durability of the glass products. This paper discusses the results of the vitrification of high molybdenum feeds in the presence of reprocessing liquor in both the borosilicate glass formulation currently utilised on WVP and the modified formulation which contain calcium.

INTRODUCTION

Highly Active Liquor (HAL) resulting from the reprocessing of used nuclear fuel on the Sellafield site is stored in Highly Active Storage Tanks (HASTs), and then fed to the Waste Vitrification Plant (WVP) to be immobilised in a glass matrix. As vitrification of HAL progresses, the storage tanks will be emptied to a point where they only contain a residual heel of HAL, which is likely to contain a large quantity of solids. The tanks will thus have to be subjected to Post Operational Clean Out (POCO) to reduce their activity, prior to decommissioning. There are several potential routes for undertaking POCO, one of which is to wash the tanks out and direct the resultant waste stream to WVP. Such a waste stream is likely to contain high levels of zirconium and molybdenum; molybdenum is known to be a problematic element during vitrification due to its low solubility in the borosilicate glass matrix used in the UK[1]. At concentrations above the solubility limit molybdenum tends to form compounds with alkali, alkali earth and rare earth elements, which aggregate in a relatively low density phase commonly termed "yellow phase".

Yellow phase accelerates corrosion of the melting crucibles used in the vitrification process and due to its water soluble alkali component is also detrimental to the chemical durability of the vitrified product. Therefore, WVP process parameters are generally designed to minimise the chance of yellow phase formation. If a "pure" POCO waste stream were to be vitrified in the standard UK alkali borosilicate base glass crizzle (known as "MW"), the waste oxide loading would have to be limited to unacceptably low levels to completely avoid phase separation. Therefore, several alternatives are currently being explored to allow completion of POCO whilst generating the minimum number of additional High Level Waste (HLW) product containers. One of the options is to "co-vitrify" the pure POCO waste streams in combination with existing HAL stocks using standard MW crizzle. A second option is to again combine the pure POCO waste stream with existing HAL stocks, but use a base glass formulated specially for the task. This paper explores the co-vitrification of POCO and HAL waste streams in both MW and a bespoke base glass, and considers the factors affecting incorporation rates in both cases.

MOLYBDENUM INCORPORATION

The solubility of Mo and the formation of yellow phase in HLW glasses has been well documented[e.g. 2, 3], and on WVP is likely to be affected by three primary factors, namely:
- The composition of the waste stream
- The total waste oxide incorporation
- The composition of the base glass matrix

Ultimately, any one of these three primary factors can influence or dictate the two remaining parameters, and at present all three factors have yet to be defined for POCO co-vitrification.

Waste stream composition

To date, the vast majority of the HLW that has been vitrified by WVP has directly arisen from the reprocessing of spent Magnox and uranium reactor fuels, by the Magnox reprocessing plant and the Thermal Oxide Reprocessing Plant (THORP). A number of the vitrification campaigns have processed waste streams comprising solely of HAL from Magnox reprocessing operations (commonly referred to as "Magnox" waste), and other campaigns have vitrified a blend of waste (commonly referred to as "Blend" waste) from oxide and Magnox reprocessing. The typical blending ratio used to date gives a fully oxidised waste product in which 75wt% of the waste oxides arise from reprocessed oxide fuels, and the remaining 25wt% from reprocessed Magnox fuels. This is described as a 75o:25m blend, and this XXo:YYm nomenclature will be used throughout this paper. The 75o:25m blending ratio was chosen for use with the standard MW base glass composition as it maximises the reprocessed uranium oxide component of the waste stream, and produces a vitrified product with an acceptable waste oxide incorporation and chemical durability. However, the future operational strategies of Magnox Reprocessing, THORP and Highly Active Liquor Evaporation and Storage (HALES) plants may mean that alternative blending ratios will be utilised for vitrification. At present though, the operational strategies for Magnox, THORP and HALES plants are not available for the period in which POCO waste will be accessible for co-vitrification.

As the reprocessing streams from Magnox and THORP are significantly different in terms of their chemical composition, it is likely that varying the Oxide:Magnox blending ratio of the reprocessing waste that is combined with the POCO waste for co-vitrification will influence Mo solubility and yellow phase formation in the final product. To this end, a wide range of blending ratios have been investigated in laboratory studies.

Also currently undefined are the exact compositions of the waste streams likely to arise from the POCO operations. It is probable that POCO streams will consist primarily of Zr and Mo, with small quantities of Ba, Sr and Cs present. This paper will review in detail the co-vitrification of POCO with a 50o:50m blend as a comprehensive range of data are available for this formulation.

Total waste oxide incorporation

For vitrified HLW products that are to remain in the UK, there is currently no product specification for the total waste oxide loading, or the blending ratio of the waste streams. For co-vitrified (reprocessing + POCO waste) products, it is likely that this parameter will ultimately be governed either by fission product loading or yellow phase formation which, in turn, will be a factor of the Mo concentration in the WVP feed.

For reprocessing waste streams in general, it is preferable to maximise the total waste oxide loading in each product container in order to minimise the total number of containers required to immobilise a given volume of HLW. However, when considering POCO waste streams the blending ratio, and thus the vitrification strategy, may change going forwards. Therefore current vitrification studies have to be broad ranging enough to envelop all of the possibilities.

Base glass composition

To date, all of the HLW vitrified in the UK has been immobilised using a single base glass crizzle feedstock known as Mixture Windscale (MW) which was developed during the Highly Active Residue Vitrification Experimental Studies (HARVEST) programme in the 1970s[4]. This alkali borosilicate composition was chosen for its ability to tolerate a wide variation in waste stream composition, appropriate thermal characteristics (such as viscosity at target WVP melter temperatures) and the chemical durability of the fabricated waste products. Having a single base glass formulation that was applicable to all of the HLW streams predicted at the time was also beneficial, as it minimised the need for extensive laboratory testing regimes associated with changing the glass composition to suit the characteristics of the waste, such as those performed in the US.

However, the solubility of Mo in products made with the MW formulation is limited, and the yellow phase into which Mo segregates tends to be detrimental to the overall product quality, thus becoming a limiting factor on the overall waste oxide incorporation rates attainable with high Mo waste streams. If an alternative base glass formulation can offer either increased Mo solubility, or allow Mo to segregate into a secondary phase that has no negative impact on the vitrified product quality, it might offer significant benefits and allow an increase in the waste loading of high Mo HLW products. This would then reduce the number of containers required for immobilisation. To this end, a new base glass formulation, termed "Ca/Zn" has been developed[5], and some of the co-vitrification work that has been carried out with the Ca/Zn formulation is described in this paper. The compositions of the MW and Ca/Zn formulations are shown in Table I below.

Table I – Compositions of MW and Ca/Zn base glasses

	MW base glass (wt%)	Ca/Zn base glass (wt%)
SiO_2	61.20	47.63
B_2O_3	22.40	23.41
Li_2O	5.40	4.20
Na_2O	10.99	8.56
Al_2O_3	-	4.20
CaO	-	6.0
ZnO	-	6.0
Total	100.00	100.00

EXPERIMENTAL DETAILS

General approach

The data presented in this report spans a ~2 year period of research into high Mo waste vitrification, during which the assorted factors governing potential Mo incorporation described earlier were explored. The research was primarily based on co-vitrification of reprocessing and POCO waste streams in MW base glass, and this formed the basis of the initial laboratory experiments, and culminated in a full scale trial on the Vitrification Test Rig (VTR). The results of MW co-vitrification trials prompted further trials with Ca/Zn base glasses, and the laboratory phase of a selection of that work is reported here.

In all cases, the general experimental approach involved combining a simulated 50o:50m reprocessing waste stream with a simulated POCO stream in the appropriate ratios, resulting in either a specific target waste oxide incorporation, or a specific MoO_3 incorporation in the final vitrified product. The compositions of the simulated waste streams are given in Table II. The MoO_3 incorporation rates were increased in successive melts until the onset of yellow phase was observed, and/or the quantity of yellow phase became significant.

Table II – Compositions of simulated waste streams

Oxide	50o:50m (wt%)	Laboratory POCO (wt%)	Full scale POCO (wt%)
Na$_2$O	0.2		
MgO	10.9		
Al$_2$O$_3$	11.2		
PO$_4$	1.4		0.3
SO$_4$	0.2		
Cr$_2$O$_3$	2.1		
Fe$_2$O$_3$	9.6		
NiO	1.3		
SrO	1.6	1.0	1.2
Y$_2$O$_3$	1.0		
Zr$_2$O$_3$	8.8	26.0	23.1
MoO$_3$	8.1	60.0	60.8
RuO$_2$	1.5		
TeO$_2$	0.9		
Cs$_2$O	5.7		1.7
BaO	3.3	13.0	12.9
La$_2$O$_3$	2.7		
CeO$_2$	5.4		
Pr$_2$O$_3$	2.6		
Nd$_2$O$_3$	8.8		
Sm$_2$O$_3$	1.8		
Gd$_2$O$_3$	10.9		

N.B. There were slight differences in the laboratory and full scale POCO simulant compositions reflecting possible variations in the waste stream

Laboratory Scale Production

Glasses for the laboratory trials were made from oxides, carbonates and calcined reprocessing waste simulant (see "Full Scale Production" section). The required quantities of each component were weighed, combined and mixed in a rotary powder mixer, then loaded into a silica or platinum crucible and placed into a furnace at 1050°C. After 3 hours the melts were manually stirred for approximately 5 seconds using a silica rod, then returned to the furnace for a further hour prior to casting into a brass mould. The resulting monoliths were annealed for 3 hours at 500°C, then cooled to room temperature at 0.5°C/min. Batches were calculated to yield ≤80g of vitrified product, depending on the final use (e.g. exploratory trial for yellow phase limit determination or viscometry test specimen). The range of target glass compositions manufactured during the laboratory trials are shown in Table III.

Table III – Compositions of MW and Ca/Zn base glasses

Base glass type	MW			Ca/Zn			
50o:50m (wt% of product)	25	28	32	0	5	15	25
POCO (wt% of product)	5.42 – 6.25	5.03 – 5.87	4.08 – 5.75	16.67 – 23.33	11.83 – 16.00	11.33 – 14.67	7.50 – 10.83
Total MoO$_3$ (wt% of product)	5.25 – 5.75	5.25 – 5.75	5.00 – 6.00	10.00 – 14.00	7.50 – 10.00	8.00 – 10.00	6.50 – 8.50
Total waste oxide incorporation rate of product (wt%)	30.42 – 31.25	33.03 - 33.87	36.08 – 37.75	16.67 – 23.33	16.83 – 21.00	26.33 – 29.67	32.5 – 35.83

Full Scale Production

Full scale trials were performed on the VTR, a full scale waste vitrification plant designed to process non-active HAL simulants and used as a research tool to support operation of the WVP at Sellafield [e.g. 6]. Highly active liquid reprocessing waste was simulated by dissolving non-active metal nitrates and oxides in nitric acid, and this "HAL simulant" was then fed to a rotating tube furnace (calciner) in which it dried to form a solid granular "calcine" consisting of metal nitrates and oxides. At this point, the calcine was either collected for laboratory use (see "Laboratory Scale Production" section above) or quality assurance purposes, or fed into an induction heated melting crucible along with base glass crizzle, and heated to ~1050°C for vitrification.

During vitrification, batches targeting 190kg of final vitrified product were fed to the melter, then poured via a freeze valve into demountable containers. Once cool, the containers were dismantled to reveal the glass monolith, which was inspected, photographed, then sampled for laboratory analysis.

Table IV below indicates the target compositions of those co-vitrified products made on the VTR. A 38wt% incorporation 50o:50m product with no POCO component was used as the baseline.

Table IV – Target compositions of full scale co-vitrified products

	Baseline	Co-vitrification trials
MW base glass (wt% of product)	62	62
50o:50m (wt% in product)	38	33.86 – 31.95
POCO (wt% in product)	0	4.14 – 6.05
Total MoO$_3$ (wt% in product)	3.08	5.25 – 6.0
Total waste oxide incorporation rate of product (wt%)	38	38

In this study, visual inspection of the vitrified products was an essential component of the analysis process. Quantitative methods for assessing yellow phase in vitrified product are difficult to implement consistently, and thus a qualitative evaluation by highly experienced technical staff was deemed a more reliable indicator of the onset and relative quantity of yellow phase in the glasses.

Samples for SEM/EDS were mounted in resin, ground to 1000 grit using SiC paper and polished to 1 μm using diamond paste, then cleaned with methanol. Conductive silver paint was added and then they were gold coated to provide a conductive sample necessary for SEM analysis. Examination was performed with a JEOL JSM-5600 SEM fitted with a Princeton Gamma-Tech Prism Digital Spectrometer. Digital image processing in backscattered mode was used to determine the percentage of crystalline material in the relevant samples.

XRD analysis was carried out using an INEL Equinox 1000 X-ray diffractometer with a CuKα source and a curved positive sensitive detector ($2\theta = 5 - 110°$). The diffractometer was operated at 35 kV and 40 mA. The XRD patterns obtained for each material were compared with standard patterns obtained from the Joint Committee for Powder Diffraction Studies (JCPDS) X-pert database using Match software for phase identification.

Viscosity measurements were performed using a Theta Rheotronic II high temperature rotating viscometer. Measurements were made every 10 seconds as the melt was cooled from 1200°C to 900°C at a rate of 2°C/min.

Glass monoliths for soxhlet testing, approximately 1x1x1cm, were cut from specimens using a low speed diamond saw. The monoliths were cleaned in methanol, dried and weighed, and the surface area accurately measured. The specimen was then placed in the Soxhlet apparatus and leached with deionised water at ~97°C for 28 days. Afterwards, the leached monolith was removed, dried and weighed, and the mass loss calculated to give a bulk leach rate. The volume of the liquor was recorded and the leachate was analysed in-house using an Vista-MPX CCD simultaneous Inductively Coupled Plasma Optical Emission Spectrometer (ICP-OES) to obtain an elemental leach rate.

Static powder durability tests were carried out to the Product Consistency Test (PCT) ASTM standard[7] in deionised water at 90 °C for 7, 28, 56 and 112 days. Some of the tests carried out were only intended to be indicative, hence only single tests were performed as opposed to the triplicate testing detailed in the standard. Powders for the test were prepared by crushing, grinding and sieving to obtain a 75 – 150 μm size distribution. Adherent fines were removed using deionised water and absolute ethanol as in the standard. Finally, the beaker containing the cleaned powder was dried in an oven at 90 °C overnight. Each test used 4g of powder and 40 ml deionised water (18 MΩ). Assuming a glass density of 2.7 gcm^{-3}, this gave a surface area to volume (S/V) ratio of ~2000 m^{-1}. The powder was weighed into clean perfluoroalkoxy (PFA) leach vessel containers, the leachant added and the lids screwed on tightly. The containers were then weighed and placed in a Memmert UFE500 oven at 90 ± 1 °C. The lids were re-tightened several hours after the start of the test to prevent evaporative losses of the leachant. After the required test duration, the vessels were removed and allowed to cool. They were weighed to determine the evaporative losses of the leachant, which was <5 % in all cases. A sample of the leachate was filtered through a 0.45 μm syringe filter and transferred to a clean plastic specimen bottle, acidified with concentrated nitric acid then analysed by ICP-OES to determine the elemental concentrations. The normalized mass losses (NL_i) were calculated according to equation (1) below:

$$NL_i = \frac{c_i}{f_i (S/V)} \qquad (1)$$

where c_i is the concentration of element i in the leachate in ppm, and f_i is the mass fraction of element i in the unleached waste form.

Some glass specimens were subjected to a heat treatment programme to simulate a thermal excursion event. The selected samples were heated to either 650°C or 750°C for 2 weeks.

RESULTS

Laboratory trials with standard MW base glass

Table V indicates the general findings of the laboratory trials performed on co-vitrified products made with 50o:50m reprocessing waste, POCO simulant and MW base glass. Where available, equivalent data is included for a typical 25wt% incorporation Magnox product (with no POCO component) made at full scale on the VTR for comparative purposes[8]. For the co-vitrified products, the waste loading of the reprocessing component of the waste stream (25-32wt%) did not influence the amount of MoO_3 that could be tolerated in the product before the onset of significant yellow phase formation, which in all cases was 5.25wt% MoO_3. Unfortunately, elemental analysis data was not available for all of the vitrified products discussed, and so all reported values refer to target values. Figure 1 shows the difference in appearance of typical laboratory scale product glasses below and above the yellow phase threshold.

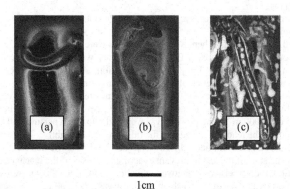

1cm

Figure 1 – Simulated laboratory scale HLW co-vitrified products made from MW base glass and reprocessing + POCO waste streams
(a) product contains ≤4.5wt% MoO_3, no phase separation
(b) product contains 4.5 - 5.25wt% MoO_3, some minor $BaMoO_4$ phase separation
(c) product contains >5.25wt% MoO_3, significant yellow phase formation (seen here as light grey "spots")

SEM and XRD analysis of the products containing 5.25wt% MoO_3 indicated that RuO_2 ([01-075-4303]) and $BaMoO_4$ ([01-089-4570]) were the only significant crystalline phases present. Soxhlet leach testing of the 5.25wt% MoO_3 products gave Bulk Leach Rates (BLRs) of 4-5gm^{-2}d^{-1} and normalised elemental leach rates for Mo of 5.3 - 5.7gm^{-2}d^{-1}. These values did not change significantly at higher MoO_3 loadings, despite the presence of significant yellow phase in the glass. After heat treatment of the co-vitrified product containing 28wt% 50o:50m and 5.25wt% MoO_3, various crystalline molybdate phases were observed under SEM and XRD. The soxhlet BLR and Mo elemental leach rates decreased to 2.3 and 1.7gm^{-2}d^{-1} respectively.

Table V – Laboratory scale co-vitrification results with MW base glass

	Co-vitrification products with maximum MoO3 loading prior to appearance of significant yellow phase			"Standard" 25wt% incorporation Magnox glass
50o:50m waste oxide incorporation in product (wt%)	25	28	32	N/A
POCO waste oxide incorporation in product (wt%)	5.40	5.00	4.50	N/A
MoO3 incorporation in product (wt%)	5.25	5.25	5.25	1.62
Phases identified under SEM	Not analysed	RuO2	Not analysed	RuO2 Spinel
Phases identified under XRD	RuO2 BaMoO4	RuO2	RuO2 BaMoO4	RuO2 Spinel
Soxhlet bulk leach rate (gm⁻²d⁻¹)	4.4	4.9	4.4	3.0
Phases identified under SEM in heat treated product	Not analysed	RuO2 CeOx Spinel Rare-earth molybdate	Not analysed	RuO2 Spinel
Phases identified under XRD in heat treated product	Not analysed	RuO2 CeZrO2 BaMoO4 Rare-earth molybdate	Not analysed	RuO2 Spinel
Soxhlet bulk leach rate of heat treated product (gm⁻²d⁻¹)	Not analysed	2.3	Not analysed	2.5

The normalised mass losses of B and Mo obtained from PCT leach testing of the co-vitrified product samples containing 5.25wt% MoO_3 are shown in Figure 2. There was a general trend of decreasing mass loss with increasing waste loading, but no evidence of preferential leaching of Mo. A standard 25wt% incorporation Magnox glass is shown for comparative purposes.

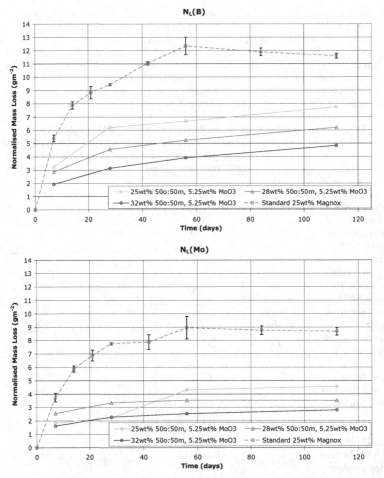

Figure 2 – Static PCT results for co-vitrified MW products, with standard 25wt% Magnox as comparison (1σ errors). No error data was available for co-vitrified products as the tests were not carried out in triplicate

The viscosity profile of the co-vitrified product with 5.25wt% MoO_3 loading and 28wt% reprocessing waste was very similar to that of standard Magnox and 75o:25m glasses at 28wt% waste oxide incorporation rates, indicating that the increased MoO_3 level in the co-vitrified product did not have a significant effect on the viscosity response.

Full scale trials with standard MW base glass

Table VI highlights the major findings of the full scale co-vitrification trials carried out with 50o:50m reprocessing waste and POCO simulant. Where a range of results is given in the table, the data covers those products made over the range of processing parameters investigated, from standard melting temperatures and sparge rates ($1050°C$ and 465 l/h global sparge flow rate), through to low temperature and sparge rates ($1016°C$ and 175 l/h global sparge flow rate). As was found in the laboratory scale trials, the maximum amount of MoO_3 that could be incorporated in the VTR co-vitrified product prior to the onset of significant yellow phase formation was 5.25wt%, although the total waste oxide loading in the full scale co-vitrified products was slightly higher than those of the laboratory scale trials. The as cast properties of the co-vitrified product at 5.25wt% MoO_3 were typical of 38wt% incorporation products made at full scale in the absence of POCO simulants, with the soxhlet BLRs falling into the $1-6gm^{-2}d^{-1}$ range. After heat treatment, additional molybdate phases were observed in the samples, in both the POCO and non-POCO containing products. The soxhlet BLRs also increased after heat treatment, with the highest BLRs found in those samples that were produced at the low melter temperature and low sparge conditions.

The normalised B and Mo mass losses from PCT leach testing of the full scale samples are shown in Figure 3. In general, the normalised mass losses of the co-vitrified products at 5.25wt% MoO_3 incorporation were comparable to, or even less than, the baseline condition of 38wt% 50o:50m incorporation with zero POCO content. The mass loss of the full scale product produced at low melter temperatures and low sparge conditions was greater than the same product manufactured at standard operating conditions.

Table VI – Full scale co-vitrification results with MW base glass

	38wt% 50o:50m vitrified product made under optimum processing conditions	38wt% 50o:50m co-vitrified product
50o:50m waste oxide incorporation in product (wt%)	38	33.9
POCO waste oxide incorporation in product (wt%)	0	4.1
MoO_3 incorporation in product (wt%)	3.06	5.25
Phases identified under SEM in as-cast product	RuO_2 $CeZrO_x$ Spinel	RuO_2 $CeZrO_x$ Spinel Alkali-RE-molybdates
Surface area of as-cast product covered in crystalline material (%)	0.7	1.0 – 3.0
As-cast soxhlet bulk Leach rate $(gm^{-2}d^{-1})$	3.5	4.2 – 4.9
Phases identified under SEM in heat treated product	RuO_2 $CeZrO_x$ Spinel Rare-earth molybdates	RuO_2 REO_x Spinel Rare-earth molybdates BaSr-molybdates
Surface area of heat treated product covered in crystalline material (%)	8.5	7.5 – 20.1
Heat treated soxhlet Bulk Leach rate $(gm^{-2}d^{-1})$	4.9	5.2 – 8.0

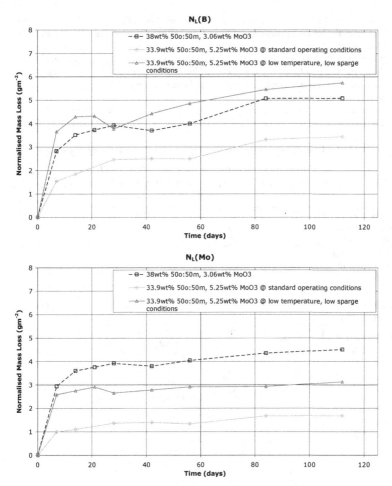

Figure 3 – PCT results from full scale MW co-vitrified products. The majority of errors are <0.2gm^{-2} at 1σ

<u>Laboratory trials with standard Ca/Zn base glass</u>

Table VII indicates the findings to date from the co-vitrification laboratory trials with Ca/Zn base glass. As the incorporation of the 50o:50m blend reprocessing waste was increased, the maximum amount of MoO$_3$ that could be tolerated in the product without the formation of significant phase separation decreased from 10wt% MoO$_3$ with zero reprocessing waste, to 7.5wt% MoO$_3$ with 25wt% reprocessing waste. CaMoO$_4$ and BaMoO$_4$ were the primary crystalline forms of Mo found in the products up to the onset of yellow phase formation, and Figure 4 indicates the relative proportions of these crystalline phases in the XRD data. As the amount of reprocessing waste in the product was increased, the amount of Ca and Ba molybdate decreased significantly, until at 25wt% 50o:50m reprocessing waste loading and 7.5wt% MoO_3, no crystalline molybdate phases could be detected by XRD. The relatively small reduction in MoO_3 content in those products indicates that the reduction in $Ca/BaMoO_4$ was due to the increase in reprocessing waste loading.

The soxhlet leach testing results indicate a trend of decreasing BLR with increasing waste loading, but all results are within the expected range for HLW vitrified products of $1 - 6gm^{-2}d^{-1}$.

Table VII – Laboratory scale co-vitrification results with Ca/Zn base glass

	Co-vitrification products with maximum MoO_3 loading prior to appearance of significant yellow phase				"Standard" 25wt% incorporation Magnox glass
50o:50m waste oxide incorporation in product (wt%)	0	5	15	25	Not analysed
POCO waste oxide incorporation in product (wt%)	16.7	15.2	12.2	8.3	Not analysed
MoO_3 incorporation in product (wt%)	10	9.5	8.5	7	1.62
Phases identified under SEM	$CaMoO_4$ $BaMoO_4$	RuO_2 $CaMoO_4$ $BaMoO_4$	RuO_2 $CaMoO_4$ $BaMoO_4$	RuO_2 Spinel	RuO_2 Spinel
Phases identified under XRD	$CaMoO_4$ $BaMoO_4$	$CaMoO_4$ $BaMoO_4$	RuO_2 $CaMoO_4$ $BaMoO_4$	RuO_2	RuO_2 Spinel
Soxhlet bulk leach rate $(gm^{-2}d^{-1})$	5.5	5.1	3.3	3.3	3.0

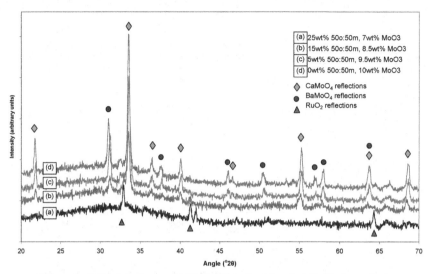

Figure 4 – XRD data for co-vitrified products made using Ca/Zn base glass

The normalised B and Mo mass losses from PCT leach testing of the Ca/Zn glasses is shown in Figure 5. General network dissolution (as indicated by N_LB) was comparable to those products made with 50o:50m waste and MW glass at laboratory scale, and significantly less than for standard 25wt% Magnox glass. Mo release was also comparable to both laboratory scale and full scale products made using 50o:50m waste and MW glass, despite the MoO_3 loading in the Ca/Zn glasses being significantly greater. Again, the Mo release for the Ca/Zn glasses at all MoO_3 loadings was significantly less than for standard 25wt% Magnox glass.

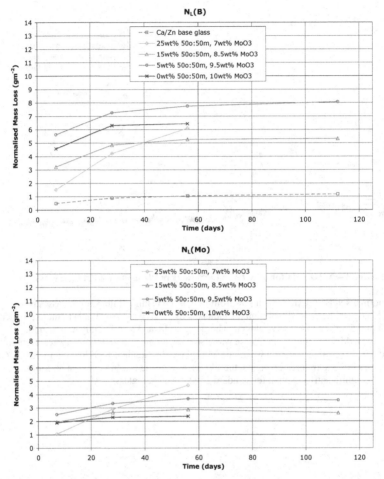

Figure 5 – Static PCT results for co-vitrified products, made using Ca/Zn base glass. No N_L(Mo) data is shown for Ca/Zn base glass as there was no Mo in the formulation, hence levels were <LoD

Viscosity data for the co-vitrified Ca/Zn products is shown in Figure 6, along with data for the Ca/Zn base glass for comparative purposes. As the temperature was decreased, the rate of increase of the viscosities of the co-vitrified Ca/Zn products changed sharply at temperatures between ~1020 and 990°C, when compared to the base glass formulation.

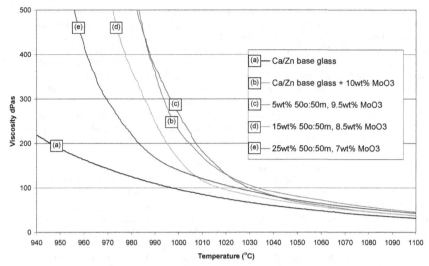

Figure 6 – Viscosity data for co-vitrified products made using Ca/Zn base glass

DISCUSSION

The initial laboratory trials indicated that 5.25wt% was the maximum MoO_3 loading that could be tolerated in a co-vitrified product made with 50o:50m reprocessing waste and MW base glass prior to the onset of significant yellow phase formation. The reprocessing waste loading did not influence the maximum MoO_3 loading, although correspondingly less POCO waste could be incorporated in the product as the reprocessing waste loading was increased. This was because the reprocessing stream supplied a greater fraction of the total MoO_3 content. At the 5.25wt% MoO_3 level, the co-vitrified products retained similar characteristics to standard HLW vitrified products (i.e. those made solely with Magnox or 75o:25m blend reprocessing waste), especially in terms of their chemical durability and viscosity/temperature response. Prior to the full scale trials on VTR it was expected that the superior mixing and homogenisation that has generally been witnessed on the full scale plant compared to laboratory scale trials would yield a higher MoO_3 limit prior to the onset of significant yellow phase formation.

The full scale trials were undertaken with a slightly different philosophy, in that the reprocessing waste oxide loading was not fixed as in the laboratory scale trials, but the reprocessing:POCO ratio was varied in order to attain a 38wt% total waste oxide loading at a given MoO_3 target. Nonetheless, the results were the same in that the maximum MoO_3 incorporation that could be tolerated prior to the onset of significant yellow phase formation was 5.25wt%. As for the laboratory glasses, the properties of the full scale products made at and below this MoO_3 loading displayed comparable properties to the baseline condition of 38wt% incorporation of 50o:50m blend with zero POCO product. The soxhlet and PCT leaching results indicated that the Mo was generally not segregating into water soluble phases such as alkali molybdates. However the BLR and normalised mass loss was slightly higher for the products manufactured at low melting temperatures and sparge rates, possibly as a result of incomplete mixing and homogenisation of the melt. Upon heat treatment, molybdate phases did develop in both the POCO free and POCO dosed products, but the soxhlet BLRs indicated that those new molybdate phases were not highly soluble ones. The highest BLR values for the heat treated products containing 5.25wt% MoO_3 were higher than the usual range for vitrified HLW product (typically $1 - 6$ $gm^{-2}d^{-1}$), although the higher values were all from products made at the lower end of the operating parameters. Thus, the results confirmed that it should be possible to co-vitrify POCO waste with 50o:50m reprocessing waste on WVP at various waste loadings up to 38wt% total waste oxide incorporation without any significant changes to the current base glass formulation, or significant yellow phase formation, as long as the total MoO_3 incorporation level remains $\leq$5.25wt%.

However, the Ca/Zn glass results indicated that higher MoO_3 incorporation rates (and thus POCO concentrations in co-vitrified products) were attainable with the alternative base glass formulation. The increased yellow phase resistance of the Ca/Zn formulation stems from the formation of $CaMoO_4$ and $BaMoO_4$ phases within the product, which "mop up" a significant amount of the Mo. As alkali earth molybdates have a much lower solubility in water than alkali molybdates, the $CaMoO_4$ and $BaMoO_4$ phases can be tolerated in the product if they prevent (or impede) the formation of alkali molybdates. The XRD results showed that the addition of reprocessing waste retarded the formation of crystalline Ba and Ca molybdates in the glasses, although the mechanism for that observation has not yet been determined. The leach testing data for the Ca/Zn co-vitrified products containing $\geq$7wt% MoO_3 confirmed the durability was comparable to the MW co-vitrified products with $\leq$5.25wt% MoO_3. The viscometry data from the Ca/Zn glasses indicated that, at normal processing temperatures ($\sim$1050°C), all of the glasses were fluid enough to allow good mixing and homogenisation. The sharp viscosity increase in the Ca/Zn products at $\sim\leq$1020°C may have been due to the formation of the $CaMoO_4$ secondary phase, as this transition to higher viscosity appeared to happen at higher temperatures when the amount of MoO_3 in the product was higher. This would have to be considered if processing these formulations at full scale. However, as the glasses on the active plant would only cool to those temperatures after pouring into the vitrification canister, it should not affect plant operation. In addition, increasing the reprocessing waste component of the total waste oxide loading decreased the temperature at which the viscosity step change occurred, thus processing flowsheets for the full scale process could be adjusted to take this into account if required. Full scale co-vitrification trials on the VTR with Ca/Zn base glass have been scheduled for early 2013.

CONCLUSIONS

Co-vitrification of POCO with a 50o:50m reprocessing waste stream in MW base glass could be achieved without significant yellow phase formation, as long as the total MoO_3 incorporation rate in the product remained $\leq$5.25wt%. Varying the 50o:50m content of the co-vitrified product between 25 and 33.9wt% did not affect the MoO_3 incorporation at which the onset of yellow phase formation occurred.

Using the Ca/Zn base glass formulation for co-vitrification of POCO and 50o:50m reprocessing waste increased the MoO_3 tolerance of the product, via the formation of a durable $CaMoO_4$ phase. With the Ca/Zn base glass and a 25wt% 50o:50m incorporation rate, the MoO_3 tolerance increased from 5.25 to 7.5wt% of the product, without significant yellow phase formation. Decreasing the 50o:50m loading in the Ca/Zn co-vitrified product from 25wt% to 5wt% increased the MoO_3 tolerance further, allowing 9.5wt% MoO_3 incorporation before yellow phase formation became significant.

REFERENCES

[1] W. Lutze and R. Ewing, *Radioactive Wasteforms for the Future*, (North Holland, Amsterdam), 1988
[2] E. Schiewer, H. Rabe, S. Weisenburger, Mater. Res. Soc. Symp. Proc. Vol. 11, pp. 289-297
[3] I. L. Pegg, H. Gan, K. S. Matlack, Y. Endo, T. Fukui, A. Ohashi, I. Joseph and B. W. Bowan, *Mitigation of Yellow Phase Formation at the Rokkasho HLW Vitrification Facility*, Waste Management Symposium WM-2010 Conference, March 7-11, 2010, Phoenix, AZ
[4] C. Magrabi, W. Smith, M. J. Larkin, *Radioactive Waste Management and the Nuclear Fuel Cycles*, 19(1-3), (Ed J. R. Grover, 1987) pp. 85-106
[5] B. F. Dunnett, N. R. Gribble, R. Short, E. Turner, C. J Steele & A. D. Riley, Glass Technol.: Eur. J. Glass Sci. Technol. A, 53 (4), pp. 166–171
[6] K. Bradshaw, N. Gribble, P. Mayhew, M. Talford, A. Riley, Full scale non-radioactive vitrification development in support of UK highly-active waste vitrification", Waste Management Symposium WM '06 Conference, Feb 26-Mar 2, 2006, Tuscon, AZ
[7] ASTM C1285-02: Standard Test Methods for Determining Chemical Durability of Nuclear, Hazardous, and Mixed Waste Glasses and Multiphase Glass Ceramics: The Product Consistency Test (PCT)
[8] C R Scales, NNL report 10929 Issue 6, *Characterisation of simulated vitrified Magnox Product manufactured on the VTR*, 2011

Mater. Res. Soc. Symp. Proc. Vol. 1518 © 2013 Materials Research Society
DOI: 10.1557/opl.2013.203

The Effect of γ-radiation on Mechanical Properties of Model UK Nuclear Waste Glasses

Owen J. McGann, Amy S. Gandy, Paul A. Bingham, Russell J. Hand, and Neil C. Hyatt

Immobilisation Science Laboratory, Department of Materials Science and Engineering,
The University of Sheffield, Mappin Street, Sheffield S1 3JD, UK.

ABSTRACT

The effect of γ-radiation on the mechanical properties of model UK intermediate and high level nuclear waste glasses was studied up to a dose of 8 MGy. It was determined that γ-irradiation up to this dose had no measurable effect upon the Young's modulus, shear modulus, Poisson's ratio, indentation hardness, or indentation fracture toughness. The absence of measurable radiation induced changes in mechanical properties was attributed to redox mediated healing of electron-hole pairs generated by γ-irradiation by multivalent transition metal ions, in particular the Fe^{3+} - Fe^{2+} couple.

INTRODUCTION

Vitrification of high level radioactive wastes (HLW), arising from the reprocessing of nuclear fuels, in an alkali borosilicate glass matrix has been adopted as the most appropriate waste treatment technology by USA, UK, Russia, France and Germany [1, 2] and has been deployed on an industrial scale since at least 1977 [3]. The vitrification of intermediate level radioactive wastes (ILW) arising from nuclear fuel cycle operations is not widespread, as summarized in a recent IAEA state of the art review [4]. However, vitrification of ILW offers the combined benefit of reduced packaged waste volume (due to the destruction of combustible components, removal of entrained water, and minimisation of void space) and enhanced passive safety (arising from passivation of reactive components and improved compatibility with the host matrix). [4]. These advantages could yield significant cost savings associated with waste storage and emplacement costs in a geological disposal facility, as well as a more robust disposal system safety case, compared to conventional cement encapsulation technology applied to ILW treatment. Here, we apply the UK definition of HLW and ILW, for which the activity exceeds the threshold of 4 GBq α / tonne or 12 GBq β,γ / tonne for Low Level Waste (LLW) designation; and for which heat generation is either a definitive consideration for the design of storage or disposal facilities (in the case of HLW), or may be neglected (in the case of ILW). IAEA guidance is that ILW thermal power should be below approximately 2 kW m^{-3}.

The purpose of these vitrified wasteforms is to function as the primary barrier to the release of radionuclides into the environment, within a multi-barrier geological disposal facility. The decay of fission products and actinides in radioactive waste glasses, over the service lifetime of 10^2-10^6 years results in the emission of ionising radiation which may adversely impact the long term physical and chemical stability of the wasteform [5]. In alkali-borosilicate glasses, exposure to ionizing radiation has been shown to induce a small (but measurable) increase in density, with a concomitant effect on mechanical properties [6]. Furthermore, exposure to ionising radiation has been shown to lead to the formation of oxygen bubbles; various defect centres [6]; and, at very high radiation doses, even phase separation [7]. These radiation induced changes in physical and chemical properties are attributed to radiation induced changes in the structure of the glass network.

Ionising radiation can interact with the vitrified wasteform materials through several mechanisms. For example, β-decay of fission products leads to the emission of γ-ray photons which interact with the glass matrix primarily through Compton Scattering and the photoelectric effect [5]. Absorption of γ-ray photons results in the formation of mobile electron and hole pairs which can lead to metastable defects associated with a change in element valence state and coordination. In borosilicate glass compositions, similar to those used for radioactive waste vitrification, it was observed that γ-radiation induced the formation of boron-oxygen hole centre (BOHC) defect sites, which were detectable by electron paramagnetic resonance (EPR) spectroscopy [8]. Typically, in borosilicate glasses, a range of effects have been reported: the formation of colour centres occurs at low doses of γ-radiation (0.018 MGy) [9]; and at larger doses (>1 MGy) increases in silicate network polymerisation, and increases in glass density connected with an increase in mechanical strength, occurs [10]. γ-radiation might therefore be expected to cause a range of detectable effects on borosilicate glasses due to the induced defects and changes in bonding of the borate and silicate networks caused by electron and hole pairs.

This paper examines the effect of γ-irradiation on the physical properties of a variety of model vitreous wasteforms applied to, or conceived for, immobilisation of UK intermediate and high level radioactive wastes.

EXPERIMENT

A selection of glass samples were prepared, which were inactive model glasses of materials applied to the vitrification of UK HLW (glass MW-25%) or under consideration for application to the vitrification UK ILW (glasses G11, G73 and G78). All of the glasses were loaded with appropriate inactive simulated wastes. Glasses were melted in re-crystallised alumina crucibles using an electrical muffle furnace. Melting temperatures ranged from 1060 °C to 1200 °C depending on glass composition. Glasses were melted for 2 hours before pouring into a steel mould and annealed in a muffle furnace, set close to the glass transition temperature (T_g) for the glass composition. Glass compositions, as well as melting and annealing temperatures, are given in Table I.

Annealed glass prisms, measuring approximately 10mm x 10mm x (100 – 200) mm, were subsequently exposed to a γ-radiation dose of 4 or 8 MGy, a control specimen (0 MGy) was retained for each composition. The γ-radiation source was [60]Co (2.824 MeV) and irradiation was carried out at the UKAEA Harwell Site. The highest does of 8 MGy γ-radiation represented the anticipated accumulated dose acquired over 8.7 years for the wasteforms G11, G73 and G78 and approximately 0.13 years for the wasteform MW-25%. Post irradiation, the samples were stored in a dark and refrigerated environment (approximately −15 °C) to reduce the potential for the annealing of radiation induced defects. A resonant frequency technique was applied to determine the effect of γ-radiation on Young's modulus, shear modulus and Poisson's ratio, as described elsewhere [11]. Due to sample scarcity, non-destructive techniques were applied to examine potential effects of γ-irradiation on physical properties of the wasteforms. Bending and mechanical strength were, therefore, not measured, and examination of the Young's modulus was considered sufficient in determining potential changes due to γ-radiation. Vickers indentation, utilising a diamond tipped indenter, was applied to determine the effect of γ-radiation on glass hardness and fracture toughness.

Table I. Nominal Glass compositions (mol%) and glass melting temperatures

Component	G11	G73	G78	MW-25%
SiO$_2$	58.2	50.2	58.8	48.4
B$_2$O$_3$	9.1	2.2	17.8	17.2
Al$_2$O$_3$	2.5	0.4	-	5.1
Fe$_2$O$_3$	5.2	2.9	3.9	3.4
CaO	-	10.7	-	-
BaO	-	17.6	-	-
Li$_2$O	10.6	7.6	10.4	4.2
Na$_2$O	13.3	3.0	8.0	8.7
MgO	-	-	-	8.4
Other waste components	-	-	-	4.6
Melting Temperature	1200 °C	1200 °C	1200 °C	1060 °C
Annealing Temperature	450 °C	480 °C	480 °C	500 °C

RESULTS AND DISCUSSION

The measured Young's and shear moduli of the sample prisms as determined by the resonant frequency method, is compared with that of the unirradiated control samples, in Figures 1a and 1b, respectively. Both the measured Young's modulus and shear modulus of the control sample are in agreement with the range of values (78 – 81 GPa and 31 – 32 GPa, respectively) reported for similar glass compositions in previous research [12, 13]. The measured moduli of the γ-irradiated specimens when compared with the un-irradiated control specimens show no significant change, within the precision of the measurements. The error bars shown in Figure 1 represent the propagated errors associated with measurement of the dimensions, density and resonant frequency of the glass prisms, based on at least three independent measurements. They were derived from the accumulated systematic errors of the individual measurements involved in making the measurement using the resonant frequency technique.

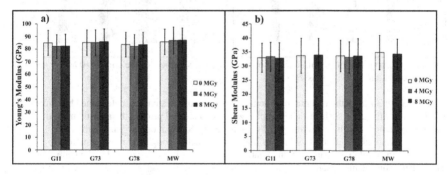

Figure 1. Comparison of a) Young's modulus and b) shear modulus of control (0 MGy) and γ-irradiated (4 and 8 MGy) glasses determined by the resonant frequency technique.

Figure 2a compares Poisson's ratio, v, for the γ-irradiated and control specimens, determined from the Young's and shear moduli using Equation 1:

$$v = E/2G - 1 \qquad\qquad \text{(Equation 1)}$$

where E is the Young's modulus and G is the shear modulus. For the unirradaited control samples, the Posisson ratio of $v \approx 0.25$, is similar to that reported for related glass compositions in previous research [12, 13]. Given that there was no measured impact of γ-radiation on the Young's modulus and shear modulus, the Poisson's ratio of the irradiated and control samples were identical within precision, as shown in Figure 2a. The error bars quoted in Figure 2a represent the propagated errors associated with measurement of the Young's modulus and the shear modulus of the prism. Data for MW-25% and G73 4 MGy are not shown in Figures 1b and 2a, respectively, due to the absence of satisfactory shear modulus data.

Figure 2b compares the hardness values for the control and irradiated glass samples, as determined from micro-indentation of the glass surfaces. The measured hardness values determined for the control samples were in good agreement with the range of values, 5.9 – 7.1 GPa, reported in previous research for similar glass compositions [13]. The hardness values determined for the γ-irradiated glass samples showed no significant variation from the values obtained for the control samples within the estimated precision of the measurements. These results are consistent previous γ-irradiation studies of alkali-borosilicate glasses, albeit at lower doses (1 MGy) [5, 10]. The errors bars reported in Figure 2b were derived from the standard deviation of the 30 – 50 results obtained for each sample.

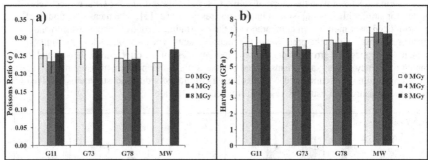

Figure 2. Comparison of a) Poisson's ratio and b) hardness values for the control (0 MGy) and γ-irradiated (4 and 8 MGy) glasses.

Figure 3 shows the indentation fracture toughness of the samples, determined by micro-indentation of the glass surface. The samples were not annealed prior to the measurement of fracture toughness, as this was anticipated to heal any defects induced by γ-irradiation. Therefore, although the absolute values of indentation fracture toughness may be subject to systematic error, relative changes in indentation fracture toughness are considered meaningful. The indentation fracture toughness values obtained for the control glass samples were in good agreement with the range of values, $0.81 - 1.06$ N m$^{-3/2}$, reported in previous research [13]. Within the estimated precision of the measurements, the γ-irradiated glass samples presented no significant variation in fracture toughness when compared with the results determined for the control samples. These results are again broadly consistent with previous studies of the effect of γ-radiation on alkali-borosilicate glasses, albeit at lower doses (1 MGy) [5, 10]. The errors bars reported in Figure 3 were derived from the standard deviation of the $30 - 50$ results obtained for each sample.

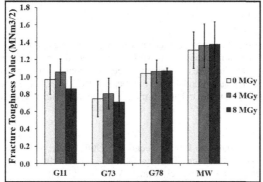

Figure 3. Comparison of indentation fracture toughness values in control (0 MGy) and γ-irradiated (4 and 8 MGy) glasses.

Comparison of the mechanical properties of unirradiated control glass samples against samples subject to γ-irradiation to a dose of 4 MGy or 8 MGy, showed no significant change in Young's modulus, shear modulus, hardness or fracture toughness. This behaviour is not generally observed in simplified alkali-borosilicate compositions with low Fe contents, where γ-irradiation typically results in increased density with concomitant change in the mechanical properties [14]. According to Shelby et al., such an increase in density arises from an increase in the degree of network polymerisation induced by γ-irradiation, although the mechanism has not been fully elucidated [14]. The results presented here support past observations of Fe rich alkali-borosilicate exposed to γ-irradiation [8, 10, 15] and parallel observations made in β-radiation experiments [16]. It has been suggested that the presence of redox active transition metal and actinide species in silicate glasses, such as Fe, Cr, and U, may prevent the accumulation of radiation induced defects, and hence deleterious impact on the chemical and physical properties [8]. The mechanism of this effect, proposed by Debnath et al. [17], is linked to the ability of these elements to effectively 'trap' the excitons (electron and hole pairs) produced by the irradiation via changes in oxidation states, such as the Fe^{2+} and Fe^{3+} [18].
We hypothesise that the substantial Fe content of the glasses studied, which is present for either waste loading or for waste compatibility reasons, may assist in healing g-radiation induced electron – hole pair defects, in accordance with the mechanism described above, leading to negligible impact on mechanical properties below 8 MGy.

CONCLUSIONS

This work has shown that γ-radiation produces no significant change in mechanical properties of the wasteform materials studied, up to 8 MGy, within the precision of the techniques applied. This has been attributed to the presence of a significant proportion of iron within the glass compositions, which has been identified providing a mechanism to heal radiation induced defects in glasses of similar compositions [12, 15].

ACKNOWLEDGMENTS

N.C. Hyatt is grateful to The Royal Academy of Engineering and the Nuclear Decommissioning Authority for funding. This work was supported in part by an EPSRC Case award co-funded by Magnox Ltd. We also acknowledge part support from the Engineering and Physical Sciences Research Council under grant number EP/I012214/1. The authors gratefully acknowledge useful discussion with Dr Tony Burnett in connection with this research.

REFERENCES

1. N. D. Hutson, C. A. Herman, and J. R. Zamecnik, US DOE Report No.WSRC-MS-2000-00884, http://sti.srs.gov/fulltext/ms2000884/ms2000884.html, (2000).
2. J. Sheng, Glass Technol Eur J Glass Sci Technol A. 45, 153 (2004).
3. V. A. Morozov, IAEA Bull. 21, 17 (1979).
4. "Application of Thermal Technologies for Processing of Radioactive Waste, IAEA-TECDOC-1527, December (2006).
5. W. J. Weber, Nucl Inst Meth Phys Res B. 32, 471 (1998).
6. R. C. Ewing, W. J. Weber, F. Jr. Clinard, Prog Nucl Energy. 29, 63 (1995).
7. K. Sun, L. M. Wang, R. C. Ewing, W. J. Weber, Nucl Inst Meth Phys Res B. 218, 368 (2004).
8. G. J. Brown, J Mater Sci. 10, 1841 (1975).
9. E. A. Vanina, M. A. Chibisova, S. M. Sokolova, Glass Ceram. 63, 11 (2006).
10. G. J. Brown, J Mater Sci. 10, 1481 (1975).
11. W. R. Davis, Trans Brit Ceram Soc. 67, 515 (1968).
12. H. Matzke, E. Toscano, J Am Ceram Soc. 69, C138 (1986).
13. A. J. Connelly, R. Hand, P. A. Bingham, N. C. Hyatt, J Nucl Mater. 408, 188 (2011).
14. J. Shelby, J Appl Phys. 51, 2561 (1980).
15. D. L. Griscom, C. I. Merzbacher, R. A. Weeks, R. A. Zuhr, J Non-Cryst Solids. 258, 34 (1999).
16. F. Y. Olivier, B. Boizot, D. Ghaleb, G. Petite, J Non-Cryst Solids. 351, 1061 (2005).
17. R. Debnath, J Mater Res. 6, 127 (2001).
18. E. Malchukova, B. Boizot, G. Petite, D. Ghaleb, Eur Phys J Appl Phys. 45, 10701 (2009).

Mater. Res. Soc. Symp. Proc. Vol. 1518 © 2013 Materials Research Society
DOI: 10.1557/opl.2013.95

Structural Characterization and Analysis of Glasses in the Al_2O_3-B_2O_3-Fe_2O_3-Na_2O-SiO_2 System

S.V. Stefanovsky,[1,2] B.S. Nikonov,[3] B.I. Omelyanenko,[3] K.M. Fox,[4] J.C. Marra[4]

[1] SIA Radon, 7th Rostovskii lane 2/14, Moscow 119121 Russia, profstef@mtu-net.ru
[2] Institute of Physical Chemistry and Electrochemistry RAS, Leninskii av. 31, Moscow 119071 Russia
[3] Institute of Geology of Ore Deposits RAS, Staromonetniy 35, Moscow 119117 Russia
[4] Savannah River National Laboratory, Building 773-A, Aiken 29808 U.S.A.

ABSTRACT

Glasses in the Al_2O_3-B_2O_3-Fe_2O_3-Na_2O-SiO_2 system were produced at a temperature of 1150 °C, annealed, and examined using XRD and SEM/EDX. Surfaces of same samples were additionally heat-treated and etched with HCl. The pristine samples were X-ray amorphous and rather homogeneous except the B1 sample that contained trace crystalline phases of carnegieite/nepheline and spinel. Corrosion of these glasses via an etching treatment proceeds by a conventional mechanism with damage of their sur-face layers, however, the B2 glass exhibits a "drop-type" microstructure after etching that suggests oc-currence of liquid-liquid phase separation.

INTRODUCTION

Borosilicate glasses are used as forms for various types of radioactive wastes due to their relatively low melting temperature (1050-1200 °C), capability to accommodate ions with wide range of charges and radii, high chemical durability, mechanical integrity, and good radiation resistance [1]. The Defense Waste Processing Facility (DWPF) [2] vitrifies high level waste (HLW) at the Savannah River Site in Aiken, SC. To facilitate operation of the vitrification facility, development work on HLW glass formula-tions is performed. The wide variety of elements comprising the HLW dictates that the glass composi-tions are very complex [3], and therefore structural characterization becomes difficult. A study of simpli-fied glass compositions would be useful in providing insight into the performance of the more complex glasses while making characterization data easier to interpret. For example, the formation of nepheline crystals in complex glass compositions, which reduces the chemical durability of the glass waste form [4], may be easier to understand by characterizing a simplified composition.

The structure of Al_2O_3-B_2O_3-Na_2O-SiO_2 glasses has been reported in the literature using nuclear magnetic resonance (NMR) and optical characterization techniques [5,6]. The published work has been successful in identifying compositionally driven coordination changes in aluminum and boron, changes in the fraction of non-bridging oxygens, and the tendency for sodium to preferentially associate with alumi-num. However, the addition of iron (a major component of HLW glass) makes NMR studies difficult. The objective of this task was to characterize the chemistry and structure of Al_2O_3-B_2O_3-Na_2O-SiO_2 glasses with the addition of Fe_2O_3 using different spectroscopic techniques (infrared – IR, Raman, X-ray absorp-tion fine structure – XAFS) as well as X-ray diffraction (XRD) and scanning electron microscopy (SEM). The present paper describes XRD and SEM characterization of the as-prepared glasses and the effect of etching on the microstructure of the glasses. Comparisons were made with the literature data, with any additional influences of Fe_2O_3 being identified and described.

EXPERIMENTAL

Samples of glasses were prepared from the proper proportions of reagent-grade Al_2O_3, Fe_2O_3, SiO_2, Na_2CO_3 and H_3BO_3 in 150 g batches. The batches were placed in Pt/Rh crucibles, heated to 1150 °C, kept at this temperature for 1 hr, and heat-treated according to Canister Centerline Cooling (CCC) regime to simulate conditions during pouring into a DWPF canister and subsequent cooling [7]. The target and actual chemical compositions of the glasses are given in Table I. The polished surfaces of the samples A1, A2, B1, B2, D1 and D2 were etched with 0.1M HCl for 1 day and studied by SEM/EDS.

Table I. Chemical Compositions of Borosilicate Glasses.

Oxides	A1 (SB6) target mol %	A1 (SB6) target wt. %	A1 (SB6) actual wt. %	A2 (SB6 w/o Fe) target mol %	A2 (SB6 w/o Fe) target wt. %	A2 (SB6 w/o Fe) actual wt. %	B1 (SB19) target mol %	B1 (SB19) target wt. %	B1 (SB19) actual wt. %	B2 (SB19 w/o Fe) target mol %	B2 (SB19 w/o Fe) target wt. %	B2 (SB19 w/o Fe) actual wt. %	D1 (high B$_2$O$_3$ frit) target mol %	D1 (high B$_2$O$_3$ frit) target wt. %	D1 (high B$_2$O$_3$ frit) actual wt. %	D2 (high B$_2$O$_3$ frit w/o Fe) target mol %	D2 (high B$_2$O$_3$ frit w/o Fe) target wt. %	D2 (high B$_2$O$_3$ frit w/o Fe) actual wt. %
Al$_2$O$_3$	8.5	12.5	12.5	9.0	14.1	14.3	15.0	21.1	21.0	16.0	24.1	24.1	15.0	20.9	20.6	16.0	23.9	23.5
B$_2$O$_3$	5.0	5.0	4.9	5.5	5.9	5.6	4.2	4.0	3.8	4.5	4.6	4.4	8.5	8.1	7.7	9.0	9.2	9.0
Fe$_2$O$_3$	5.0	11.5	10.4	-	-	0.1	5.3	11.7	11.1	-	-	0.1	5.5	12.0	12.6	-	-	0.1
Na$_2$O	25.0	22.3	21.5	26.5	25.3	23.8	29.5	25.2	24.1	31.0	28.3	26.8	25.0	21.2	20.3	26.5	24.1	23.0
SiO$_2$	56.5	48.8	48.3	59.0	54.6	54.3	46.0	38.1	38.0	48.5	43.0	43.3	46.0	37.8	37.4	48.5	42.8	42.4
Total	100	100	97.6	100	100	98.1	100	100	99.0	100	100	98.7	100	100	98.6	100	100	98.0
ψ_B	3.3			3.2			3.5			3.3			1.2			1.2		
ψ_B (Fe)	3.0						3.1						1.0					
K	11.3			10.7			11.0			10.8			5.5			5.4		

Actual chemical compositions of the glasses were determined at SRNL using a Perkin-Elmer 403 ICP-OES spectrometer on chemically digested glasses. Samples were examined by X-ray diffraction (XRD) using a Rigaku D / Max 2200 diffractometer (Cu Kα radiation, 40 keV voltage, 20 mA current, stepwise 0.02 degrees 2θ), scanning electron microscopy with energy dispersive spectrometry using a JSM-5610LV+JED-2300 analytical unit, and optical microscopy (OM) using an Olympus unit.

THEORETICAL REMARKS

There are several structural parameters characterizing some features of the structure of glasses. The most important among them are degree of connectedness of the silica-oxygen network (f_{Si}) for silicate-based and relative fraction of four-coordinated boron (ψ_B) for borate and borosilicate glasses [8]:

$$\psi_B=\{(Na_2O+K_2O+BaO)+[0.7(CaO+SrO+CdO+PbO)]+[0.3(Li_2O+MgO+ZnO)]-Al_2O_3\}/B_2O_3 \qquad (1)$$

and K = SiO$_2$/B$_2$O$_3$, where oxide concentrations are given in mol.%.

Formally, boron is three-coordinated if $0 \leq \psi_B \leq 1/3$ and forms boron-oxygen triangles. At $1/3 < \psi_B < 1$ both three- and four-coordinated boron atoms co-exist and are present in complex borate groups. At $\psi_B \geq 1$ all the boron is four-coordinated and forms boron-oxygen tetrahedra associated with alkali and, to a less extent, alkali earth cations as, for example, Me$^+$[BO$_{4/2}$] units. Actually, a significant excess of alkali or/and alkali earth oxides as oxygen donors is required to form BO$_4$ tetrahedra and convert all the boron into four-coordinated state.. This process depends also on the silica content in the glass or factor K = [SiO$_2$]/[B$_2$O$_3$] (Table I). At relatively high silica content (60-80 wt.%), for fully four-coordinated boron to exist, the ψ_B value is estimated to be 1.5-2. At lower silica contents (44-60 wt.%), this value should be much higher. We have demonstrated using IR and electron paramagnetic resonance (EPR) spectroscopy that in borosilicate glasses for high-sodium intermediate level waste immobilization containing 40-50 wt.% SiO$_2$, a minor fraction of trigonally coordinated boron is present even at $\psi_B \cong 4$ to 5 [9].

Because Al$_2$O$_3$ is a stronger acceptor of oxygen than B$_2$O$_3$, oxygen introduced with alkali and alkali earth oxides initially results in transformation of Al into a four-coordinated state. Generalized parameters for coordination transformations in aluminoborosilicate glasses are shown in Table II [8].

Table II. Coordination of boron and aluminum in glasses at various ψ_B values [8].

φ_B	Coordination			
$\psi_B>1$	[AlO$_4$]	[BO$_4$]	-	-
$1/3<\psi_B<1$	[AlO$_4$]	[BO$_4$]	[BO$_3$]	-
$0<\psi_B<1/3$	[AlO$_4$]	-	[BO$_3$]	-
$\psi_B<0$	[AlO$_4$]	-	[BO$_3$]	[AlO$_6$]

As seen from Table II, aluminum concentration negatively influences the transformation of trigonally coordinated boron into tetragonally coordinated one. Stability of a coordination state depends on a value of cationic to anionic radii ratio (r_c/r_a). For tetrahedral oxygen coordination in solids this ratio ranges between 0.22 and 0.41 [10] (radius of O^{2-} anion is assumed to be 1.36 Å [11]). Deviation from the average value of the r_c/r_O^{2-} ratio expressed as $|\Delta| = [r_c/r_O^{2-}]_{av} - [r_c/r_O^{2-}]_{calc}$ may be considered as a measure of stability of tetrahedral coordination for the given cation. The values calculated from two different reference data are given in Table III. Due to higher stability of [AlO$_4$] tetrahedra as compared to [BO$_4$] tetrahedra, oxygen provided by alkali oxides is initially applied to formation of AlO$_4$ tetrahedra and at relatively high Al$_2$O$_3$ concentrations the majority of boron remains three-coordinated. Nevertheless, this does not result in a negative effect on chemical durability of glasses, because aluminosilicate glass networks built from SiO$_4$ and AlO$_4$ tetrahedra with associated alkali ions are generally leach resistant [12].

Table III. Cationic radii, cationic to oxygen anion radii ratios and their deviations from average value (0.315 Å) for tetrahedrally coordinated ions.

Cation	B^{3+}		Si^{4+}		Al^{3+}		Fe^{3+}	
Refs	[10]	[11]	[10]	[11]	[10]	[11]	[10]	[11]
r_c, Å	0.20	0.11	0.39	0.26	0.57	0.39	0.67	0.49
r_c/r_O^{2-}	0.15	0.08	0.29	0.19	0.42	0.29	0.49	0.36
$|\Delta|$	0.165	0.235	0.025	0.125	0.105	0.025	0.175	0.045

The effect of iron oxides on boron coordination in borosilicate glasses is more complicated. Fe^{3+} ions being network-formers act similarly to Al^{3+} ions forming FeO$_4$ tetrahedra and suppressing B$^{III} \rightarrow$ B^{IV} transformation. However, since stability of [Fe^{3+}O$_4$] tetrahedron is lower then that of [AlO$_4$], the effect of Fe$_2$O$_3$ on B$^{III} \rightarrow$ B^{IV} transformation is weaker. Molar concentration of Fe$_2$O$_3$ should be subtracted like Al$_2$O$_3$ in eq. (1) but with a coefficient less than 1. Taking into account that the energy of Fe^{3+}—O bond is lower than that of Al—O bond by ~3 times, this coefficient may be suggested to be ~3 as well. Therefore, in the numerator of the eq. (1) we have to subtract additionally ~0.3×[Fe$_2$O$_3$]. Thus, in the presence of Fe$_2$O$_3$, the ψ_B values for the glasses studied are somewhat lower (Table I). At high iron concentrations, Fe^{3+} ions may become network-modifiers with higher coordination number (CN=6) or form separate crystalline phases such as hematite (Fe$_2$O$_3$) or spinels (especially in the presence of Fe^{2+} ions or different transition metal ions (Mn^{2+}, Ni^{2+}, Co^{2+}, Cu^{2+}, Zn^{2+}) as well as Mg^{2+} and Al^{3+} ions). Fe^{2+} ions, if present, are network-modifiers, but often form spinel type phases.

RESULTS AND DISCUSSION

Pristine glasses

Target and actual chemical compositions of glasses were similar (Table I). Trace of Fe$_2$O$_3$ was present in Fe free glasses due to impurity in the batch chemicals. All the glasses were found to be X-ray amorphous except glass B1 that contained a minor amount of carnegieite (Fig. 1 and 2), which is a high temperature variety of nepheline. Features of the chemical composition of this phase will be discussed later. Glasses A2, B2, D1, and D2 had uniform texture, homogeneous distribution of major elements over the bulk of the glass and did not contain any inclusions.

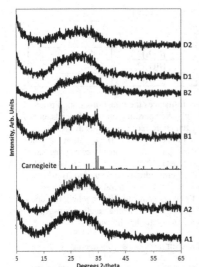

Figure 1. XRD patterns of glass samples.

OM showed that thin section of the glass A1 is predominantly composed of transparent glass but minor brown-colored glass also occurs. Brown color is due to saturation of these areas with the finest (submicron-sized) spinel crystals. Local aggregations of fine gas bubbles are observed (Fig. 2, *1*).

Areas of partly devitrified glass B1 have symmetric zoned structure (Fig. 2, *2-4*). The rim is uniform and composed of a carnegieite structure crystalline phase. OM data in transparent cross-section at single Nichol shows that unlike the glass having light-brown color, carnegieite is colorless and has double reflection in the crossed Nichols. It has also contraction cracks being characteristic of glass devitrification. This zone with a thickness of 50-70 μm is darker than the glass in backscattered electron mode (Fig. 2, *2*). The chemical composition of carnegieite is similar to that of the glass but does not contain boron (Table IV).

The next zone (Fig. 2, *2-4*) has a thickness of about 25 μm and clear borders and brown color at single Nichol. It is isotropic in crossed Nichols and has light-gray color and uniform texture in backscattered electrons (Fig. 2, *2*). It is different in chemical composition from carnegieite (Table IV). At similar sodia and silica contents it is depleted with alumina and enriched with ferrous oxides. This phase is also satisfactory recalculated to a formula with four oxygen ions and may be referred to as ferrous carnegieite.

The core is dark-brown and nearly opaque in transparent cross-sections (Fig. 2, *4*). This zone is composed of two phases. Phase-1 having light-gray color on SEM images is an aggregate of micro-sized

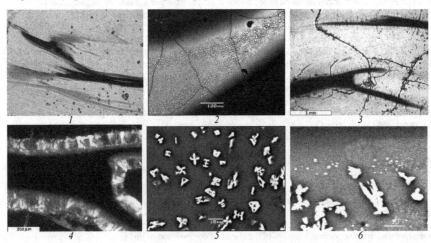

Figure 2. OM (*1,3,4*) at single (1), parallel (*3*) and crossed Nichols (*4*) and SEM (*2,5,6*) images of A1 (*1*) and B1 (*2-6*) glasses.

Table IV. Chemical compositions and formulae of the clear and brown areas and spinel in Sample B1.

Oxides	Concentration, wt.%				Ions	Formula units		
	Glass	Clear	Brown	Spinel		Clear	Brown	Spinel
Na_2O	23.5	21.9	19.5	7.0	Na^+	1.1	1.1	-
Al_2O_3	23.3	23.8	18.0	11.2	Al^{3+}	0.8	0.6	0.2
SiO_2	37.4	37.1	30.7	11.2	Si^{4+}	1.0	0.9	-
					Fe^{2+}			1.0
Fe_2O_3	10.8	11.0	23.0	73.2	Fe^{3+}	0.2	0.5	1.8
B_2O_3	ND	ND	ND	-	Total	3.1	3.1	3.0
Sum	95.0	93.8	91.2	102.6	O^{2-}	4.0	4.0	4.0

crystals of carnegieite (Fig. 2, 4). Chemical composition (Table IV) and structure of this area are similar to the previous ones. Brown-colored areas of the core are represented by both individual and dendrite crystals of spinel up to 10 μm in size (Fig. 2, 5,6) distributed in the glass and the chemical composition given in Table IV corresponds to spinel with captured glass. The spinel phase is not clearly observed in the XRD patterns due to the low concentration in the glass.

Effect of Etching on the Structure of the Surface of Glasses

The surface of the glasses after etching for all but the B2 glass is indicative of general corrosion (Fig. 3). Chemical compositions of the corroded layer and unaltered bulk were determined in the most strongly corroded areas (Table V).

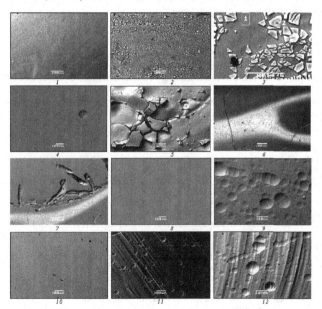

Figure 3. BSE SEM Images of the surface of the glasses A1 (1-3), A2 (4,5), B1 (6,7), B2 (8,9), D1 (10,11), and D2 (12) before (1,4,8,10) and after (2,3,5,7,9,11,12) etching in 0.1 M HCl.

Table V. Chemical composition (wt.%) by EDX of various areas on the etched surfaces of the glasses.

Oxides	Glass A1		Glass A2		Glass D1	
	Core	Rim	Core	Rim	Actual	Average
Na$_2$O	19.1	0.5	24.4	4.7	20.3	19.5
Al$_2$O$_3$	14.1	2.3	14.5	17.2	20.6	22.9
SiO$_2$	50.7	73.6	55.2	76.6	37.4	40.8
Fe$_2$O$_3$	8.9	22.5	0.1	1.2	12.6	14.4
B$_2$O$_3$	ND	ND	ND	ND	7.8	ND
Total	92.8	98.9	94.2	99.7	98.7	77.6

ND – not determined by EDX

As follows from theoretical consideration, glasses A1, A2, B1 and B2 should not have a tendency to liquid-liquid phase separation because the ψ_B values for their composition significantly exceed 1 at high K-factor. Corrosion of these glasses proceeds by conventional mechanism with a damage of their surface layers (Fig. 3) and formation of a rim enriched with silica and depleted with alkali ions. The strongest corrosion takes place along the fractures on the glass surface (Fig. 3, *3,5,7*).

Nevertheless, a "drop-type" microstructure was observed after etching of the surface of glass B2 (Fig. 3, *9*) that suggests occurrence of liquid-liquid phase separation: the structure of this glass consists of areas with higher and lower durability phases that react differently to acid attack and, therefore, differ in chemical composition. The diameter of the drops is widely varied and reaches ~100 μm.

The glasses D1 and D2 have both the lowest ψ_B and K factors and, therefore, may be susceptible to phase separation. The etched surfaces of the heat-treated glasses demonstrate the drop-type microstructure due to selective leaching of the lower durability phase (Fig. 3, *11,12*).

CONCLUSION

Glassy materials A1, A2, B2, D1, D2 are X-ray amorphous (glasses); the material B1 is also amorphous but contains trace concentrations of carnegieite/nepheline and spinel phases. No liquid-liquid phase separation in glasses A1, A2, B1, B2 was found. Etching of the surface of these glasses with HCl resulted in typical surface damage under acid attack with formation of a high-silica layer. The glasses D1 and D2 have some tendency for phase separation. The etched surfaces of the heat-treated glasses demonstrate the drop-type microstructure due to selective leaching of the lower durability phase.

REFERENCES

1. W. Lutze, "Silicate Glasses," *Radioactive Waste Forms for the Future*, ed. W. Lutze and R.C. Ewing (Elsevier, 1988) pp. 1-159.
2. S.L. Marra, R.J. O'Driscoll, T.L. Fellinger, J.W. Ray, P.M. Patel, and J.E. Occhipinti, in: *Waste Management '99*. Proc. Int. Conf. (Tucson, AZ, 1999) ID 48-5. CD-ROM.
3. K.M. Fox, D.K. Peeler, Mater. Res. Soc. Symp. Proc. **1193** (2009) 229-237.
4. H. Li, P. Hrma, J.D. Vienna, M. Qian, Y. Su, D.E. Smith, *J. Non-Cryst. Solids*, **331** (2003) 202-216.
5. J.G. Darab, J.C. Linehan, and B.P. McGrail, *Mat. Res. Soc. Symp. Proc.* 556 (1999) 337-344.
6. D. Holland, B.G. Parkinson, M.M. Islam, A. Duddridge, J.M. Roderick, A.P. Howes, and C.R. Scales, *Mater. Res. Soc. Symp. Proc.* **1107** (2008) 199-206.
7. S.L. Marra and C.M. Jantzen, Characterization of Projected DWPF Glasses Heat Treated to Simulate Canister Centerline Cooling (U). WSRC-TR-92-142, 1993.
8. A.A. Appen, *Chemistry of Glass* (Russ., Khimiya, Leningrad, 1974).
9. S.V. Stefanovskii, O.A. Knyazev, T.N. Lashchenova, and S. Merlin, *J. Adv. Mater.* 3 [6] (1996) 479-487.
10. G.B. Bokiy, *Crystal Chemistry*, (Russ., Moscow State University Publ., 1960).
11. R.D. Shannon, *Acta Cryst.* **A32** (1976) 751-767.
12. A.P. Kobelev, S.V. Stefanovsky, V.V. Lebedev, M.A. Polkanov, O.A. Knyazev, J.C. Marra, *Glass Tech.: Eur. J. Glass Sci. Technol.* A. **50** [1] (2009) 47-52.

Mater. Res. Soc. Symp. Proc. Vol. 1518 © 2013 Materials Research Society
DOI: 10.1557/opl.2013.143
Infrared and Raman Spectroscopic Study of Glasses in the Al$_2$O$_3$-B$_2$O$_3$-Fe$_2$O$_3$-Na$_2$O-SiO$_2$ System

S.V. Stefanovsky,[1,2] K.M. Fox,[3] J.C. Marra[3]

[1] SIA Radon, 7th Rostovskii lane 2/14, Moscow 119121 Russia, profstef@mu-net.ru
[2] Institute of Physical Chemistry and Electrochemistry RAS, Leninskii av. 31, Moscow 119071 Russia
[3] Savannah River National Laboratory, Building 773-A, Aiken 29808 U.S.A.

ABSTRACT

Glasses in the Al$_2$O$_3$-B$_2$O$_3$-Fe$_2$O$_3$-Na$_2$O-SiO$_2$ system were produced at a temperature of 1150 °C, poured onto a metal plate and annealed. The nature of the structural units and their bonding in the structure were studied by infrared and Raman spectroscopic techniques. The structural network of all the glasses studied is built from major [SiO$_4$] tetrahedra with 2-3 non-bridging oxygens (NBO). Incorporation of Fe$_2$O$_3$ offers a destructive effect in the glass network.

INTRODUCTION

Major glass properties, among them key characteristics of nuclear waste glasses such as chemical durability and radiation resistance, depend on the structure of glass, nature of chemical bonding and local structure around waste elements. The structure of Al$_2$O$_3$-B$_2$O$_3$-Na$_2$O-SiO$_2$ glasses has been reported in the literature using nuclear magnetic resonance (NMR) and optical characterization techniques (see, for example [1,2]). The published work has been successful in identifying compositionally driven coordination changes in aluminum and boron, changes in the fraction of non-bridging oxygens (NBO) and the tendency for sodium to preferentially associate with aluminum. However, the addition of iron (a major component of HLW glass) makes NMR studies difficult. The objective of this task was to characterize the chemistry and structure of Al$_2$O$_3$-B$_2$O$_3$-Na$_2$O-SiO$_2$ glasses with the addition of Fe$_2$O$_3$ using infrared (IR) and Raman spectroscopic techniques.

EXPERIMENTAL

Samples of glasses were prepared from the proper proportions of reagent-grade Al$_2$O$_3$, Fe$_2$O$_3$, SiO$_2$, Na$_2$CO$_3$ and H$_3$BO$_3$ in 150 g batches. The batches were placed in Pt/Rh crucibles, heated to 1150 °C, kept at this temperature for 1 hr, and heat-treated according to Canister Centerline Cooling (CCC) regime to simulate conditions in a canister produced in full-scale operations in the Defense Waste Processing Facility (DWPF) [3]. The target and measured chemical compositions are given in Table I. Samples that were heat-treated at 500 °C for 10 hrs were also studied.

Table I. Chemical Compositions of Borosilicate Glasses.

Oxides	A1 (SB6)			A2 (SB6 w/o Fe)			B1 (SB19)			B2 (SB19 w/o Fe)			D1 (high B$_2$O$_3$ frit)			D2 (high B$_2$O$_3$ frit w/o Fe)		
	target		actual	target		actual	target		actual	target		actual	target		actual	target		actual
	mol %	wt. %	wt. %	mol %	wt. %	wt. %	mol %	wt. %	wt. %	mol %	wt. %	wt. %	mol %	wt. %	wt. %	mol %	wt. %	wt. %
Al$_2$O$_3$	8.5	12.5	12.5	9.0	14.1	14.3	15.0	21.1	21.0	16.0	24.1	24.1	15.0	20.9	20.6	16.0	23.9	23.5
B$_2$O$_3$	5.0	5.0	4.9	5.5	5.9	5.6	4.2	4.0	3.8	4.5	4.6	4.4	8.5	8.1	7.7	9.0	9.2	9.0
Fe$_2$O$_3$	5.0	11.5	10.4	-	-	0.1	5.3	11.7	11.1	-	-	0.1	5.5	12.0	12.6	-	-	0.1
Na$_2$O	25.0	22.3	21.5	26.5	25.3	23.8	29.5	25.2	24.1	31.0	28.3	26.8	25.0	21.2	20.3	26.5	24.1	23.0
SiO$_2$	56.5	48.8	48.3	59.0	54.6	54.3	46.0	38.1	38.0	48.5	43.0	43.3	46.0	37.8	37.4	48.5	42.8	42.4
Total	100	100	97.6	100	100	98.1	100	100	99.0	100	100	98.7	100	100	98.6	100	100	98.0
ψ$_B$	3.3			3.2			3.5			3.3			1.2			1.2		
ψ$_B$ (Fe)	3.0						3.1						1.0					
K	11.3			10.7			11.0			10.8			5.5			5.4		

ψ$_B$={(Na$_2$O+K$_2$O+BaO)+[0.7(CaO+SrO+CdO+PbO)+[0.3(Li$_2$O+MgO+ZnO)]-Al$_2$O$_3$}/B$_2$O$_3$; K=SiO$_2$/B$_2$O$_3$ (mol.%)

The infrared and Raman spectra were recorded using a Perkin-Elmer 2000 Fourier spectrophotometer (compaction of glass powders in pellets with KBr), and a Jobin Yvon U1000 spectrophotometer operated at an excitation wavelength of 532 nm, respectively.

RESULTS AND DISCUSSION

Infrared spectra

IR spectra of glasses (Fig. 1) consist of the bands due to stretching (3100-3600 cm^{-1}) and bending modes (1600-1800 cm^{-1}) in the molecules of absorbed and structurally bound water, weak bands due to hydrogen bonds in the structure of glasses and numerous bands lower than 1600 cm^{-1} due to stretching and bending modes in the units forming an anionic motif of the structure of glasses.

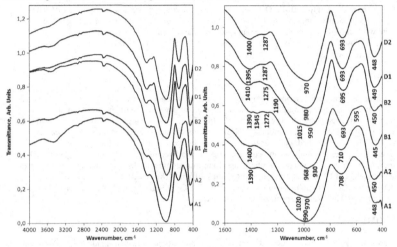

Figure 1. IR spectra of pristine glasses (right is a fragment of the left).

IR spectra of all the glasses within the range of 400-1600 cm^{-1} (Fig. 1) consist of the bands due to stretching and bending modes in silicon-oxygen, boron-oxygen, aluminum-oxygen and iron-oxygen (in spectra of glasses A1, B1 and D1 only) structural groups. The wavenumber ranges of 1550-1300 cm^{-1} and ~1260-1270 cm^{-1} are typical of vibrations in the boron-oxygen groups with trigonally coordinated boron (boron-oxygen triangles BO$_3$) [4]. These bands were attributed to components of twice degenerated asymmetric valence ν_3 O–B–O vibrations (stretching modes). The band with components ~710-730 and 650-670 cm^{-1} may be associated with twice degenerated asymmetric deformation δ (ν_4) O–B–O vibrations (bending modes) [4]. Strong absorption in both IR and Raman spectra within the range of 1150-850 cm^{-1} is caused by asymmetric ν_3 vibrations (stretching modes) in silicon-oxygen units bound to zero (900-950 cm^{-1}), one (~950-1000 cm^{-1}), two (~1000-1050 cm^{-1}), three (~1050-1100 cm^{-1}) and four (~1100-1150 cm^{-1}) neighboring SiO$_4$ tetrahedra (Q^0, Q^1, Q^2, Q^3, Q^4, respectively) [5] and, to a lesser extent, BO$_4$ tetrahedra (1000-1100 cm^{-1}) [4]. In IR spectra of all the glasses the broad band within the range of ~800-1200 cm^{-1} is multicomponent due to superposition of vibrations (stretching modes) in SiO$_4$ and BO$_4$ tetrahedra. Stretching modes of Al—O bonds in AlO$_4$ tetrahedra and Fe—O bonds in FeO$_4$ tetrahedra are positioned at 700-800 cm^{-1} and 550-650 cm^{-1}, respectively [6]. Bending modes of Si—O—Si bonds in SiO$_4$ tetrahedra are positioned within the range of 350-550 cm^{-1}.

Glasses A1, A2, B1, and B2 are characterized by similar values of the ψ_B (3.0-3.3) and K (10.7-11.3) factors. As follows from theoretical suggestions (see our previous paper [7]), the majority of boron in these glasses must be four-coordinated on oxygen. In the structure of these glasses BO$_4$ tetrahedra are built in the network or chains of SiO$_4$ tetrahedra forming [BO$_{4/2}$]Na$^+$ units. Therefore, these glasses should have similar structure and their IR and Raman spectra should be similar as well.

IR spectra of glasses A1 and A2 are actually very similar. Two strong bands within the ranges of 850-1200 cm^{-1} and 400-550 cm^{-1} are due to stretching and bending modes of Si—O—Si and Si—O$^-$ bonds. In the IR spectra of both glasses, the broad band at 850-1200 cm^{-1} consists of a shoulder at ~1160 cm^{-1}, three narrower bands centered at 1085, 1055-1056, and 1002-1006 cm^{-1} and a shoulder at ~890 cm^{-1} due to asymmetric stretching modes in Q^4, Q^3, Q^2, and Q^1 units, including valence vibrations of Si—O—Si(Al) bridging bonds (970-1100 cm^{-1}) and non-bridging Si—O$^-$ bonds (850-950 cm^{-1}). As it is seen from the relative intensity of the bands, Q^2 units dominate. However, it should be taken into account that vibrations (stretching modes) of Si—O—B bridges linking SiO$_4$ and BO$_4$ tetrahedra are also positioned at 950-1000 cm^{-1} and may make a contribution to this band, although due to the low content of B$_2$O$_3$ in glasses this contribution should be minor. There are also weak absorption bands within the ranges of 1300-1550 cm^{-1}, 1250-1300 cm^{-1} and a stronger band within the range of 650-800 cm^{-1}. Because these bands are associated with trigonally coordinated boron, it may be concluded that the amount of ternary coordinated boron is low and is consistent with theoretical representations. Nevertheless, major contribution to the band at 650-800 cm^{-1} is made by Al—O vibrations in AlO$_4$ tetrahedra and symmetric stretching modes of Si—O$^-$ bonds. Incorporation of 5 mol.% Fe$_2$O$_3$ in glass does have a significant effect on the IR spectra with the exception of very weak absorption near 575 cm^{-1} which may be assigned to vibrations of Fe—O bonds in FeO$_4$ tetrahedra [6].

In general, the structure of A1 and A2 glasses is formed by a network of SiO$_4$ and AlO$_4$ tetrahedra with minor contribution of BO$_4$ tetrahedra; some of which enter complex borate groups with BO$_3$ triangles.

The IR spectra of glasses B1 and B2 are also similar (Fig. 1) but somewhat different from those of glasses A1 and A2. There is a weak band centered at 1390-1410 cm^{-1} and a shoulder at ~1272-1275 cm^{-1} due to vibrations in the units with trigonally coordinated boron, two broad strong bands within the range of 800-1200 cm^{-1} and 400-550 cm^{-1} due to stretching and bending modes of Si—O—Si bridges and Si—O$^-$ bonds and a moderate intensity band at 700-800 cm^{-1}. IR spectra of glasses B1 and B2 within the range of ~800-1200 cm^{-1} differ in intensities of higher and lower wavenumber components. In the spectrum of glass B1 this band has a maximum at 968 cm^{-1} and a shoulder at ~930 cm^{-1} whereas in the spectrum of the glass B2 maximum of this band is positioned at 950 cm^{-1} while the lower intensity shoulder is located at ~1015 cm^{-1}. Thus, incorporation of 5.3 mol.% Fe$_2$O$_3$ in glass increases the fraction of Q^2 and Q^1 units and, therefore, the number of non-bridging oxygen ions or decreases the degree of polymerization of the structural network although this effect is rather minor.

By comparison of IR spectra of glass B1 and nepheline [8], it can be shown that no combination of their spectra could result in the spectrum observed for the carnegieite/nepheline containing glass B2. An increase of nepheline content in the glass should result in growth in absorption intensity within the range of 950-1000 cm^{-1} and formation of additional bands at lower wavenumber values. However, this does not occur. Therefore, carnegieite/nepheline content in glass is expected to be low, i.e. $\leq$ 5 vol.%.

IR spectra of glasses D1 and D2 are somewhat different from those of the other glasses. A major difference is much stronger absorption in wavenumber range of 1250-1550 cm^{-1}. In the spectra of glasses A1, A2, B1 and B2 a weak shoulder at ~1280-1285 is present. However in the spectra of glasses D1 and D2 a well-formed band centered at ~1285-1300 cm^{-1} occurs. Because the bands centered at ~1400 and ~1270-1300 cm^{-1} are due to ternary coordinated boron, it may be concluded that the fraction of trigonally coordinated boron in D1 and D2 glasses is much higher than in the other glasses. This is in a good agreement with theoretical suggestions – the ψ_B factor for these glasses is about 1 and K$\approx$5.5, whereas for A and B glasses it is about 3 and K$\approx$11. The edge of the broad band at 850-1200 cm^{-1} is shifted to higher wavenumbers pointing to a markedly smaller contribution of four-coordinated boron as compared to A and B glasses. The high wavenumber region of this band is similar to that of A1 and A2 glasses.

Overall, glasses D1 and D2 have a lower degree of connectedness (polymerization) of the silicon-oxygen network as follows from lower wavenumber maxima of the band due to vibrations in SiO$_4$ tetrahedra and low intensity of its higher wavenumber component.

IR spectra of the glasses heat-treated at 500 °C for 10 hrs were also recorded (Fig. 2). The major difference in spectra of heat-treated glasses is the splitting of the bands due to vibrations in SiO$_4$ tetrahedra. This normally occurs at structural ordering in the glass network due to formation of pre-crystallization areas followed by devitrification [9]. Glasses A1, A2 and D2 seem to be structurally ordered to the most extent. The band at 800-1200 cm^{-1} in spectra of glasses B1, B2 and D1 is less resolved than in the spectra of glasses A1, A2 and D2. In the spectrum of iron-containing glass B1, band components are located at ~1160 (shoulder), 1045, 1005, 975, ~885, and 808 cm^{-1} (shoulders). Some of them may be assigned to the contribution of nepheline (~1085, 1045, ~1000, ~700, ~520 and ~480 cm^{-1} [4,5,8]). In the spectrum of iron free glass B2, the band contains components ~1082 (shoulder), 1045, 970, and 885 cm^{-1} (shoulder). The difference in location of the bands may be assigned to the effect of iron.

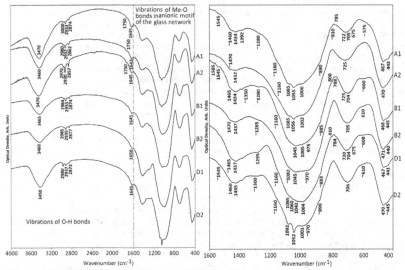

Figure 2. IR Spectra of Glasses Heat-Treated at 500 °C (right is a fragment of left).

Shoulders at ~808 and 788 cm^{-1} are probably symmetric stretching modes of Si—O—Al or Si—O—B bridges whereas a weak band at ~600 cm^{-1} in the spectrum of glass B1 is due to contribution of Fe—O bonds in FeO$_4$ tetrahedra. Weak splitting of the band at 650-750 cm^{-1} is possibly due to formation of AlO$_4$ tetrahedra with various number of non-bridging oxygen ions.

Raman spectra

Raman spectra of the glasses studied are shown on Fig. 3. Similarly to the IR spectra, Raman spectra consist of strong bands within the ranges of 300-600 and 850-1200 cm^{-1} and weaker bands at 700-800 and 1250-1550 cm^{-1}. But unlike in IR spectra, different locations of major bands due to stretching and bending modes in various structural units of glasses in the Raman spectra are observed. There is a significant difference between Raman spectra of iron-containing and iron-free glasses especially within the range of 850-1200 cm^{-1}. The band at 850-1200 cm^{-1} consists of two major components at lower (850-1000 cm^{-1}) and higher (1000-1200 cm^{-1}) wavenumber ranges. In the spectra of iron free glasses the second component is stronger than the first one whereas in the spectra of iron-containing glasses the opposite occurs. Since these components correspond to stretching modes in $Q^0+Q^1+Q^2$ units and Q^3+Q^4 SiO$_4$ units, respectively, it may be concluded that Fe^{3+} (and Fe^{2+} ions if present) offer a destructive effect on the structural network of these glasses.

For an understanding of the structure of glasses in more details, computer simulation is required. Glasses A1 and A2 are similar in chemical composition, but unlike glass A2, glass A1 contains 5 mol.% Fe$_2$O$_3$. Raman spectrum of iron free glass A2 may be represented as a superposition of the bands centered at 449, 511, 578, 631, 695, 771, 941, 1042, 1074, 1333, and 1556 cm^{-1} (Fig. 3). Strong bands centered at 449 (broad) and 511 cm^{-1} (narrower) may be attributed to bending modes in the SiO$_4$ units with various number of non-bridging oxygen ions. Much weaker bands centered at 578, 695 and 771 cm^{-1} are due to stretching modes of Al—O$^-$ bonds in AlO$_6$ (first band) and AlO$_4$ units. The band centered at 631 cm^{-1} is due to symmetric valence vibrations (v_1 stretching modes) of Si—O—Si(Al) bridges in depolymerized

56

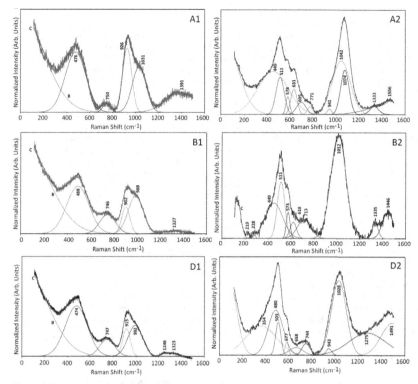

Figure 3. Raman spectra of glasses.

silicate or aluminosilicate anions. Weak bands centered at 941, 1333 and 1556 cm^{-1} may be attributed to vibrations of B—O bonds in BO_4 (first band) and BO_3 units with various number of non-bridging oxygen ions. The bands centered at 1042 and 1074 cm^{-1} are due to v_3 stretching modes in SiO_4 tetrahedra with two (Q^2) and three (Q^3) bridging oxygen ions. The broad band centered at 1042 cm^{-1} may be also given a contribution due to $Q^4(mAl)$ structural units [10].

The Raman spectrum of glass A1 is simpler than that of glass A2 (Fig. 3). Computer simulation gives the following set of overlapping bands centered at 476, 750, 926, 1031, and 1391 cm^{-1}. It is apparent that the band centered at 476 cm^{-1} is a superposition of narrower lines due to bending modes in SiO_4 units, various number of non-bridging oxygen ions (350-550 cm^{-1}), and stretching modes of Fe—O bonds in FeO_4 tetrahedra (550-600 cm^{-1}) as well as v_1 stretching modes of Si—O—Si(Al) bridges. The band centered at 750 cm^{-1} is due to stretching modes of Al—O$^-$ bonds in AlO_4 tetrahedra. The strong broad asymmetric band within the range of 850-1150 cm^{-1} is a superposition of the bands centered at 926 and 1031 cm^{-1} (v_3 stretching modes in Q^1 and Q^2 tetrahedra, respectively). Thus, Raman spectra unambiguously demonstrate the depolymerizing effect of Fe_2O_3 on the glass network resulting in formation of pyro-groups $Si_2O_7^{4-}$ or short chains of SiO_4 tetrahedra. The broad band centered at 1391 cm^{-1} is due to vibrations in BO_3 units. Like previously considered glasses, B1 and B2 have similar chemical composition but glass B1 contains 5.3 mol.% Fe_2O_3. These glasses should have B^{III}/B^{IV} ratio similar to glasses A1 and A2 due to close ψ_B and K values (Table I).

As follows from computer simulation of the Raman spectrum of glass B2 its low wavenumber range is similar to that of glass A2 (Fig. 3) with the exception of two weak bands centered at 219 and 228 cm^{-1}. The broad band within the range of 300-600 cm^{-1} may be resolved into components 449, 511, 573,

618 and 713 cm^{-1}. As discussed previously, the bands centered at 449 and 511 cm^{-1} are due to bending modes of Si—O bonds in SiO$_4$ tetrahedra. The band centered at 573 cm^{-1} may be identified as the band at 578 cm^{-1} in glass A2 and assigned to stretching modes of Al—O$^-$ bonds on AlO$_6$ octahedra, whereas the band centered at 713 cm^{-1} is due to stretching modes of Al—O$^-$ bonds on AlO$_4$ tetrahedra. Unlike the Raman spectrum of glass A2, the spectrum of glass B2 contains a band within the range of 850-1150 cm^{-1} that is nearly Gaussian symmetric with a maximum at 1012 cm^{-1} pointing to predominant Q^2 tetrahedra in the structure. Two bands in the high wavenumber range centered at 1335 and 1446 cm^{-1} are due to BO$_3$ triangles predominantly isolated and bound three-dimensionally, respectively.

Incorporation of 5.3 mol.% Fe$_2$O$_3$ in glass B2 yields a spectrum very similar to that of glass A1 (Fig. 3). In the low wavenumber range, the spectrum of glass B2 exhibits a strong broad band centered at 488 cm^{-1} due to bending modes of Si—O$^-$ and Si—O—Si bonds. The band centered at 746 cm^{-1} is due to stretching modes of Al—O bonds in AlO$_4$ tetrahedra. The strong band due to vibrations in SiO$_4$ tetrahedra becomes asymmetric and is composed of components at 902 and 988 cm^{-1} which may be attributed to ν_3 stretching modes of Si—O bonds in Q^1 and Q^2 units, respectively. Despite that Q^2 units remain the major constituent of the structure of glass B2, the formation of Q^1 units show the effect of Fe^{3+} ions to disrupt the glass network. The occurrence of BO$_3$ triangles is responsible for the weak band centered at 1327 cm^{-1}. A minor contribution by BO$_4$ tetrahedra into the edge of a major band within the range of 900-1000 cm^{-1} may also occur.

For the spectrum of iron free glass D2, the best fit was achieved at a set of lines centered at 354, 480, 503, 577, 658, 744, 943, 1028, 1275, and 1491 cm^{-1} (Fig. 3). As discussed above, the bands in the low wavenumber range are due to bending modes in SiO$_4$ tetrahedra and stretching modes in AlO$_4$ tetrahedra. In the spectrum of iron-containing glass D1, this range is approximated by the only band with a maximum at 474 cm^{-1} due to superposition of several bands including stretching modes of Fe—O bonds. However their contribution is minor due to low Fe$_2$O$_3$ content in the glass. The major band in the high wavenumber range is resolved into components at 915 and 992 cm^{-1}. These may be attributed to ν_3 stretching modes of Si—O bonds in Q^1 and Q^2 tetrahedra, respectively. Because the band centered at 992 cm^{-1} is broader, it may be suggested that it contains some contribution of stretching modes of B—O bonds in BO$_4$ tetrahedra. Two weak overlapped bands with maxima at 1246 and 1323 cm^{-1} are due to vibrations in BO$_3$ triangles. In all the glasses studied the incorporation of Fe$_2$O$_3$ results in a disruption of the glass network.

CONCLUSION

IR spectra show the minor effect of Fe$_2$O$_3$ on the structure of the anionic motif of the glasses studied. As follows from Raman spectra, incorporation of Fe ions causes a disruption of the glass network increasing the number of non-bridging oxygen ions and, thus, the fraction of SiO$_4$ tetrahedra with lower degree of connectedness.

ACKNOWLEDGMENT

The work was funded by Savannah River National Laboratory under Subcontract No. AC69549N.

REFERENCES

1. J.G. Darab, J.C. Linehan, B.P. McGrail, *Mat. Res. Soc. Symp. Proc.* **556** (1999) 337-344.
2. D. Holland, B.G. Parkinson, M.M. Islam, A. Duddridge, J.M. Roderick, A.P. Howes, C.R. Scales, *Mater. Res. Soc. Symp. Proc.* **1107** (2008) 199-206.
3. S.L. Marra and C.M. Jantzen, Characterization of Projected DWPF Glasses Heat Treated to Simulate Canister Centerline Cooling (U). WSRC-TR-92-142, 1993.
4. V.A. Kolesova, *Glass Phys. Chem.* (Russ.) **12** [10] 4-13 (1986).
5. V.N. Anfilogov, V.N. Bykov, and A.A. Osipov, Silicate Melts (Russ., Nauka, Moscow, 2005).
6. I.I. Plyusnina, *Infrared Spectra of Minerals* (Russ., MGU, Moscow, 1977).
7. S.V. Stefanovsky, B.S. Nikonov, K.M. Fox, and J.C. Marra, *Mater. Res. Soc. Symp. Proc.* (2013).
8. http://rruff.info/nepheline/display=default/R040025
9. Infrared Spectra of Inorganic Glasses and Crystals, A.G. Vlasov and V.A. Florinskaya (eds.) (Russ., Khimiya, Leningrad, 1972).
10. P.F. McMillan, B. Piriou, A. Navrotsky, *Geochim Cosmochim. Acta.* **46** (1982) 2021-2037.

Mater. Res. Soc. Symp. Proc. Vol. 1518 © 2013 Materials Research Society
DOI: 10.1557/opl.2013.204
XAFS Study of Fe K Edge in Al_2O_3-B_2O_3-Fe_2O_3-Na_2O-SiO_2 Glasses

S.V. Stefanovsky,[1,2] A.A. Shiryaev,[2] Y.V. Zubavichus,[3] K.M. Fox,[4] J.C. Marra[4]

[1] SIA Radon, 7th Rostovskii lane 2/14, Moscow 119121 Russia, profstef@mtu-net.ru
[2] Institute of Physical Chemistry and Electrochemistry RAS, Leninskii av. 31, Moscow 119071 Russia
[3] NRC "Kurchatov Institute", Kurchatov sq. 1 123182, Moscow 119117 Russia
[4] Savannah River National Laboratory, Building 773-A, Aiken 29808 U.S.A.

ABSTRACT

Valence state and local environment of Fe in complex glasses related to the system Al_2O_3-B_2O_3-Fe_2O_3-Na_2O-SiO_2 were studied. In all the glasses, the major fraction of Fe exists as Fe^{3+} ions but a minor fraction of Fe^{2+} ions especially in the glass with the lowest $K=[SiO_2]/[B_2O_3]$ ratio was also present. Average Fe—O distance in the first shell is 1.80-1.85 Å and coordination number is 4-6. The intensity due to the second sphere is rather weak demonstrating homogeneous distribution of Fe ions in the glass.

INTRODUCTION

Iron is a typical component of radioactive wastes occurring as both activated corrosion product and component of chemicals used in waste pre-treatment processes [1]. During waste vitrification, iron may have a significant effect on phase composition of glassy materials (spinel formation), their structure and properties (viscosity, rheology, electric resistivity of glassmelts and chemical durability of glasses) [2-4]. Fe_2O_3 in the presence of MgO, MnO, FeO, CoO, NiO forms spinel-type phases which have no impact on chemical durability of borosilicate glasses [4-7]. In the glass structure Fe_2O_3 can act similarly to Al_2O_3 and retard conversion of tri- to four-coordinated boron (but typically to a less extent than Al_2O_3). An extreme excess of Fe_2O_3 may result in phase separation and reduction in chemical durability of borosilicate glasses.

Structural parameters $R = R_2O/B_2O_3$ and $K = SiO_2/B_2O_3$ (oxide concentrations are given in mol.%) for evaluation of relative fractions of three- and four-coordinated boron were proposed based on investigation of the structure of alkali borosilicate glasses [8]. For complex borosilicate glasses the R parameter is calculated from the expression [9]:

$$R=\{(Na_2O+K_2O+BaO)+[0.7(CaO+SrO+CdO+PbO)+[0.3(Li_2O+MgO+ZnO)]-Al_2O_3\}/B_2O_3$$

For the Fe_2O_3 bearing glasses this expression may be modified as [10]:

$$R^{Fe} =\{(Na_2O+K_2O+BaO)+[0.7(CaO+SrO+CdO+PbO)+[0.3(Li_2O+MgO+ZnO)]-Al_2O_3-0.3Fe_2O_3\}/B_2O_3,$$

where all oxide concentrations are also given in mol.%, and for the glasses studied in the current work the expreessions are the following: $R = (Na_2O-Al_2O_3)/B_2O_3$ and $R^{Fe} = (Na_2O-Al_2O_3-0.3Fe_2O_3)/B_2O_3$, respectively (Table I).

The methods of determination of Fe valence state and local environment in glass structure include techniques such as Mössbauer effect, electron paramagnetic resonance (EPR), and X-ray absorption spectroscopy (XAS). The latter is being extensively applied during recent years (see, for example, [11-16]). This technique was used to study Fe speciation in complex Al_2O_3-B_2O_3-Fe_2O_3-Na_2O-SiO_2 glasses considered as models for nuclear waste glasses for immobilization of Savannah River Site (SRS) waste.

EXPERIMENTAL

Glasses were prepared from the proper proportions of reagent-grade Al_2O_3, Fe_2O_3, SiO_2, Na_2CO_3 and H_3BO_3 in 150 g batches. The batches were placed in Pt/Rh crucibles, heated to 1150 °C, kept at this temperature for 1 hr, and heat-treated according to the Canister Centerline Cooling (CCC) regime to simulate conditions during melt pouring and cooling of a canister in the Defense Waste Processing Facility (DWPF) [17]. The target and measured chemical compositions of the glasses are given in Table I. The glasses were characterized by X-ray diffraction, electron microscopy, infra-red (IR) and Raman spectroscopy in our previous studies [18,19].

Table I. Chemical Compositions of Borosilicate Glasses.

Oxides	A1 (SB6)			B1 (SB19)			D1 (high B_2O_3 frit)		
	target		actual	target		actual	target		actual
	mol%	wt. %	wt. %	mol%	wt. %	wt. %	mol%	wt. %	wt. %
Al_2O_3	8.5	12.5	12.5	15.0	21.1	21.0	15.0	20.9	20.6
B_2O_3	5.0	5.0	4.9	4.2	4.0	3.8	8.5	8.1	7.7
Fe_2O_3	5.0	11.5	10.4	5.3	11.7	11.1	5.5	12.0	12.6
Na_2O	25.0	22.3	21.5	29.5	25.2	24.1	25.0	21.2	20.3
SiO_2	56.5	48.8	48.3	46.0	38.1	38.0	46.0	37.8	37.4
Total	100.0	100.0	97.6	100.0	100.0	99.0	100.0	100.0	98.6
R	3.3			3.5			1.2		
R^{Fe}	3.0			3.1			1.0		
K	11.3			11.0			5.5		

X-ray absorption fine structure (XAFS) spectra were recorded at the Structural Materials Science (STM) beamline of the synchrotron source at NRC "Kurchatov Institute" [20]. The glass samples were measured at room temperature either as dispersed powder or as pellets pressed from powder mixed with sucrose in transmission mode using a Si(220) channel-cut monochromator and two air-filled ionization chambers. Fluorescence spectra were also acquired. Powders of reagent-grade oxides Fe_2O_3, Fe_3O_4 and FeO were used as standards and measured under identical conditions. Experimental XAFS spectra were fitted in R-space using an IFEFFIT package [21]; FEFF8 [22] was used for calculation of phase shift from crystal structures of corresponding oxides and silicates.

Additional information was obtained using Wavelet transform (WT) approach. As shown in refs [23,24], WT is easily adapted to analysis of EXAFS spectra, and the expression of the WT of the k^n-weighted EXAFS data takes the form:

$$W_\chi^\psi(k,r) = (2r)^{1/2} \int_{-\infty}^{\infty} \chi(k')k'^n \psi^*[2r(k'-k)]dk', \tag{1}$$

where $\chi(k)$ is the EXAFS signal and $\psi^*[2r(k'-k)]$ is the complex wavelet function.

The WT is able to resolve the k dependence of the absorption signal, which potentially allows separation of contributing backscattering atoms even situated at the same distances from the core. One of the advantages of the wavelet analysis is the visualization of the WT modulus in a k-R plot, which provides an easy way to interpret the results. Our analysis of EXAFS data for Fe were performed using the FORTRAN program HAMA employing Morlet wavelet algorithm [23,24]. The Morlet wavelet is well-suited for the EXAFS signal since it consists of a slowly varying amplitude term and a fast oscillating phase term. Its mathematical description is broadly analogous to the Fourier transform. The Morlet wavelet is obtained by taking a complex sine wave with frequency η (as in FT) and by confining it with a Gaussian envelope with the half width σ,

$$\psi(k) = \frac{1}{(2\pi)^{1/2}\sigma} \exp(i\eta k)\exp(-k^2/2\sigma^2). \tag{2}$$

The choice of the η and σ parameters is important for data analysis since, besides other issues, it determines resolution in k-R space. Various combinations of these parameters were used in an attempt to resolve contributions from atoms at similar distances from the central atom. As shown in ref.[24] use of higher k-weighting decreases resolution in the k-space, since backscattering amplitudes become flattened and shifted to higher values. Note, that in all plots of the WT modulus the interatomic distances are given without phase shift correction.

RESULTS

X-ray absorption near-edge structure (XANES) spectra of Fe K edge in glasses and their first derivatives are shown on Fig. 1. As follows from comparison with reference data [11-16], spectra may be attributed to Fe^{3+} ions in mixed major tetrahedral and minor octahedral oxygen environments. Spectrum of Fe K edge in the D1 glass is slightly different from spectra of the A1 and B1 glasses. In particular the pre-edge peak in the XANES spectrum of the D1 glass is positioned at 7110.9 eV whereas the same peak in the spectra of A1 and B1 glasses is positioned at 7111.4 eV.

Comparison of first derivatives (FD) of XANES spectra with reference data [12,13] demonstrates that FD of Fe K XANES spectra of glasses are similar to that of glass with $NaFeSi_2O_6$ composition. FD of spectra of the glass D1 suggests minor contribution due to Fe^{2+}.

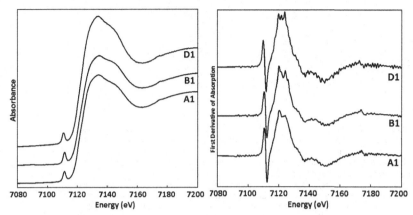

Figure 1. XANES spectra of Fe K edge in glasses (left) and their first derivatives (right).

Magnitudes of Fourier transformed (FT) EXAFS spectra of Fe K edge in glasses are shown on Fig. 2. It is evident that only the first coordination shell is well-pronounced. Therefore, Fe ions are quite homogeneously distributed within the bulk glass.

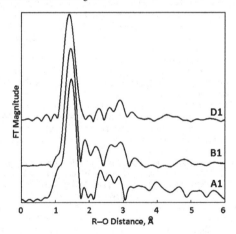

Figure 2. FT of Fe K-edge EXAFS spectra in glasses (no phase shift correction)

Fit of the experimental spectra (Table II) shows that in glass A1 average coordination number of Fe ions (CN) is ~4.6. The nearest oxygen is positioned at an average distance of 1.89 Å. In the structure of glass B1, the average Fe CN = 4.4 and Fe—O distance in the first shell is 1.92 Å. In the D1 glass, the

61

average CN = 4.0 pointing to tetrahedral oxygen environment of Fe ions. Fe—O distance in the first shell for the glass D1 is ~1.86 Å.

The second coordination shell actually exists even though it is weakly manifested. The second-nearest neighbor (oxygen or another element) is located at a distance of 3.5-3.8 Å with CN = 1.5-1.9. The second coordination sphere is better pronounced on Morlet-Wavelet transforms (Fig. 3).

Table II. Results of fitting of the Fe K Edge EXAFS Spectra of Glasses.*

Glass	A1		B1		D1	
Bond	CN	R—O, Å	CN	R—O, Å	CN	R—O, Å
Fe—O	4.6±0.2	1.89±0.04	4.4±0.2	1.92±0.04	4.0±0.2	1.86±0.04
Fe—M	1.5±0.2	3.7±0.1	1.9±0.2	3.6±0.1	1.6±0.2	3.6±0.1

* Amplitude (S_0^2) was fixed at 0.9, Debye-Waller factor was fixed at 0.005 $Å^{-1}$.

DISCUSSION

XRD and SEM characterization showed that the A1 and D1 materials were fully amorphous whereas the B1 material contained a minor content of carnegieite crystals ($Na_{1.13}Al_{0.75}Fe_{0.22}Si_{0.99}O_{4.00}$) and a very low concentration of magnetite-type spinel crystals [18,19]. Due to the low content of these phases in the B1 glass their effect on Fe partitioning is negligible and the majority of Fe in the glass enters the vitreous phase similar to the A1 and D1 glasses.

As seen from similar values of R and K structural factors, the A1 and B1 glasses have similar structure in particular similar contents of tetrahedrally and trigonally coordinated boron. High R and K values indicate that the ^{IV}B to ^{III}B ratio in the structure of these glasses should favor tetrahedrally coordinated boron. Fe_2O_3 and Fe^{3+} ions in borosilicate glasses play a role similar to Al_2O_3 and Al^{3+} ions by interfering conversion of ^{III}B to ^{IV}B. As a result, even at high R and K values, occurrence of Fe^{3+} ions in glasses reduces fraction of B^{3+} ions in tetrahedral coordination and maintains significant amount of [BO_3] units in the glass structure as it is seen from infrared and Raman spectroscopic data [19].

XANES spectra of Fe K edge in the A1 and B1 glasses and their first derivatives are typical of Fe^{3+} ions and their major oxygen environment is tetrahedral but minor concentrations of [FeO_6] octahedra are also present that is consistent with FT EXAFS data showing average CN≈4.5, i.e. ~75% Fe ions have CN = 4 and ~25% Fe – CN = 6. The nearest oxygens are positioned at a distance of ~1.89 Å (Table II). For the B1 glass, the first coordination shell consists of major [FeO_4] and minor [FeO_6] polyhedra. CN=4.4 corresponds to ~80% Fe ions occupying tetrahedral sites and ~20% - octahedral sites with nearest oxygens at an average distance of 1.92 Å. In the structure of the D1 glass essentially all the Fe^{3+} ions have tetrahedral oxygen environment with the nearest oxygens at a distance of ~1.86 Å. It is seen that the Fe—O distance in the first shell for the B1 glass is the longest among all the glasses and this may be due to contribution of spinel phase observed in the glass.

In the structure of the D1 glass with the highest Al_2O_3 content among all the glasses and relatively low R and K values, Al^{3+} and Fe^{3+} ions compete for tetrahedral sites. Because the energy of $^{IV}Al^{3+}$—O bond is higher than energy of $^{IV}Fe^{3+}$—O bond, Al^{3+} ions preferentially form [AlO_4] tetrahedra whereas a minor fraction of Fe^{3+} ions may be reduced to Fe^{2+} ions as follows from the shift of pre-edge peak to lower energies (by ~0.5 eV) remaining four-coordinated.

In various minerals and glasses in IR and Raman spectra, the absorption bands due to Fe^{3+}—O bonds in [$Fe^{3+}O_4$] tetrahedra are positioned at 550-650 cm^{-1} whereas in the octahedral units [$Fe^{3+}O_6$] they are positioned at lower wavenumbers – 300-400 cm^{-1} [25]. In our paper [19] we described and discussed IR and Raman spectra of the A1, B1 and D1 glasses. We found weak absorption at ~575-595 cm^{-1} in IR spectra of all three glasses which may be attributed to vibrations of Fe^{3+}—O bonds in [$Fe^{3+}O_4$] tetrahedra. Raman spectra demonstrate broad strong absorption within the range of 300-600 cm^{-1} caused by overlapping of several bands including the bands due to vibrations of Fe^{3+}—O bonds in [$Fe^{3+}O_4$] tetrahedra with possible minor contribution due to Fe^{3+}—O bonds in [$Fe^{3+}O_6$] octahedra. Rather low molar concentration of Fe_2O_3 in these glasses does not allow us to make definitive conclusions on iron speciation from IR and Raman spectra but the data obtained generally confirm the XAFS data on occurrence of Fe^{3+} ions in primarily tetrahedral oxygen environment.

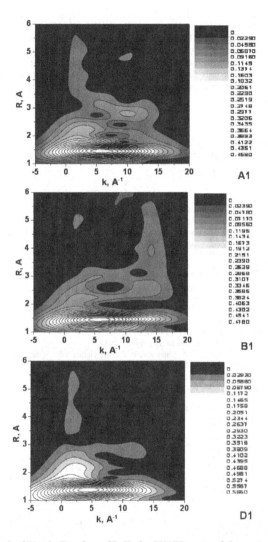

Figure 3. Magnitude of Wavelet Transform of Fe K edge EXAFS spectra of glasses. A strong oxygen-related peak corresponding to the first coordination sphere is observed at a (uncorrected) distance of ~1.4 Å and $k \sim 5$ Å^{-1}. Indication of the second coordination sphere around 2.9 Å is apparent for the A1 sample. However, unique identification of the backscattering atom is barely possible.

CONCLUSION

In the glasses studied, Fe occurs predominantly in a trivalent form as Fe^{3+} ions but a minor fraction of Fe^{2+} may be also present especially in the D1 glass. A1 and B1 glasses with similar values of ψ_B and K structural factors have similar structure and Fe speciation. In the structure of the A1 glass, Fe^{3+} ions have CN = 4.6 (70% Fe ions occupy tetrahedral and 30% -octahedral sites), in the B1glass, the average CN = 4.4 (80 Fe ions occupy tetrahedral sites and 20% - octahedral), in the D1glass, the average CN = 4 (essentially all the Fe ions are tetrahedrally coordinated). The second coordination shell of Fe ions is not well defined in the spectra; therefore, Fe ions are quite homogeneously distributed within the bulk glass. Morlet wavelet analysis showed the nearest neighbor (oxygen or heavier element) to Fe ions is located at a distance of 3.5-3.8 Å with CN = 1.5-1.9.

ACKNOWLEDGEMENT

The work was funded by Savannah River National Laboratory under Subcontract No. AC69549N. XAFS spectra were recorded at Synchrotron source at NRC "Kurchatov Insitute"funded by the State Contract N16.552.11.7003.

REFERENCES

1. S.A. Dmitriev, S.V. Stefanovsky, *Management of Radioactive Wastes* (Russ., RUCT, 2000).
2. J.G. Reynolds, P. Hrma, *Mater. Res. Soc. Symp. Proc.* **465**, 65 (1997).
3. Y. Inagaki, H. Sakata, H. Furuya, K. Idemitsu, T. Arima, T. Banba, T. Maeda, S. Matsumoto, Y. Tamura, S. Kikkawa, *Mater. Res. Soc. Symp. Proc.* **506**, 177 (1998).
4. C.M. Jantzen, K.G. Brown, *J. Amer. Ceram. Soc.* **90**, 1866 (2007).
5. A.P. Kobelev, S.V. Stefanovsky, O.A. Knyazev, T.N. Lashchenova, E.W. Holtzscheiter, J.C. Marra, *Mater. Res. Soc. Symp. Proc.* **932**, 351 (2006).
6. A.P. Kobelev, S.V. Stefanovsky, V.V. Lebedev, M.A. Polkanov, O.A. Knyazev, J. C. Marra, *Glass Tech.: Eur. J. Glass Sci. Technol. A*, **49**, 307 (2008).
7. A.P. Kobelev, S.V. Stefanovsky, V.V. Lebedev, D.Y. Suntsov, M.A. Polkanov, O.A. Knyazev, J.C. Marra, *Ceram. Trans.* **222**, 91 (2010).
8. W.J. Deli, P.J. Bray, *J. Non-Cryst. Solids*, **58**, 1 (1983).
9. A.A. Appen, Chemistry of Glass (Russ., Khimiya, Leningrad, 1974).
10. S.V. Stefanovsky, K.M. Fox, J.C. Marra, A.A. Shiryaev, Y.V. Zubavichus, *Phys. Chem. Glasses: Eur. J. Glass Sci. Technol. B*, **53**, 158 (2012).
11. G. Paschina, G. Piccaluga, G. Pinna, M. Magini, G. Cocco, *J.Non-Cryst.Solids*, **72**, 211 (1985).
12. G.E. Brown, Jr., G.A. Waychunas, C.W. Ponader, W.E. Jackson, and D.A. McKeown, *J. Phys. Colloques*, **47**, C8-661 (1986).
13. N. Binsted, G.N. Greaves, and C.M.B. Henderson, *J. Phys. Colloques*, **47**, C-841 (1986).
14. Z. Wu, M. Bonnin-Mosbah, J.P. Duraud, N. Métrich, and J.S. Delaney, *J.Synchrotron Rad.* **6**, 344 (1999).
15. G. Giuli, E. Paris, K.-U. Hess, D.B. Dingwell, M.R. Cicconi, S.G. Eeckhout, K.T. Fehr, and P. Valenti, *Amer.Miner.*, **96**, 631 (2011).
16. M. Wilke, G.M. Partzsch, R. Bernhardt, and D. Lattard, *Chem.Geology*, **220**, 143 (2005).
17. S.L. Marra and C.M. Jantzen, Characterization of Projected DWPF Glasses Heat Treated to Simulate Canister Centerline Cooling (U). WSRC-TR-92-142, 1993.
18. S.V. Stefanovsky, B.S. Nikonov, K.M. Fox, and J.C. Marra, *Mater. Res. Soc. Symp. Proc.* (2013) LL10.6.
19. S.V. Stefanovsky, K.M. Fox, and J.C. Marra, *Mater. Res. Soc. Symp. Proc.* (2013) LL13.11.
20. A.A. Chernyshov, A.A. Veligzhanin, and Y.V. Zubavichus, *Nucl. Instrum. Meth. Phys. Res.* A, **603**, 95 (2009).
21. B. Ravel and M. Newville, *J. Synchrotron Rad.*, **12**, 537 (2005).
22. A.L. Ankudinov and J.J. Rehr, *Phys.Rev.* B, **56**, 1712 (1997).
23. H. Funke, A.C. Scheinost and M. Chukalina, *Phys. Rev. B*, **71**, 094110 (2005).
24. H. Funke, M. Chukalina and A.C. Scheinost, *J. Synchrotron Radiat.* **14**, 426-432 (2007).
25. I.I. Plyusnina, *Infrared Spectra of Minerals* (Russ., MSU, Moscow, 1977).

Ceramic Wasteforms - Beta Decay

Mater. Res. Soc. Symp. Proc. Vol. 1518 © 2013 Materials Research Society
DOI: 10.1557/opl.2013.202

Thermal Conversion of Cs-exchanged IONSIV IE-911 into a Novel Caesium Ceramic Wasteform by Hot Isostatic Pressing

Tzu-Yu Chen[1], Joseph A. Hriljac[1], Amy S. Gandy[2], Martin C. Stennett[2], Neil C. Hyatt[2] and Ewan. R. Maddrell[3]

[1]School of Chemistry, University of Birmingham, Edgbaston, Birmingham, B15 2TT, U.K.,
[2]Department of Materials Science and Engineering, University of Sheffield, Sheffield, S1 3JD, U.K.,
[3]National Nuclear Laboratory, Sellafield, Seascale, Cumbria, CA20 1PG, U.K.

ABSTRACT

Hot Isostatic Pressing of Cs-exchanged IONSIV IE-911 samples is shown to produce a mixture of ceramic phases, the nature and mass fractions of these have been determined by Rietveld analysis of powder X-ray diffraction data. The main Cs phase that forms is $Cs_2TiNb_6O_{18}$, after this reaches approximately 30% of the total crystalline content the remaining Cs is partitioned into $Cs_2ZrSi_6O_{15}$. Durability tests using the PCT-B method for 7 days at 90 °C with deionised water lead to Cs leach rates of 0.032 and 0.038 $g \cdot m^{-2} \cdot day^{-1}$ for samples exchanged to 6 and 12 wt% Cs, respectively, indicating a durable wasteform is produced.

INTRODUCTION

IONSIV IE-911 is a commercial mixture of a crystalline silicotitanate (CST) with a formula of $(H_3O)_xNa_y(Nb_{0.3}Ti_{0.7})_4Si_2O_{14} \cdot zH_2O$, where x~2, y~1 and z~4, and an amorphous $Zr(OH)_4$ binder in a 4:1 ratio. It has been widely used in the nuclear industry as an inorganic ion-exchanger to separate ^{137}Cs[1-3] and shows excellent selectivity even in the presence of large amounts of Na^+, K^+, Mg^{2+}, Ca^{2+} or Ba^{2+} and over a broad pH range.[4] After use, however, the options for long term storage or disposal are limited and warrant further investigation. Particular points to consider are large volumes of ILW due to the very low bulk density of the material as used in pellet form, 1.1 kg m^{-3}, potential back-exchange or leaching especially at higher temperatures, and large fraction of Ti which limits the amount that can be incorporated into most glasses.

The thermal behaviour of IONSIV has been previously reported,[5] CST starts to lose crystallinity at around 500 °C and with further heating new ceramic phases begin to form until the onset of melting at around 1000 °C. The reported crystalline phases that form from Cs-exchanged IONSIV are $Cs_2ZrSi_3O_9$, $Na(Ti,Nb)O_3$ and $Na_2Ti_6O_{13}$.

Hot isostatic pressing (HIP) is a more advanced ceramic processing technique that involves heating a powder sample in a sealed and evacuated metal container whilst simultaneously applying pressure. It has been widely studied for nuclear waste treatment for more than fifteen years[6-10] and offers several advantages over heating powders or pressed pellets. After HIPing, internal pores and defects within a solid body collapse, as a consequence of this a homogeneous material with a uniform grain size and a nearly 100% density is achieved.[11, 12] The waste is enclosed in a sealed can and processed at relatively low temperature, therefore there are no high temperature volatility losses and no expensive off-gas emission processing system is required. Here we report the results of HIPing IONSIV samples with a range of Cs-exchanges up to the maximum possible level of 12 weight percent.

EXPERIMENT

Samples of Cs-exchanged IONSIV IE-911 were prepared by stirring portions of the solid with aqueous caesium nitrate solutions for 3 days to give Cs loadings of 2, 4, 6, 8, 10 and 12 wt%. Samples were recovered by filtration and washed with deionized water. These were dried and then calcined in air at 800 °C overnight before filling into 2 mm thick mild steel HIP cans, welding these shut and then hot isostatically pressing at 1100 °C and 190 MPa for 2 hours under an argon atmosphere. A sample of unexchanged IONSIV was also prepared.

Fused borate glass beads were analysed by X-ray fluorescence spectrometry (Bruker S8 Tiger WDXRF) to determine bulk elemental compositions. Circular glass beads with flat surfaces were prepared by mixing ground samples with lithium tetraborate flux in a 1:10 ratio and heating in a platinum crucible at 1050 °C. Ammonium iodide was added to help the bead exfoliate from the crucible.

The identification and relative ratios of the crystalline phases in the HIPed Cs-IONSIV samples were investigated using Rietveld analysis of X-ray powder diffraction data collected using a Bruker D8 Advance diffractometer operating in transmission mode using Ge-monochromatised Cu $K_{\alpha 1}$ radiation and a Lynxeye detector. Diffraction patterns were collected between 5 and 90° 2θ. Rietveld refinements were performed using the General Structure Analysis System (GSAS) program[13] with EXPGUI software suite. The starting values of the atomic positions, lattice parameters and displacement parameters for each phase were taken from the literature listed in Table I.

Table I. Starting models for the phases used in Rietveld refinement

Phase	Space group	Reference
$(Ti_{0.83}Nb_{0.17})O_2$	$P4_2/mnm$	Okrusch et al.[14]
$ZrSiO_4$	$I4_1/amd$	Torres et al.[15]
$NaNbO_3$	$Pbma$	Hewat[16]
$(Zr_{0.5}Ti_{0.5})O_2$	$Pbcn$	Troitzsch et al.[17]
SiO_2	$P3_221$	Kihara[18]
$Cs_2TiNb_6O_{18}$	$P-3m1$	Desgardin et al.[19]
$Cs_2ZrSi_6O_{15}$	$C2/m$	Jolicart et al.[20]

Microstructure characterisation using scanning electron microscopy (SEM) on polished samples was carried out on a Philips XL30 ESEM-FEG with an Oxford Inca 300 EDS system operating at 10 kV. Backscattering electron image mode was selected to observe the elemental distribution in HIP samples. Microstructure of the selected HIPed sample was also examined on a TEM (Tecnai, operating at 200 kV) with an EDX system. The TEM sample was prepared by mechanical polishing using SiC paper and a rotating polishing wheel, followed by dimpling and ion milling using a Precision Ion Polishing System (PIPS) to achieve electron transparency.

The durability tests of powdered HIPed IONSIV 6 wt% and 12 wt% samples were carried out in deionised water at 90 °C for 7 days using 75-150 μm sized particles according to the ASTM standard PCT-B method[21]. The leachate liquid was analysed by ICP-MS (Agilent 7500ce). The normalised elemental leach rates were calculated by equation 1

$$L_i = M_i / (A_s \times t \times F_i) \qquad (1)$$

where L_i is the leach rate (g·m^{-2}·day^{-1}) of element i, M_i is the mass of element i in the leachate (g), A_s is the surface area (m^2) of the sample, t is the leaching time (day) and F_i is the mass fraction of element i in the original sample.

RESULTS AND DISCUSSION

SEM images of some of the HIPed samples are shown in Figure 1. The microstructure reduces in grain size with an increase in the Cs loading. It is also clear that bright grains with rectangular shape, suggesting Cs containing phases, form with increasing Cs loading. At higher Cs content, above 6 wt%, a second bright phase with a hexagonal shape is also observed. Attempts at direct phase identification using EDS were not successful as the incident beam was much larger than any of the individual grains.

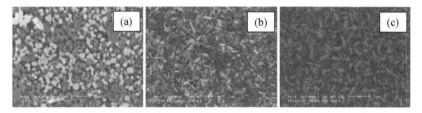

Figure 1. SEM images of HIPed samples with (a) 2 wt%, (b) 6 wt% and (c) 12 wt% Cs.

Powder X-ray diffraction patterns of the samples were analysed to determine which phases are present and from multi-phase Rietveld analyses it was possible to estimate the weight fractions of the crystalline components. In addition, TEM analysis indicates that some small amounts of glassy areas rich in Si also form. Examples of the fits for the 0, 6 and 12 wt% samples are given in Figure 2 and a summary of the phases present in Table II. An excellent fit to the data for the sample without Cs shows it consists of a majority of Nb-doped TiO$_2$ rutile with significant quantities of ZrSiO$_4$ and NaNbO$_3$. The level of doping of Nb in the rutile phase is difficult to know exactly, but the refined lattice parameters match well with those reported for (Ti$_{0.83}$Nb$_{0.17}$)O$_2$. This phase is found to be the predominant one for all samples. Zircon is present as a significant phase with low Cs dopings, but above 6 wt% is no longer found. NaNbO$_3$ is also only observable with low Cs dopings, after the ion exchange of Cs for Na a new phase, Cs$_2$TiNb$_6$O$_{18}$, forms instead. This is the only Cs-containing phase present up to 6 wt%, above this level Cs$_2$ZrSi$_6$O$_{15}$ is also found. Finally, zirconium titanate is observed for all of the Cs-containing samples. It is a disordered solid solution and known to have a compositional range from 43 to 67% Ti, but from the refined lattice parameters and TEM/EDX analysis of several particles the composition was fixed as 50:50.

From the Rietveld refinements, bulk elemental compositions have been calculated and these are compared to those found using XRF analysis in Table III. These are generally in very good agreement with a few exceptions. The calculated Zr contents of the 2 to 8 wt% samples are all high, this is probably due to an overestimation of the amount of zircon in the 2% sample and Zr content in the zirconium titanate solid solution. Increasing the Ti content of the latter and the Nb content of the Nb-doped rutile would bring all 3 elements closer in line. This would not, however, change the phases present nor alter the key aspect of the analysis regarding the nature

69

and content of the two Cs-containing phases. The other element that is not well accounted for in the Rietveld analysis is Si, but as mentioned previously TEM analysis does indicate some Si-containing glass is present. A more detailed analysis and reconciliation of the results requires higher quality diffraction data such as that obtained using synchrotron X-rays.

In early work by Su et al.[5] it was reported that a number of new phases appear when IONSIV is heat-treated. NMR, XRD and TEM results indicated that major phases in thermally converted CST are $Cs_2ZrSi_3O_9$, $Na(Ti,Nb)O_3$ and $Na_2Ti_6O_{13}$. However, the reported Cs host phase, $Cs_2ZrSi_3O_9$, was not found in any of the HIPed IONSIV samples in this work and the only Na containing phase was $NaNbO_3$. The reason for this discrepancy is not yet clear and is under further investigation. One possibility is the temperature to which the samples have been heated, in the earlier study it was 900 °C whereas in this work it was 1100 °C.

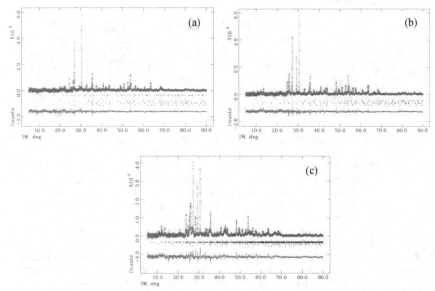

Figure 2. Simulated (top line), experimental (crosses) and difference (bottom line) diffraction patterns for (a) to (c) the HIPed Cs-IONSIV 2, 6 and 12 wt% samples, respectively.

The standard method for assessing long term durability of a possible waste form is the PCT-B test. Samples with 6 and 12 wt% caesium loadings were tested for 7 days and the results for all elements are given in Table IV. These show excellent values for all elements, most importantly Cs, and indicate that a good wasteform can be produced directly from IONSIV even at the maximum Cs loading. We believe the main Cs phase present, $Cs_2TiNb_6O_{18}$, warrants further attention as it not only appears to have excellent chemical durability but also should retain Ba^{2+} produced by radioactive decay of Cs^+ by reduction of some of the Nb^{5+} that is present to Nb^{4+}.

Table II. Phases and weight percentages in the HIPed Cs-IONSIV samples based on the Rietveld analysis of the powder X-ray diffraction results.

Phase	Cs loading						
	0 wt%	2 wt%	4 wt%	6 wt%	8 wt%	10 wt%	12 wt%
$(Ti_{0.83}Nb_{0.17})O_2$	62.6%	28.4%	28.4%	31.9%	25.5%	25.4%	27.4%
$ZrSiO_4$	30.7%	37.6%	11.6%	3.4%	–	–	–
$NaNbO_3$	6.8%	9.7%	–	–	–	–	–
$(Zr_{0.5}Ti_{0.5})O_2$	–	16.2%	36.4%	40.2%	33.4%	29.8%	17.6%
SiO_2	–	5.1%	6.8%	–	–	–	–
$Cs_2TiNb_6O_{18}$	–	3.1%	16.9%	24.5%	26.3%	28.0%	24.2%
$Cs_2ZrSi_6O_{15}$	–	–	–	–	14.9%	16.9%	30.8%
Goodness of fit parameters							
χ^2	2.42	2.25	3.26	2.33	2.34	3.17	3.53
R_{wp}	4.72	3.96	4.86	4.10	4.20	4.91	5.61
R_p	3.23	2.81	3.45	2.89	3.05	3.59	3.92

Table III. Bulk elemental compositions based on Rietveld analyses of the crystalline phases (top line) and from XRF analyses (second line, italics).

Element	0 wt.%	2 wt.%	4 wt.%	6 wt.%	8 wt.%	10 wt.%	12 wt.%
Nb	3.8	4.3	8.1	11.8	12.6	13.5	11.6
	14.9	*15.0*	*15.8*	*15.6*	*15.4*	*16.2*	*15.0*
Zr	15.3	21.7	22.1	19.8	16.8	15.4	11.6
	13.5	*13.2*	*14.2*	*14.6*	*13.7*	*14.9*	*13.9*
Ti	21.5	18.8	19.0	21.5	17.7	16.9	14.6
	20.8	*20.5*	*21.2*	*20.6*	*20.9*	*22.0*	*20.2*
Si	4.7	9.1	5.0	0.5	3.3	3.7	16.2
	8.0	*8.1*	*8.9*	*8.8*	*8.6*	*9.2*	*8.3*
Na	1.0	0.7	0.0	0.0	0.0	0.0	0.0
	2.4	*2.0*	*1.6*	*0.9*	*0.5*	*0.3*	*0.3*
Cs	0.0	0.7	3.9	5.6	11.2	12.3	6.8
	0.0	*2.6*	*4.5*	*7.9*	*10.6*	*13.8*	*12.9*

Table IV. Leach rates at 7 days (g m^{-2} day^{-1})

Sample	Cs	Na	Si	Ti	Zr	Nb
6 wt% Cs IONSIV	0.032	0.12	0.042	1.5×10^{-4}	5.0×10^{-4}	5.3×10^{-4}
12 wt% Cs IONSIV	0.038	0.11	0.041	8.5×10^{-5}	1.8×10^{-4}	1.3×10^{-4}

ACKNOWLEDGMENTS

We acknowledge technical assistance from M. Glynn S. Baker and Dr. J. Deans. This work was part funded by the Nuclear Decommissioning Authority via an award administered by the National Nuclear Laboratory. NCH is grateful to the Royal Academy of Engineering and NDA for funding. We are grateful to EPSRC for support of MCS and ASG under grant numbers EP/F055412/1, EP/I012214/1 and EP/G037140/1.The XRD and XRF systems used in this research were obtained, through the Science City Advanced Materials project: Creating and Characterising Next Generation Advanced Materials project, with support from Advantage West Midlands (AWM) and part funded by the European Regional Development Fund (ERDF).

REFERENCES

1. T.J. Tranter, R.D. Tillotson and T.A. Todd, Separ. Sci.Tech. **40**, 157 (2005).
2. D.T. Bostick, S.M. DePaoli and B. Guo, Separ. Sci. Tech. **36**, 975 (2001).
3. N.R. Mann and T.A. Todd, Separation Sci. Tech. **39**, 2351 (2004).
4. R.G. Anthony, R.G. Dosch, D. Gu and C.V. Philip, Ind. Eng. Chem. Res. **33**, 2702 (1994).
5. Y.L. Su, M.L. Balmer and B.C. Bunker, Scientific Basis for Nuclear Waste Management XX **465**, 457 (1997).
6. H. Li, Y. Zhang, P.J. McGlinn, S. Moricca, B.D. Begg and E.R. Vance, J. Nucl. Materials **355**, 136 (2006).
7. Y. Zhang, H. Li and S. Moricca, J. of Nucl. Materials **377**, 470 (2008).
8. M.L. Carter, A.L. Gillen, K. Olufson and E.R. Vance, J. Am. Ceram. Soc. **92**, 1112 (2009).
9. E.R. Vance, D.S. Perera, S. Moricca, Z. Aly and B.D. Begg, J. Nucl. Mater. **341**, 93 (2005).
10. S.V. Raman, J. Mater. Sci. **33**, 1887 (1998).
11. A.B. Harker, P.E.D. Morgan and J.F. Flintoff, J. Amer. Ceram. Soc. **67**, C26 (1984).
12. H.V. Atkinson and B.A. Rickinson, eds., *Hot Isostatic Processing*, Adam Hilger, Bristol, 1991.
13. A.C. Larson and R.B. von Dreele, GSAS program, Los Alamos National Lab. 523, 1994.
14. M. Okrusch, R. Hock, U. Schussler, A. Brummer, M. Baier and H. Theisinger, Am. Mineral. **88**, 986 (2003).
15. F.J. Torres, M.A. Tena and J. Alarcon, J. Eur. Ceram. Soc. **22**, 1991 (2002).
16. A.W. Hewat, Ferroelectrics **7**, 83 (1974).
17. U. Troitzsch, A.G. Christy and D.J. Ellis, Phys. Chem. Mineral. **32**, 504 (2005).
18. K. Kihara, Eur. J. Mineral. **2**, 63 (1990).
19. G. Desgardin, C. Robert and B. Raveau, Mater. Res. Bull. **13**, 621 (1978).
20. G. Jolicart, M. Leblanc, B. Morel, P. Dehaudt and S. Dubois, Eur. J. Sol. State Inorg. Chem. **33**, 647 (1996).
21. ASTM International, Standard Test Methods for Determining Chemical Durability of Nuclear, Hazardous, and Mixed Waste Glasses and Multiphase Glass Ceramics: The Product Consistency Test (PCT), 2002.

Mater. Res. Soc. Symp. Proc. Vol. 1518 © 2013 Materials Research Society
DOI: 10.1557/opl.2013.926

Aging Studies of Pu-238 and -239 Containing Calcium Phosphate Ceramic Waste-forms

Shirley K. Fong[1], Brian L. Metcalfe[1], Randall D. Scheele[2], Denis M. Strachan[2]
[1]Atomic Weapons Establishment, Aldermaston, Berkshire RG7 4PR, UK
[2]Pacific Northwest National Laboratories, Richland, WA, USA

ABSTRACT

A calcium phosphate ceramic waste-form has been developed at AWE for the immobilisation of chloride containing wastes arising from the pyrochemical reprocessing of plutonium. In order to determine the long term durability of the waste-form, aging trials have been carried out at PNNL. Ceramics were prepared using Pu-239 and -238, these were characterised by PXRD at regular intervals and Single Pass Flow Through (SPFT) tests after approximately 5 yrs.

While XRD indicated some loss of crystallinity in the Pu-238 samples after exposure to 2.8×10^{18} α decays, SPFT tests indicated that accelerated aging had not had a detrimental effect on the durability of Pu-238 samples compared to Pu-239 waste-forms.

INTRODUCTION

A two stage process immobilisation route has been developed to immobilise a number of halide and plutonium containing wastes arising from the pyrochemical reprocessing of plutonium at AWE. Initially the wastes are reacted with calcium hydrogen phosphate to produce a substituted β-tricalcium phosphate (β-TCP) and apatite type phases, which have been shown to be resistant to radiation damage [1-3] and to accommodate a range of actinides [4-5]. The waste is then sintered with an aluminophosphate glass to convert into a monolithic waste-form.

Development work was carried out at AWE using Hf and Sm as surrogates for Pu and Am. Trials with non-radioactive materials were carried out at Pacific Northwest National Laboratories (PNNL) before active waste-forms were synthesised. Samples were characterised by XRD at regular intervals and Single Pass Flow Through (SPFT) tests after aging for approximately 5 yrs to determine suitability for long term waste disposal. Waste-forms were prepared using both Pu-239 and Pu-238 to accelerate aging by a factor of 274.9. However, in order to reduce operator dose, the majority of the Am was replaced by Sm in these waste-forms.

EXPERIMENTAL

Ceramics were prepared containing (a) 94% Pu-239, 6% Pu-240 and (b) 78% Pu-238, 22% Pu-239, these are referred to as Pu-239 and Pu-238 samples. Simulant waste materials were prepared using HfO_2 in the correct mol% ratio.

Pu-239 containing ceramics was prepared by milling Pu(IV) oxalate and Am(II) oxalate with the remaining waste-form constituents (Table 1) and $CaHPO_4$ reagent. Oxalates were added in the correct proportions to yield the oxides compositions below. Pu-238 loaded ceramics were prepared in a similar manner, however a Pu(III) oxalate precursor was used. Mixtures were ball milled with zirconia grinding media for over 12 hrs.

Table 1 Waste composition

Constituent	Mass %
PuO_2	4.85
Am_2O_3	0.05
Ga_2O_3	6.45
Al_2O_3	2.26
Sm_2O_3	1.00
MgO	1.46
Fe_2O_3	0.31
Ta_2O_5	0.30
NiO	0.30
CaF_2	2.40
KCl	3.76
$CaHPO_4$	76.86

Ceramic waste-forms were prepared by mixing the ceramic and GTI/168 glass ($40Na_2O$. $19Al_2O_3$. $39P_2O_5$. $2B_2O_3$) in a 75:25 ceramic:glass ratio. Mixtures were ball milled for at least 48 h, then were sintered in alumina crucibles using the following heating regime, heat to 400 °C at 150 °Ch^{-1}, holding for 4 h, heat to 750 °C at 100 °Ch^{-1}, hold for 4 hr then cool to room temperature at 100 °Ch^{-1}.

Waste-forms were polished with silicon carbide papers to improve contact between adjacent specimens during storage and maximize radiation damage to the adjoining surfaces.

The majority of the active XRD data at PNNL was collected on a Scintag PAD V X-ray diffractometer, however the final dataset was measured using a Riga Ultima, both using Cu K_α radiation. For SPFT testing, specimens were ground down and the 74-149 μm particle fraction size utilized. In order to remove the majority of the fine particles adhering to the required particle size fraction, the sample was washed for one minute, three times each in deionised water, ethanol, deionised water in an ultrasound bath and finally ethanol in ultrasound bath. Samples were leached at 90 °C in titanium vessels. The leachate was a 0.03 M solution of tris(hydroxymethyl)aminomethane (TRIS) buffer adjusted to a pH of 6.9 with concentrated HNO_3.

Apparent dissolution rates were normalised to the mass fraction of element present with the following formula:-

$$r_i = \frac{\left(C_i - \overline{C}_{i,b}\right)q}{f_i S}$$

r_i	Normalised dissolution rate for element gm^{-2}d^{-1}
C_i	Concentration in leachate (gL^{-1})
$\overline{C}_{i,b}$	Average background concentration of element (gL^{-1})
q	Flow rate (Ld^{-1})
f_i	Mass fraction of element in waste-form
S	Surface area (m^2)

Electron microscopy on inactive samples was carried out using a FEI QUEMSCAN 650F equipped with two Bruker XFlash 5030 detectors.

RESULTS AND DISCUSSION

Non-active XRD studies indicated that waste-forms are composed of apatite, substituted β-TCP and β-$Ca_2P_2O_7$ and $AlPO_4$ [6]. Additional electron microscopy studies [7] have shown that these phases are embedded in a matrix phase composed of a sodium deficient GTI/168 glass as shown in Fig 1 where the thin needle like crystallites are composed of apatite, larger crystallites (3-15 μm) are substituted β-TCP and the darker regions $AlPO_4$ containing Ga.

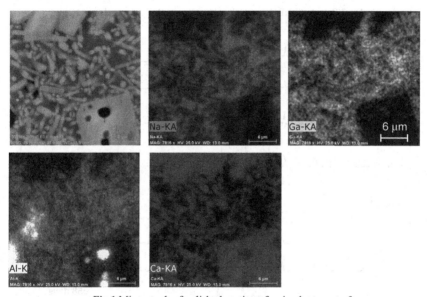

Fig 1 Micrograph of polished section of a simulant waste-form

Active waste-form-samples were indexed to the same phases, although there was some change in proportions of phases in the active waste-forms [6]. These studies have also shown that no PuO_2 was observed when added in the form of oxalates. After 2053 days Pu-238 and Pu-239 samples had been subjected to 4.6×10^{17} and 3.2×10^{15} α-decays respectively and Pu-238 samples had undergone the equivalent of 1235 yrs of accelerated aging. After this time no ingrowth of additional phases was observed (Fig 2), however comparison of the ratio of maximum XRD peak intensities for Pu-239 to Pu-238 (Table 2) indicated some loss of crystallinity of the Pu-238 samples (Table 2).

Table 2 Ratio of maximum XRD peak intensities for Pu-239 and Pu-238 samples by age and α-decays

Sample age (days)		α-decays (αg⁻¹)		Pu-239:Pu-238
Pu-239	Pu-238	Pu-239	Pu-238	I_{max} ratio
358	329	3.2×10^{15}	4.6×10^{17}	0.9
1309	1275	1.2×10^{16}	1.8×10^{18}	1.27
1567	1528	1.4×10^{16}	2.1×10^{18}	1.31
2085	2053	1.8×10^{15}	2.8×10^{18}	1.4

Fig 2 Selected PXRD spectra for Pu-239 and Pu-238 samples

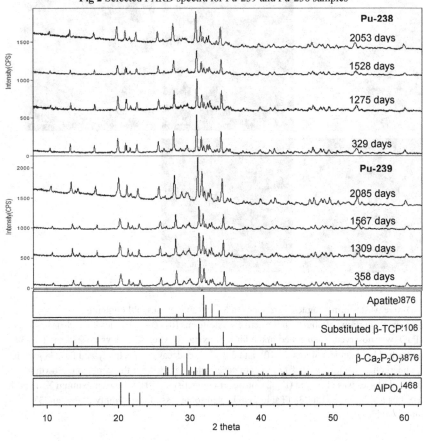

SPFT tests were run on Pu-238 and Pu-239 samples on days 1550 and 1652 respectively. Initially, up to 6 aliquots were taken at regular time intervals each day for analysis; however once steady state was reached at around 6-9 days only 2 samples per day were withdrawn. No increase in the dissolution rates for Pu 238 compared to Pu-239 samples was observed (Fig 3). The differences in release rates, once steady state was reached, are thought to arise due to errors in measurement of the surface areas, i.e. the damage caused by accelerated aging has not had a detrimental effect on the durability of the Pu-238 samples.

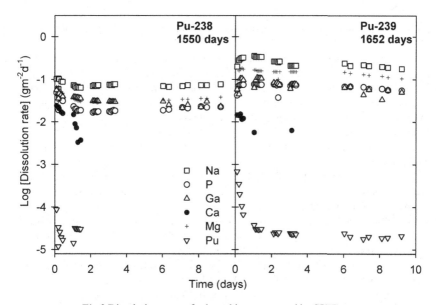

Fig 3 Dissolution rates of selected ions measured by SPFT tests

Although previous SEM/EDX studies [8], have shown some exchange of sodium and calcium between the mineral phases and glass, the bulk of the Na content is retained in the glass matrix, whilst the majority of the Ca is retained in the apatite, β-$Ca_2P_2O_7$ and substituted β-TCP mineral phases. The majority of the Ga is also present in the glass matrix (Fig 1) while P is common to all phases present so its release rates will reflect the dissolution of all phases present.

Release rates for Na, P, Ga and Mg are relatively similar. However, although the dataset is incomplete, rates for Ca and Pu indicate that dissolution of the waste-form is incongruent. Previous studies have shown that the calcium phosphate mineral phases have release rates at least an order of magnitude lower than those of the consolidated waste-form [9], and given that the bulk of the Na content is in the glass matrix, the release rates indicate that the glass matrix has a higher dissolution rate compared to the mineral phases.

CONCLUSIONS

Whilst XRD data showed some lose of crystallinity in the Pu-238 samples after exposure to 2.8×10^{18} α decays, SPFT tests indicated that accelerated aging had not had a detrimental effect on the durability of Pu-238 samples compared to Pu-239 waste-forms.

Release rates from SPFT testing, indicate that the glass matrix has a higher dissolution rate compared to the mineral phases.

REFERENCES

1. J. Chaumont, S. Soulet, J.C. Krupa and J. Carpena, *Journal of Nuclear Materials*, **301**, 122 (2002)
2. S. Utsunomiya, S. Yudintsev, L.M. Wang , R. C. Ewing, *Journal of Nuclear Materials*, **322**, 180 (2003)
3. B. L. Metcalfe, S. K. Fong, L. A. Gerrard, I. W. Donald, D. Strachan and R. Scheele, in Scientific Basis for Nuclear Waste Management XXX, edited by D. Dunn, C. Poinssot and B Begg, (Materials Research Society, Warrendale, (2007), p157-162
4. E. R. Vance and C. J. Ball *J. Am. Ceram. Soc.*, **86**, 1223, (2003)
5. K. Horie, H. Hidaka, and F. Gauthier-Lafaye, *Geochimica et Cosmochimica Acta*, **68**, (1), 115–125, (2004)
6. B. L. Metcalfe, S. K. Fong, L. A. Gerrard, I. W. Donald, in Scientific Basis for Nuclear Waste Management XXX, edited by D. Dunn, C. Poinssot and B. Begg, (Materials Research Society, Warrendale (2006), Vol 985, p157-162
7. P. M. Mallinson, S. K. Fong, E. R .Vance, J. D. Phillips, in Scientific Basis for Nuclear Waste Management XXXV, edited by R. M. Carranza, G. S. Duffo and R. B. Rabak, (Cambridge University Press, Cambridge, (2011) Vol 1475, p233-238
8. S. K. Fong & B. L. Metcalfe in Scientific Basis for Nuclear Waste Management XXIX, edited by Van Iseghem, Materials Research Society, Warrendale, (2006), Vol 932, p209-214
9. B. L. Metcalfe, S. K. Fong, I. W. Donald, in Scientific Basis for Nuclear Waste Management XXIX, edited by p. Van Iseghem, (Materials Research Society, Warrendale (2006), Vol 932, p727-734

Mater. Res. Soc. Symp. Proc. Vol. 1518 © 2012 Materials Research Society
DOI: 10.1557/opl.2012.1714

Development of the Synthetic Rock Technique for the Immobilization of Iodine: Kinetics of the Alumina Matrix Dissolution under High Alkaline Conditions

Hideaki Miyakawa[1], Tomofumi Sakuragi[1], Hitoshi Owada[1], Osamu Kato[2], Kaoru Masuda[3]
[1] Radioactive Waste Management Funding and Research Center, 1-15-7, Tsukishima, Chuo-ku, Tokyo 104-0052, Japan.
[2] Kobe Steel Ltd., 4-2-7, Iwaya-Nakamachi, Nada-ku, Kobe 657-0845, Japan.
[3] Kobelco Research Institute Inc., 1-5-5, Takatsukadai, Nishi-ku, Kobe 651-2271, Japan.

ABSTRACT

Iodine filters expended after nuclear fuel reprocessing contain radioactive iodine (I-129), almost all of which exists as silver iodide (AgI). The synthetic rock technique is a solidification treatment technique using hot isostatic press (HIP), in which the alumina adsorbent base material is synthesized to form a dense solidified material (synthetic rock), and I-129 is physically confined in the form of AgI in the alumina matrix. Thus, it is necessary to understand the matrix dissolution behavior to evaluate the iodine release behavior.

Experiments involving the dissolution of the matrix were carried out under various temperatures (35–70 °C) and pH values (10–12.5) that reflect the disposal conditions. The results of the experiments showed that the dissolution rate of Al visibly increases with temperature and pH. The dissolution rate constant was calculated from the initial data assuming the dissolution of the matrix as a primary reaction. The logarithmic rate constant showed a good linear correlation with the pH and the reciprocal of temperature. The ^{27}Al-NMR analysis of the solutions of the dissolved matrix showed that the major chemical species present in the solutions was $Al(OH)_4^-$. This indicated that the dissolution of the matrix can be described by the following equation: $Al_2O_3 + 2OH^- + 3H_2O \rightarrow 2Al(OH)_4^-$. Subsequently, the empirical equation of the rate of dissolution of the matrix as a function of the temperature and pH was derived. It will be used to evaluate the iodine release behavior from the synthetic rock.

INTRODUCTION

Radioactive iodine (I-129) is in expended iodine filters (occurring as remnants of the silver absorbent) generated after nuclear fuel reprocessing is typically present as AgI. Since I-129 shows a significantly long half-life ($t^{1/2} = 1.57 \times 10^7$ y) and a low sorption performance to rock and engineering barrier materials, I-129 exerts the largest effect on radiation exposure in the TRU radioactive safety assessment [1]. At present, several techniques have been developed to stabilize iodine release over a long period for decreasing the radiation exposure [1]. Among these, the synthetic rock technique is a solidification process that treats I-129 by hot isostatic press (HIP), without allowing its desorption from the silver-absorbent waste. By this technique, alumina, which is the base material used in the silver-absorbent waste, forms a dense matrix and the iodine is physically confined in the matrix in the form of AgI.

To evaluate the long-term release of iodine from the synthetic rock, it is necessary to understand the properties of the solidified material and the dissolution behaviors of the matrix and AgI, which are essential to comprehend iodine release. With regard to the dissolution behavior of AgI, it is assumed that AgI is readily reduced to Ag during contact with groundwater, which is its disposal environment, and I-129 is assumed to be released in the groundwater [1]. Recent studies have indicated that dissolution is independent of the pH but are dependent mainly on the concentrations of Fe^{2+} and HS^-, which are considered as the major dissolved reductant in groundwater [2, 3].

Therefore, in this study, we carried out dissolution experiments on the matrix of the synthetic rock at various pH values and at different temperatures that reflect the geological disposal environment. We have also examined the matrix dissolution in the disposal environment in terms of the kinetic parameters derived from our results.

EXPERIMENTAL DETAILS

Specimen

Specimens were prepared by the synthetic rock solidification process using simulated silver-absorbent waste prepared by loading nonradioactive iodine on unused iodine filters. HIP treatment (at 900 °C, with a pressure of 175 MPa) was carried out after a pretreatment of the adsorbent at 480 °C to remove unnecessary gaseous components. X-ray diffraction (XRD, RAD-RU300 by Rigaku Corp.) analysis of the samples showed that the synthetic rock contains mainly α-Al_2O_3, AgI and Ag. The matrix of the synthetic rock consisted of the α-Al_2O_3 (Corundum).

After the HIP treatment, the contents were ground with a 20 ton press and subsequently, with a vibrating mill. The specimens were separated according to grain sizes into samples with grain sizes ranging from 0.15 to 0.3 mm and those with grain sizes ranging from 0.3 to 0.6 mm by using a sieve. The specific surface area of each of the specimens was calculated by assuming the grains of the pulverized samples as spheres based on the average grain sizes (grain sizes 0.15-0.3 mm: 9.4×10^{-4} m^2/g, grain sizes 0.3-0.6 mm: 3.3×10^{-3} m^2/g).

Dissolution tests

We have assumed that cement is used as the typical material around the waste during geological disposal [1]. Therefore, during the dissolution tests, a solution of $Ca(OH)_2$ was used under an environment with low levels of oxygen (with concentration of O_2 <1 ppm, parts per million) in a glove box filled with N_2. The pH values of the test solutions were adjusted to appropriate levels (10, 11 and 12.5) by adding measured quantities of $Ca(OH)_2$ into deionized water. During the experiments, the temperatures of the solutions were set to three different conditions (35, 50 and 70 °C). The ratio of the solid phase to the test solution was set to 2.5 $g/0.25$ dm^3. The testing was carried out for 30 days.

The pH measurements for each sampling were obtained using the glass electrode method (Model: D52, Electrode: 9621-D, Horiba Ltd.) and the pH values were maintained at a constant value by the addition of saturated $Ca(OH)_2$ solution when they were lowered. During the analysis, the concentrations of Al in the sampled solutions were measured using inductively coupled plasma atomic emission spectroscopy (ICP-AES, ICPS-8000, Shimadzu Corp.).

Next, the pulverized samples obtained by synthetic rock solidification (with grain sizes ranging from 38 to75 μm) were immersed in $Ca(OH)_2$ solution for 15 days in a glove box filled with N_2 in a low-oxygen environment (with concentration of O_2 <1 ppm) and were analyzed. The immersion temperature was to a constant value of 35 °C, and ionic strength of the test solution was adjusted to match the value of 0.1 mol/dm^3 NaCl solution. The pH values were adjusted to specific levels (10, 11 or 12.5) by adding Ca(OH)$_2$. The chemical forms of Al in the solutions were analyzed using ^{27}Al-NMR (AVANCE 500, Bruker Corp.)

RESULTS AND DISCUSSION

The changes in the Al concentration in the solution with time as observed in the dissolution experiments is shown in Figure 1 and 2. The Al concentration with time increased sharply during the first five days in all the cases and subsequently, the rate of increase reduced until little change was observed in the latter half and a constant value was achieved. On comparing the changes in the concentration at each pH level, the final (constant) Al concentration marginally increased with increase in pH values. With regard to the temperature differences, it was almost same that the time until a constant Al concentration was achieved. However, the constant Al concentration achieved increased with increase in the temperature. For the effect of the specific surface area of the specimen, on the difference of the constant Al concentration achieved for each temperature was insignificant, although the initial slope varied.

From the ^{27}Al-NMR, the location of the signal at around 80 ppm (reference peak was that of Al^{3+} in nitrate solution) was observed. Therefore, the chemical form of Al in the solution seemed to be Al(OH)$_4^-$.

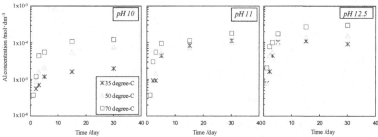

Figure 1. The concentration of Al at various pH values at different temperatures: the specimens with grain sizes ranging from 0.15 to 0.3 mm.

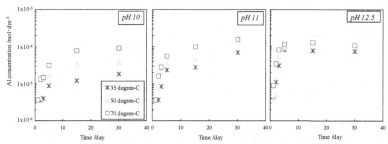

Figure 2. The concentration of Al at various pH values at different temperatures: the specimens with grain sizes ranging from 0.3 to 0.6 mm.

To evaluate the matrix dissolution reactions, the experimental results and the results of thermodynamic calculations carried out based on the ionic equilibrium state present in the solution were compared.

Previous studies reported the dissolution of Al_2O_3 in alkaline environments by reaction with OH^- [4, 5]. Therefore, it is supposed that the matrix dissolution occurred through the reaction between α-Al_2O_3 and OH^-, which resulted in the release of $Al(OH)_4^-$.

$$Al_2O_3 + 2OH^- + 3H_2O \rightarrow 2Al(OH)_4^- \qquad (1)$$

Subsequently, the experimental results and the solubility of Al from previous reports were compared. Since the Al concentration remained constant after a certain period even in experiments using different grain sizes (specific surface areas), it was assumed that Al in solution had reached the solubility limit. Thus, the solubility obtained from the experimental results (after immersion for 30 days) and those calculated from the thermodynamic equilibrium equations were compared. The equilibrium constants and the equations used were from literatures [4,6] and the equilibrium constant (K) at specific temperature was determined using van't Hoff equation. The values of the activities from the experimental results were corrected using the Debye-Hückel equation.

Figure 3 shows the results of the comparison between the experimental results and the theoretical calculations. In those cases, Al concentration was not solubility of corundum (Al_2O_3). Although concentrations close to the solubility of gibbsite ($Al(OH)_3$) are observed at pH values of 10 and 11, the experimental values departed from the solubility of gibbsite at pH 12.5. This observation indicated that the effect of Ca became more prominent with the increase in pH and that the solid phase that restricted the solubility changed from gibbsite to a complex containing Ca (for example, $Ca_3Al_2(OH)_{12}$, Hydrogarnet).

Subsequently, the matrix dissolution kinetics was examined. We assumed that kinetic equations describing phenomena in which the dissolution rate is decreasing near to the saturation concentrations range could be applied to the dissolution of the matrix in the synthetic rock solidification process. The rate constant was assumed a slope of test data until fifth day, in the initial stages of the test, during which C/C_{max} can be assumed to be approximately zero.

$$\frac{dC}{dt} = k\frac{A}{V}\left(1 - \frac{C}{C_{max}}\right) \qquad (2)$$

$$\frac{dC}{dt} = k\frac{A}{V} \text{ (in the case of } C_{max} \gg C) \qquad (3)$$

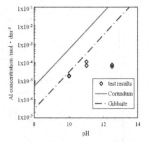

Figure 3. Concentrations of Al (experimental) and the solubilities (calculated by thermodynamics) in the pulverized matrix after immersion for 30 days, at 35 °C

Where, C is the Al concentration (mol/dm³), C_{max} is the solubility of Al (mol/dm³), k is the rate constant (mol/(cm² d)), A is the surface area (cm²), and V is the volume of the liquid phase (dm³).

Previous study reported the dissolution of Al_2O_3 in alkaline environments by reaction with OH^- [5]. And our confirmed on the chemical form of Al in solution as $Al(OH)_4^-$, it is assumed that the dissolution rate depended on OH^-. Consequently, the logarithmic rate of dissolution can be expected to be proportional to the pH in the dissolution mechanism of the matrix composed of alumina (corundum). Figure 4 shows the relationship between the rate constant and pH.

$$\log k \propto pH \text{ , i.e., } k \propto \left[OH^-\right]^\alpha \qquad (4)$$

Further, an Arrhenius-type function was assumed for the variation of rate constant with temperature. Figure 5 shows the relationship between the rate constant and the reciprocal temperature.

$$k \propto \exp\left(-\frac{E_a}{RT}\right) \qquad (5)$$

Using the above assumptions, we derived an equation for the rate constant. Equation 6 was derived by fitting the rate constant calculated for each condition based on the experiments carried out at different temperatures, pH levels, and surface areas, using multivariate least squares method for temperature and pH.

$$k = 6.28\times10^{-8} \cdot \left[H\right]^{-0.316} \cdot \exp\left(\frac{-24800}{RT}\right) \qquad (6)$$

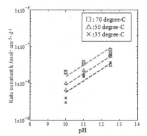

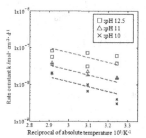

Figure 4. Variation of dissolution rate constant of matrix with pH. The dashed lines indicate the calculated values from eq. 6.

Figure 5. Variation of dissolution rate constant of matrix with reciprocal temperature. The dashed lines indicate the calculated values from eq. 6.

Based on the results shown in figure 4, which shows the dependence of the rate constant on pH, it can be seen that the experimentally observed rate increased with increase in pH, in agreement with equation 6.

The changes in the concentration observed during the dissolution tests, the values of saturation concentration obtained from the experiment at various pH values and temperatures, were compared to the calculation from the dissolution rate by equation 6. The result showed in figure 6. The concentration change after a certain period can be calculated by the integral of the rate equation.

$$C = C_{max} \cdot \left(1 - \exp\left(\frac{-kA}{VC_{max}} \cdot t\right)\right) \quad (7)$$

The values of concentration changes obtained by equation 7 showed good correlation with the experimental results, although minor deviations were observed. Hence, the our estimated equation of the dissolution kinetics we have derived for the rate constant and the dissolution chemical equation are in agreement with the results of our study.

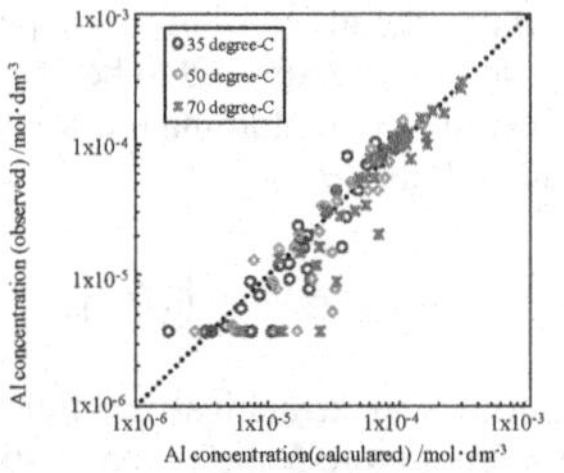

Figure 6. Comparison of the experimental results with the estimated results of the Al concentration for all the data points obtained.

CONCLUSIONS

Using specimens of pulverized products of synthetic rock solidification, we estimated the changes in Al concentration with time at various temperatures (35–70 °C) and pH levels (10–12.5) during dissolution in $Ca(OH)_2$ solution. It was found that the dissolution rate of the matrix was accelerated with increase in either the temperature or pH. Further, the matrix dissolution was restricted by the solubility of other substances such as $Al(OH)_3$ (gibbsite). Based on these results, an equation to estimate the matrix dissolution rate was proposed and a good correlation was observed between the calculated values and the experimental results.

Depending on matrix dissolution, the iodine confined in matrix as AgI is released from synthetic rock. It is therefore concluded that the long-term iodine release behavior from the solidified material can be evaluated using the dissolution behavior of the matrix.

ACKNOWLEDGMENT

This research is a part of the project on "Research and development of processing and disposal technique for TRU waste containing I-129 and C-14 (FY2011)" under a grant from the Agency of Natural Resources and Energy , the Ministry of Economy Trade and Industry of Japan (METI).

REFERENCES

1. FEPC and JAEA, "Second progress report on research and development for TRU waste disposal in Japan; Repository Design, Safety Assessment and Means of Implementation in the Generic Phase," (2007)
2. Y. Inagaki and T. Imamura, *J. Nucl. Sci. Technol.* **45**, 859 (2008).
3. M. Tada, Y. Inagaki, et al., *Proceedings of GLOBAL 2011, Paper No. 446834* (2011)
4. J. Zhang, M. Klasky, et al., *J. Nucl. Mater.* **384**, 175 (2009).
5. S. A. Carroll-Webb and J. V. Walther, *Geochim. Cosmochim. Acta* **52**, 2609 (1988).
6. D. L. Parkhurst and C. A. J. Appelo, *User's guide to PHREEQC (Version 2)—A computer program for speciation, batch-reaction, one-dimensional transport, and inverse geochemical calculations: U.S. Geological Survey Water-Resources Investigations Report* **99-4259**, 310 (1999).

Mater. Res. Soc. Symp. Proc. Vol. 1518 © 2013 Materials Research Society
DOI: 10.1557/opl.2013.68

A study on Iodine Release Behavior from Iodine-Immobilizing Cement Solid

Yoshiko Haruguchi[1], Shinichi Higuchi[1], Masamichi Obata[1], Tomofumi Sakuragi[2], Ryota Takahashi[2], Hitoshi Owada[2]

1 TOSHIBA Corporation, 4-1 Ukishima-cho, Kawasaki-ku, Kawasaki, Japan
2 Radioactive Waste Management Funding and Research Center, 1-15-7 Tsukishima, Chuo-ku, Tokyo, Japan

ABSTRACT

An iodine-immobilizing cement solidification process using calcium aluminate cement with gypsum additive was developed. Powdered cement solid was repeatedly immersed in ion-exchanged water with varying liquid-to-solid ratios (L/S) in accelerated dissolution tests simulating interaction with groundwater at waste disposal sites. The measured concentrations of iodine in the water were on the order of 10^{-5} to 10^{-3} mol·dm^{-3} in the entire L/S range. These concentration levels are extremely low compared with those in the case of ordinary Portland cement. Calculations with a solution equilibrium model for the cement immersed in ion-exchanged water showed that the observed iodine release profile versus integrated L/S ratio from the immersion test was explained by a dissolution model of minerals in the cement.

INTRODUCTION

I-129 is considered to be the most important key radionuclide in evaluating the safety of geological disposal of radioactive waste generated in reprocessing plants (TRU Waste) [1]. I-129 is a very long-lived radionuclide (half-life: 15.7 million years), and an effective approach to reducing the peak radiation dose is to solidify (immobilize) it in a solid substance having excellent long-term confinement performance and to allow the I-129 to be released gradually over a long period.

In this context, we examined a technology for immobilizing iodine using a cement material. Instead of ordinary Portland cement, we adopted calcium aluminate cement as a base material and added gypsum to form ettringite (AFt) and monosulfate (AFm) phases [2]. The iodate ion (IO_3^-) is easily absorbed in AFm and AFt compared with the iodide ion. When a solution containing IO_3^- ions is mixed with this cement, AFm and AFt with both iodine and sulfur as the oxyanion are formed in the solidification process. To employ this in TRU waste disposal, the iodine release behavior must be examined in long-term tests.

In this work, we conducted immersion tests in ion-exchanged water and evaluated the iodine release behavior using a numerical model in which the cement consists of several minerals in solution equilibrium. The iodine-containing minerals and their thermodynamic data of the numerical model were supplied by results of this work.

IODINE RELEASE BEHAVIOR RELATED TO SOLID ALTERATION

Experimental Procedure

An iodine-immobilizing cement solid was prepared by mixing $NaIO_3$ solution and calcium aluminate cement material containing gypsum as an additive. After hardening, its mineral composition and the iodine existence ratio of each mineral were evaluated by using a chemical separation method consisting of solvent extraction and dissolution in acids.

To evaluate the iodine-immobilizing capability of the cement matrix, a continuous-dissolution accelerated test was conducted by repeatedly replacing the ion-exchanged water serving as an immersing fluid, as shown in Figure 1. The concentrations of some constituent elements of the cement (Ca, Al, S, I) in each immersing fluid were measured, and the mineral composition in the recovered solid was determined quantitatively by use of a chemical separation technique based on Suzuki's method [3]. The concentrations of iodine and other cementitious elements in each sample were analyzed by the ICP method.

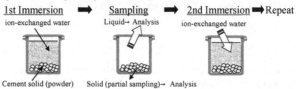

Figure 1. Set-up for continuous-dissolution accelerated test.

Results and Discussion

The initial composition of the cementitious minerals and the iodine existence ratio in the cement solid are shown in Table 1.

Table 1. Initial mineral composition and iodine existence ratio in cement solid

	AFt	AFm	HG	Al(OH)$_3$	Others
Mineral composition (%)	12	23	19	45	2
Iodine existence ratio in minerals (%)	79	6	16	N.D.	N.D.

The concentrations of the cementitious elements relative to the integrated value of the liquid-to-solid ratio (L/S) are shown in Figure 2. The iodine concentration is a sufficiently low value. Figure 3 shows the mineral compositions and iodine existence ratio of every cement solid. Alteration of the cementitious minerals started at L/S=150, where the Hydrogarnet (HG) and AFm contents decreased. The AFt content increased at L/S ratios between 300 and 500 and decreased at L/S ratios between 700 and 1200. Iodine was detected in the AFt, AFm, and HG phases. The increase in iodine concentration is related to the dissolution of these minerals. At L/S=1400, when the AFt mineral completely disappeared, the whole amount of iodine was released into liquid phase.

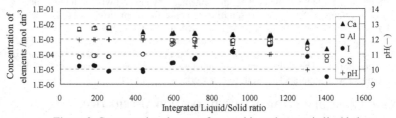

Figure 2. Concentration changes of cementitious elements in liquid phase.

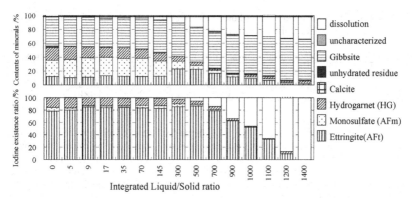

Figure 3. Mineral compositions and iodine existence ratio of every cement solid.

IODINE RELEASE MODEL FOR LONG-TERM EVALUATION

Characterization of AFt containing IO₃

AFt containing IO_3^-, having the formula $Ca_6[Al(OH)_6]_2 \cdot [x(IO_3) \cdot (1-x) (SO_4)]_3 \cdot [(26-3x)H_2O \cdot 3x(OH)]$, was synthesized by use of the chemical reagents $NaAlO_2$, $Ca(OH)_2$, $NaIO_3$, and Na_2SO_4. These chemicals, weighed as ideal molar ratio (the value of x was varied from 0.1 to 0.8), were agitated in ion-exchanged water and cured. The crystal structure and chemical composition of the synthesized minerals were estimated by XRD and SEM-EDS analysis, respectively.

Major peaks of AFt were observed in the XRD pattern of the synthesized mineral. Table 2 shows the chemical compositions evaluated by SEM-EDS. In all cases, the Ca/Al molar ratio was about 3 (Ca:Al=6:2), which is a typical composition of AFt. When x was 0.1, 0.2, and 0.5, the amount of iodine contained in the synthesized minerals increased with IO_3 input ratio (x). However, when x was 0.8, the iodine content was almost the same as that when x was 0.5, indicating that the amount of iodine taken into AFt seemed to be limited.

Table 2. Molar ratios of elements in synthesized minerals.

IO_3 Input ratio to AFt (x)		0.1	0.2	0.5	0.8
Molar ratios of elements in synthesized minerals	Ca	6.0	6.0	6.0	6.0
	Al	2.0	1.8	1.9	1.7
	I	0.4	0.6	1.8	1.7
	S	2.4	2.3	1.8	1.6

Then, immersion tests (for x=0.2 and 0.5) were performed to evaluate the solubility products. The solubility products and equilibrium constants of the dissolution reaction were evaluated based on the following reaction:

$$Ca_6[Al(OH)_6]_2 \cdot [x(IO_3) \cdot (1-x)(SO_4)]_3 \cdot [(26-3x)H_2O \cdot 3x(OH)]$$
$$\rightarrow 6Ca^{2+} + 2Al(OH)_4^- + 3x\ IO_3^- + 3xSO_4^{2-} + (4+3x)OH^- + (26-3x)\ H_2O$$

The equilibrium constant values of the synthesized minerals were evaluated and shown in Table 3. They seemed to be significantly correlated with the amount of iodine taken into the AFt.

Table 3. Values of equilibrium constant of synthesized mineral's dissolution reaction

	IO_3 Input ratio to AFt (x)=0.2			IO_3 Input ratio to AFt (x)=0.5		
Immersion temperature	4 °C	25 °C	60 °C	4 °C	25 °C	60 °C
Equilibrium constant	2.0×10^{-48}	2.8×10^{-44}	4.6×10^{-42}	3.0×10^{-52}	1.4×10^{-48}	1.8×10^{-44}

Model Description

Judging from the above discussions, it should be possible to understand the iodine release phenomenon as the dissolution of cement minerals containing iodine in their oxyanion channels. Because groundwater is eventually stationary in a waste deposit facility, the system of the minerals and the water must be described in equilibrium. Therefore, the iodine release from the cement solid is evaluated in this section by using a solubility equilibrium model of the minerals that make up the cement.

PHREEQC ver. 2[4] was used for the calculation of the equilibrium. The initial composition of minerals and the iodine distribution in the cement minerals were set using the results of chemical analysis shown in Table 1.

AFt, AFm, and HG have forms $Ca_6[Al_2(OH)_6]_2 \cdot 24H_2O \cdot (SO_4)_3 \cdot 2H_2O$, $[Ca_2Al(OH)_6]_2 \cdot (SO_4) \cdot 6H_2O$, and $Ca_3Al_2(OH)_{12}$, respectively. The findings in the previous section lead to an ideal solid solution model of AFt containing iodine that consists of two end members $Ca_6[Al(OH)_6]_2 \cdot [0.5(IO_3) \cdot 0.5(SO_4)]_3 \cdot [(26-1.5)H_2O \cdot 1.5(OH)]$ and $Ca_6[Al_2(OH)_6]_2 \cdot 24H_2O \cdot (SO_4)_3 \cdot 2H_2O$. The former is AFt, in which half of the SO_4^{2-} ions is exchanged with IO_3^- and OH^- in the oxyanion channel; thus, it is called IS_AFt for short. The latter is AFt with 100% SO_4^{2-} in the oxyanion channel (simply called AFt for short). The solubility products for the two minerals and the other minerals containing iodine, named IO_3-AFm and I-HG, are given by the dissolution tests for the synthesized minerals described in the previous section (see Table 4).

Table 4. Solubility products of IS_AFt and AFt

Mineral	Chemical Reaction	log K
IS-AFt	$Ca_6(Al(OH)_6) \cdot 2((IO_3)_{0.5}(SO_4)_{0.5}) \cdot 3(H_2O) \cdot 24.5(OH)_{1.5}$ $= 6Ca^{2+} + 2Al(OH)_4^- + 1.5IO_3^- + 1.5SO_4^{2-} + 5.5OH^- + 24.5H_2O$	-51.29
AFt	$Ca_6Al_2O_6 \cdot (SO_4)_3 : 32H_2O$ $= 6Ca^{2+} + 2Al(OH)_4^- + 3SO_4^{2-} + 4OH^- + 26H_2O$	-43.94
IO_3-AFm	$(CaO)_3 \cdot Al_2O_3 \cdot Ca(IO_3)_2 : 12H_2O$ $= 4Ca^{2+} + 2Al(OH)_4^- + 2IO_3^- + 4OH^- + 6H_2O$	-36.8
I-HG	$Ca_3Al_2(OH)_8(IO_3)_4 = 3Ca^{2+} + 2Al(OH)_4^- + 4IO_3^-$	- 33.20

IO_3-AFm is a mineral in which 100% of the SO_4^{2-} is exchanged with IO_3^-, and log K was obtained by the dissolution test for the synthesized mineral. For I-HG, a certain part of OH^- is exchanged with IO_3^-. Although such a mineral was not observed in the tests, since there is a study reporting a mineral that contains CaO and Al_2O_3 and that has Cl^- instead of OH^-, the I-HG was assumed in this study. Log K was treated as a variable to explain the iodine release in the immersion test, shown later in detail.

Dissolution of the minerals causes the release of iodine into the groundwater in solubility equilibrium. The thermodynamics database JNC-TDB (2005) [5] was used, except for the specifically defined minerals mentioned above.

Comparison with the immersion test

The initial composition of minerals in the cement solid is shown in Table 5. This table is based on the results of chemical analysis and the considerations of iodine attribution described in the previous section.

Table 5. Initial mineral composition of the cement solid [mol/kg].

IS_AFt	AFt	IO$_3$-AFm	AFm	I-HG	HG	Al(OH)$_3$	CAH10	Na$_2$O
0.0668	0.0219	0.0035	0.3672	0.0050	0.4708	5.693	0.0646	0.128

Cement alteration and iodine release with the initial minerals were calculated in equilibrium using PHREEQC, and the results were compared with the values obtained in the immersion test with ion-exchanged water. The results showed that the iodine concentrations were low compared with the test results at each L/S value up to the integrated value L/S=300. This L/S range corresponds to the interval where IS_AFt exists in the system as a solubility limiting solid phase. Very stable IS_AFt is precipitated in this L/S range. However, since the test results did not show such a tendency, it is considered that the precipitation does not occur in the test conditions. Unlike the ion-exchanged process, precipitation has such a low rate that the process could not complete within the test period. Under the assumption that complete precipitation of iodine into minerals is prohibited in the test period and only I-HG can dissolute with the proper log K value due to its relatively thermodynamically unstable properties, the iodine concentrations were in good agreement with the test. Thus, I-HG was considered to play a role of a solubility-limiting phase while it existed at relatively low L/S. After the I-HG completely dissolved, the other minerals containing iodine, namely IS_AFt and IO$_3$-AFm, determined the iodine concentration. Eventually, an iodine release transition profile versus L/S was obtained by calculation, as shown in Figure 4, in which the test results are also shown for comparison. There was good agreement between the calculated and measured element concentration in the liquid phase. Figure 5 shows the iodine distribution in the solid and liquid system, with a comparison of the calculated and measured results. It is revealed that the iodine release is governed by the dissolution of minerals containing iodine, namely I-HG at low L/S and IS_AFt at high L/S, up to complete release.

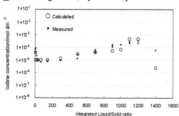 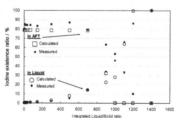

Figure 4. Comparison of iodine release tendency (calculated and measured)

Figure 5. Comparison of iodine existence ratio (calculated and measured)

CONCLUSIONS

The iodine release behavior from iodine-immobilizing cement solid was assessed by both experimental and theoretical methods. It was considered that iodine was chemically stable in the following minerals:

I-HG $(Ca_3Al_2(OH)_8(IO_3)_4)$,

IO_3-AFm$(Ca_6[Al_2(OH)_6]_2 \cdot 24H_2O \cdot (SO_4)_3 \cdot 2H_2O)$ and

IS_AFt$(Ca_6[Al(OH)_6]_2 \cdot [0.5(IO_3) \cdot 0.5(SO_4)]_3 \cdot [(26-1.5)H_2O \cdot 1.5(OH)])$.

IS_AFt was assumed to have the maximum iodine content in its oxyanion channel and to form a solid solution with AFt having 100% SO_4^{2-} in an $IO_3^- - SO_4^{2-}$ coexistent system. Despite the fact that such a treatment and other assumptions about iodine-containing minerals have not yet been directly validated, the equilibrium calculation with this mineral system immersed in ion-exchanged water shows that the iodine release profile versus integrated liquid/solid ratio in consecutive water exchange tests was explained by dissolution of these minerals.

It is worth noting that the picture of the mineral alteration and iodine release obtained in the present work does not depend on the test case used. Therefore, the evaluation method should be applicable in general to other conditions. To verify this, it will be necessary to examine other chemical conditions of groundwater in future work.

ACKNOWLEDGMENTS

This research is a part of the project on "Research and development of processing and disposal technique for TRU waste containing I-129 and C-14 (FY2011)" under a grant from the Agency of Natural Resources and Energy , the Ministry of Economy Trade and Industry of Japan (METI).

REFERENCES

1. FEPC and JAEA, "Second Progress Report on Research and Development for TRU Waste Disposal in Japan", Repository Design, Safety Assessment Means of Implementation in the Generic Phase (2007).
2. F. Tomita, et. al., "Development of Iodine Immobilization Process with Cementitious Materials", ICEM 99 (1999).
3. K. Suzuki, et. al., Concrete Research and Technology, vol.1(2), p39- (1990), (in Japanese).
4. D. L. Parkhurst and C. A. J. Appelo, "User's Guide to PHREEQC (VERSION2)," US. Geological Survey, (1999).
5. JNC TN8400, 2005-010, "Development of Thermodynamic Database for Hyperalkaline, Argillaceous Systems," July 2005.

Mater. Res. Soc. Symp. Proc. Vol. 1518 © 2013 Materials Research Society
DOI: 10.1557/opl.2013.927

Towards a Silicate Matrix for the Immobilisation of Halide-rich Wastes

M. R. Gilbert[1]
[1]AWE, Aldermaston, Reading, RG7 4PR, UK.

ABSTRACT

Single-phase calcium chlorosilicate and sodalite, two potential ceramic waste-forms for the immobilisation of $CaCl_2$-based pyroprocessing wastes, have been fabricated at temperatures below the volatilisation point of $CaCl_2$. Solid solutions doped with Sm^{3+} as an inactive analogue for trivalent actinides have been fabricated and characterised. XRD analysis shows both phases will successfully accommodate Sm^{3+}, with the sodalite in particular remaining single-phase. Fabrication of Sm-doped calcium chlorosilicate in air results in the formation of SmOCl and $Ca(Si_2O_5)$ secondary phases, however, calcination in an inert atmosphere is shown to successfully retard the formation of SmOCl allowing for higher levels of doping.

INTRODUCTION

Pyrochemical reprocessing techniques enable the recovery of Pu metal from spent nuclear material without the need to convert it to PuO_2 and back [1]. These methods utilise an electrorefining process, where the Pu is separated from the impurities in a molten chloride salt, most typically either $CaCl_2$ or an equimolar mixture of NaCl-KCl, at temperatures of between 750 – 850 °C [2]. Post-reprocessing, this chloride salt must be replaced, as it now contains a number of different waste streams which will contaminate the cathode and affect the properties of the molten salt. This contaminated salt must be disposed of in such a way as to immobilise the radionuclide chlorides contained within. However, halide-rich wastes such as these can be problematic to immobilise, as not only are their solubilities in melts very low, but even in small quantities they can seriously affect the properties of the waste-form [3,4]. In addition, processing temperatures are often severely limited in order to prevent the volatilisation of the halides.

One approach is to immobilise the radionuclide chlorides in a ceramic waste-form based upon mineral phases with naturally high chloride contents. Current research at AWE is focused on wastes arising from pyroprocessing in $CaCl_2$ salt, limiting the processing temperatures to a maximum of 800 °C to prevent substantial loss of chlorine via volatilization. Potential candidates under investigation include chloroapatite, calcium chlorosilicate, rondorfite, sodalite and quadridavyne, of which calcium chlorosilicate and sodalite are presented here.

Calcium chlorosilicate ($Ca_3(SiO_4)Cl_2$) has one of the highest Cl contents of any chlorine-containing silicate at ~ 25 wt. %. It has a monoclinic structure consisting of an approximately cubic close-packed arrangement of SiO_4^{4-} tetrahedra and Cl^- anions, with the Ca^{2+} cations occupying interspersed octahedral sites in such a way that a distorted NaCl-type structure results [5-8]. Calcium chlorosilicate has been investigated as a potential waste-form for the immobilisation of precipitated chlorine [9] and is seen as potentially highly suitable for those wastes resulting from $CaCl_2$ based processes.

Sodalite ($Na_8[AlSiO_4]_6Cl_2$) is a crystalline aluminosilicate formed of a framework of all-corner-linked tetrahedra made up by the fusing of the 4-membered rings of the β-cages, creating a microporous structure with a typical pore diameter of 4 Å [10]. Work carried out at Argonne National Laboratory on glass-bonded sodalites have demonstrated them to be effective for immobilising wastes from NaCl-KCl based reprocessing, though preparation temperatures range from 850 – 1000 °C, which leads to the loss of chlorine through volatilisation [11-13].

EXPERIMENT

Sm-doped Calcium Chlorosilicate Fabrication

Solid solutions of $(Ca_{3-1.5x}Sm_x)(SiO_4)Cl_2$, where $x = 0$, 0.05, 0.1, 0.15 and 0.2, were prepared based on the method described by Leturcq *et al* [9]. Stoichiometric quantities of $CaCl_2$, $SmCl_3$, CaO and SiO_2 were placed into a Nalgene mill pot together with yttria-stabilised zirconia (YSZ) milling media and dry milled overnight. The resulting powder mix was passed through a 212 μm sieve mesh, placed in an alumina crucible, heated to 300 °C to drive off any absorbed moisture and then calcined at 750 °C for 3 h in air. An identical set of solid solutions were also calcined at 750 °C for 3 h under flowing Ar (4 $l.min^{-1}$).

Sm-doped Sodalite Fabrication

Solid solutions of $(Na_{8-3x}Sm_x)(AlSiO_4)_6Cl_2$, where $x = 0$, 0.05, 0.1 and 0.2, were prepared via an anhydrous nepheline $(NaAlSiO_4)$ intermediate (Equations 1 and 2) based on the method described by De Angelis *et al* [14].

$$Al_2Si_2O_7.2H_2O + 2NaOH \rightarrow 2NaAlSiO_4 + 3H_2O \tag{1}$$

$$6NaAlSiO_4 + (2-3x)NaCl + xSmCl_3 \rightarrow (Na_{8-3x}Sm_x)(AlSiO_4)_6Cl_2 \tag{2}$$

Nepheline was first synthesized by weighing stoichiometric quantities of kaolinite $(Al_2Si_2O_7.2H_2O)$ and NaOH, together with a 5 wt. % excess of NaOH, into a Nalgene mill pot with YSZ milling media and dry milling overnight. The resulting powder mix was passed through a 212 μm sieve mesh, placed in an alumina crucible, heated to 300 °C to drive off any absorbed moisture and then calcined at 700 °C for 3 h in air. The calcined nepheline plug was then broken up using a pestle and mortar and passed through a 212 μm sieve mesh.
Stoichiometric quantities of $NaAlSiO_4$, NaCl and $SmCl_3$ were then placed into a Nalgene mill pot together with YSZ milling media and dry milled overnight. The resulting powder mix was passed through a 212 μm sieve mesh, placed in an alumina crucible, heated to 300 °C to drive off any absorbed moisture and then calcined at 750 °C for 48 h in air to produce the Sm-doped sodalites.

X-ray Diffraction (XRD)

Powder XRD was carried out using a Bruker D8 Advance diffractometer operating in Bragg-Brantano flat plane geometry using Cu $K_{\alpha1}$ radiation ($\lambda = 1.54056$ Å). Diffraction patterns were measured over a 2θ range of 10 – 90° using a step size of 0.025° and a collection time of 3.4 s per step.

92

RESULTS & DISCUSSION

Undoped Calcium Chlorosilicate

In the method detailed by Leturcq *et al.*, calcination of the milled $CaCl_2$, CaO and SiO_2 powders at 700 °C produces a $Ca_3(SiO_4)Cl_2$ major phase together with a second, $CaCl_2$ deficient calcium chlorosilicate phase [9]. By increasing the calcination temperature to 750 °C single-phase $Ca_3(SiO_4)Cl_2$ can be fabricated without exceeding the volatilization temperature of $CaCl_2$ (Figure 1).

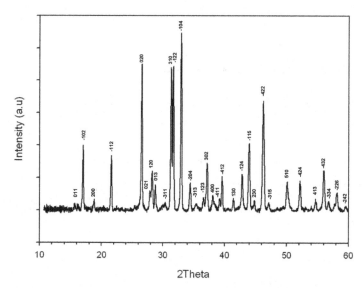

Figure 1. Indexed XRD pattern of $Ca_3(SiO_4)Cl_2$ fabricated by calcination at 750 °C for 3 h.

Sm-doped Calcium Chlorosilicate

The XRD patterns for the Sm-doped calcium chlorosilicates can be seen in Figure 2. XRD analysis of the calcined powders shows $Ca_3(SiO_4)Cl_2$ to be the major phase, together with minor phases of SmOCl and a $Ca(Si_2O_5)$ buffer phase, both of which increase with increasing levels of Sm doping.

To try and reduce the levels of SmOCl formed a series of calcinations were performed in an inert atmosphere. The XRD patterns for the $(Ca_{3-1.5x}Sm_x)(SiO_4)Cl_2$ compositions calcined under flowing Ar are shown in Figure 3. Comparison with the diffraction patterns of those chlorosilicates fabricated in air shows a significant reduction in the amount of SmOCl (and hence $Ca(Si_2O_5)$) formed when calcined under Ar, with the major 101 and 102 SmOCl reflexions now not appearing until the x = 0.1 composition.

93

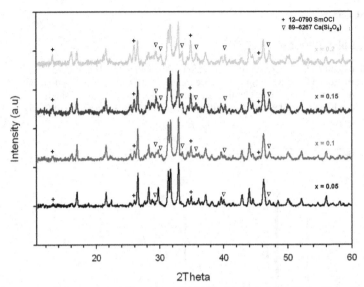

Figure 2. XRD patterns of $(Ca_{3-1.5x}Sm_x)(SiO_4)Cl_2$ calcined at 750 $^{\circ}$C for 3 h in air.

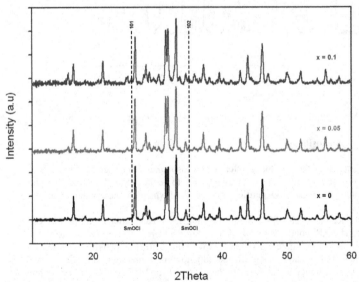

Figure 3. XRD patterns of $(Ca_{3-1.5x}Sm_x)(SiO_4)Cl_2$ calcined at 750 $^{\circ}$C for 3 h under Ar.

Sm-doped Sodalite

XRD patterns of both undoped and Sm-doped sodalite are shown in Figure 4. The XRD pattern of the undoped $Na_8(AlSiO_4)_6Cl_2$ shows that single-phase sodalite has been successfully fabricated at 750 °C, below the volatilization temperature of $CaCl_2$. Analysis of the Sm-doped sodalite patterns shows the sodalite matrix to be a superior host for the Sm^{3+} than the chlorosilicate, the $(Na_{8-3x}Sm_x)(AlSiO_4)_6Cl_2$ compositions remaining almost entirely single-phase sodalite. The first indications of a minor phase(s) can be seen appearing in the x = 0.1 and x = 0.2 compositions in the region 21 – 30° 2θ, however, at this current level of doping the intensity is too low for these to be identified.

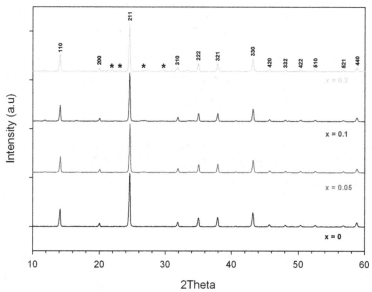

Figure 4. XRD patterns of $(Na_{8-3x}Sm_x)(AlSiO_4)_6Cl_2$ calcined at 750 °C for 48 h in air.

CONCLUSIONS

Fabrication of both single-phase calcium chlorosilicate and sodalite at temperatures below the volatilization point of $CaCl_2$ has been successful and both matrices have been shown to successfully incorporate Sm^{3+} as an inactive analogue for trivalent actinides. When calcined in air, sodalite has been shown to the be superior host, remaining almost entirely single-phase, whereas, in the case of calcium chlorosilicate, in-growth of SmOCl and $Ca(Si_2O_5)$ secondary phases can be observed, even at low doping levels. However, calcination of the calcium chlorosilicate in an inert atmosphere has been shown to significantly improve performance by retarding the formation of SmOCl (and hence $Ca(Si_2O_5)$) leading to potentially higher doping levels than when prepared in air.

© British Crown Owned Copyright 2012 /AWE. Published with the permission of the Controller of Her Britannic Majesty's Stationery Office.

REFERENCES

1. T. Nishimure, T. Koyama, M. Iizuka, H. Tanaka, *Prog. Nucl. Energy,* **32**, 381 (1998).
2. I. N. Taylor, M. L. Thompson, T. R. Johnson, *Proceedings of the International Conference and Technology Exposition on Future Nuclear,* **1,** 690 (1993).
3. W. E. Lee, R. W. Grimes, *Energy Materials,* **1,** 22 (2006).
4. T.O. Sandland, L.-S. Du, J.F. Stebbins, J.D. Webster, *Geochim. Cosmochim. Acta,* **68,** 5059 (2004).
5. E. N. Treushnikov, V. V. Ilyukhin, N. V. Belov, *Doklady Akademii Nauk SSSR,* **193,** 1048 (1970).
6. R. Czaya, G. Bissert, *Kristall und Technik,* **5,** 9 (1970).
7. R. Czaya, *Z. Anorg. Allg. Chem.,* **372,** 353 (1970).
8. R. Czaya, G. Bissert, *Acta Cryst. B,* **27,** 747 (1971).
9. G. Leturcq, A. Grandjean, D. Rigaud, P. Perouty, M. Charlot, *J. Nucl. Mater.,* **347,** 1-11 (2005).
10. J. Rouguerol, D. Anvir, C. W. Fairbridge, D. H. Everett, J. H. Haynes, N. Pernicone, J. D. Ramsay, K. S. W. Sing, K. K. Unger, *Pure Appl. Chem.,* **66,** 1739 (1994).
11. M. A. Lewis, C. Pereira, US Patent No. 5,613,240, (18 Mar 1997).
12. C. Pereira, ANL/CMT/CP--84675, (1996).
13. S. Priebe, *Nucl. Tech.,* **162,** 199 (2008).
14. G. De Angelis, I. Bardez-Giboire, M. Mariani, M. Capone, M. Chartier, E. Macerata, *Mater. Res. Soc. Symp. Proc.,* **1193,** 73-78 (2009).

Mater. Res. Soc. Symp. Proc. Vol. 1518 © 2013 Materials Research Society
DOI: 10.1557/opl.2013.162

Decontamination of Molten Salt Wastes for Pyrochemical Reprocessing of Nuclear Fuels

Martin C. Stennett[*], Matthew L. Hand, and Neil C. Hyatt.
Department of Materials Science and Engineering, The University of Sheffield, Sheffield,
S13JD, United Kingdom.

ABSTRACT

Pyrochemical reprocessing of nuclear fuels, in which electrochemical separation of actinides and fission products is mediated by a molten alkali chloride salt (typically a LiCl-KCl eutectic) is of interest for future nuclear energy cycles. A key challenge in the management of pyrochemical reprocessing wastes is decontamination and recycling of the molten salt medium to remove entrained actinides and radioactive lanthanide fission products. Since pyrochlore oxides are promising candidates for the immobilisation of lanthanides and actinides, we sought to use the "problematic" molten salt to our advantage as a reaction medium for low temperature synthesis of titanate pyrochlores. Through control of reaction time and temperature, we demonstrated the synthesis of lanthanide pyrochlores at temperatures as low as 700 °C in 1 h, compared to 1350 °C in 36 h for conventional solid state synthesis. The importance of this study is in demonstrating the potential feasibility for decontamination of pyrochemical reprocessing wastes by simple addition of TiO_2 to form lanthanide and actinide pyrochlores by rapid molten salt assisted reaction at moderate temperature.

INTRODUCTION

Conventional solid state synthesis (SSS) of mixed metal oxides is generally achieved by reaction between metals oxides and/or carbonate reagents at high temperature. SSS is controlled by diffusion of chemical species and long reaction times at high temperature, with intermediate milling steps, are often required to obtain single phase products [1, 2]. The long reaction times typically promote grain growth which may be un-favourable for certain applications such as high strength components. In addition it has been demonstrated that repeated cycling at very high temperatures can lead to decomposition of some mixed metal oxides [1] and volatilization of some reaction components, leading to product non-stochiometry, and potential environmental discharges, which are undesirable [3]. These disadvantages have lead to the development of other synthesis methods which including co-precipitation, molten salt, sol-gel, hydrothermal, liquid-phase and gas-phase reactions [1]. The prime advantage of these methods is that they reduce the diffusion distances required for phase formation which has the effect of reducing the synthesis temperature and time required. The alternative route under investigation here is the molten salt synthesis (MSS) route which utilises low melting point, water-soluble salts, as a liquid medium in which one or more of the reactant species may dissolve. This liquid salt assists rapid diffusion of reactant species at low temperature, which affords a chemically homogeneous product in relatively short reaction times [1, 2, 4, 5]. MSS generally uses a mixture of readily available and inexpensive alkali halide, sulphate or carbonate salts, making MSS an attractive process for industrial scale operations in comparison to other chemical routes such as sol-gel method, which may require costly reagents and require environmentally unfriendly organic precursors and solvents [2]. Recognising the potential of MSS, we have explored the application of this methodology to the synthesis of lanthanide titanate pyrochlores, $Ln_2Ti_2O_7$, which are of interest for immobilisation of lanthanides and actinides from nuclear fuel reprocessing [6-9]. The general composition of a pyrochlore oxide (space group Fd-3m), can be expressed as $A_2B_2O_7$, where the 8-coordinate A site is occupied by trivalent cations (e.g. Ln) and the 6-coordinate B site is occupied by tetravalent cations (e.g. Ti, Zr, Hf, *etc.*) [10].

Future advanced nuclear energy cycles are anticipated to comprise a combination of light water reactors, gas cooled high temperature reactors, and fast breeder systems, to allow multiple recycle of nuclear fuel and burning of minor actinides formed by neutron capture and decay reactions. In this scenario, alternative separations technologies for reprocessing of nuclear fuel are of interest, with the aim of increasing efficiency and throughput, whilst minimising the waste inventory, at acceptable cost. One such technology is the pyrochemical process, in which electrochemical separation of actinides and fission products is mediated by a molten alkali chloride salt (typically a LiCl-KCl eutectic) [11]. A key requirement for decontamination and recycling of the molten salt medium is a process to remove entrained actinides and radioactive lanthanide fission products. Since pyrochlore oxides are promising candidates for the immobilisation of lanthanides and actinides [6-9], the synthesis of titanate pyrochlores in a molten salt medium is clearly of interest as a route to the effective decontamination of pyrochemical molten salt waste streams.

EXPERIMENTAL

Ytterbium (III) oxide (Alfa Aesar, Ward Hill, MA, purity 99.9%) and titanium (IV) (anatase) oxide (Sigma Aldrich, St Louis, MO, purity $\geq$ 99%) were used as oxide reagents and ytterbium (III) chloride hexahydrate, $YbCl_3 \cdot 6H_2O$ (Alfa Aesar, Ward Hill, MA, purity 99.9%) as the rare earth salt reagent. The oxide materials were dried overnight at 600 °C and stored in a desiccator. The ytterbium (III) chloride was used as received; thermogravimetric analysis confirmed the hexahydrate stochiometry. The molten salt medium was composed of KCl (Sigma Aldrich, St Louis, MO, purity 99.0%) and NaCl (Arcos Organics, New Jersey, USA, reagent grade). The molten salt flux was added at an equimolar eutectic ratio (50:50% mol KCl:NaCl). A 7:1 flux to reactant molar ratio (MS: $Yb_2Ti_2O_7$) was used.

Stochiometric quantities of TiO_2, Yb_2O_3 (or $YbCl_3 \cdot 6H_2O$) and salt medium, required to form a 10 g batch, were added to a 12 ml Sialon planetary ball mill pot, containing five 10 mm spherical Sialon milling media, and milled for 5 minutes at 300 rpm. Cyclohexane was added as a carrier fluid. The resulting slurry was oven dried at $\sim$ 90 °C for 1 h and then passed through a 250 micron mesh to separate the powder from the milling media. 0.5 g samples were weighed out and pressed into pellets using a 10 mm diameter stainless steel die at a load of 1000 kg for 1 minute. The resulting green pellets were placed into a closed alumina crucible and fired in air at 700-1200 °C for 1-2 h. The heating and cooling rates were 5 °C min^{-1}. Following MSS, the pellets were broken up and washed with deionised water to remove the molten salt medium and the reactant powders were recovered by vacuum filtration using a 0.2 μm cellulose nitrate filter. For comparison $Yb_2Ti_2O_7$ samples were also prepared by conventional solid state synthesis of Yb_2O_3 and TiO_2 at air at 700-1300°C for 1-2 h and by reactions at 1350 °C for 36 h (5 °C min^{-1} ramp rate) with intermediate regrinding.

All products were analysed by X-ray powder diffraction using a STOE Stadi P diffractometer (STOE & Cie GmbH, Darmstadt, Germany) utilising a Mo Kα radiation source (λ = 0.7093 Å) operating at 40 mA and 45 kV. Data were collected over the range $5° \leq 2\theta \leq 30°$

with a step size of 0.02°. Particle size distribution of the raw materials and $Yb_2Ti_2O_7$ were determined using a VASCO-3 laser particle size analyser (Cordouan Technologies, Pessac, France). Powder morphology and microstructure were examined by scanning electron microscopy (SEM) using a JEOL JSM6400 scanning electron microscope operating with a working voltage of 20 kV. SEM samples were mixed in an acetone solution and placed into an ultrasonic bath. Each sample was withdrawn from the suspension using a disposable Pasteur pipette and dropped onto to a sticky carbon pad, fixed onto an aluminium stub; the solvent was allowed to evaporate before the specimens were carbon coated.

RESULTS AND DISCUSSION

Samples were heated for 1 h at temperatures between 700 and 1200 °C. Powder XRD, Figure 1, revealed the formation of $Yb_2Ti_2O_7$ to occur at temperatures as low as 700 °C. Single phase $Yb_2Ti_2O_7$ was formed at 1200 °C after a reaction time of only 1 h. At lower temperatures, samples were found to contain minor or trace quantities of Yb_2O_3 and TiO_2. At a constant reaction time of 1h, increasing the reaction temperature from 700 °C to 1000 °C significantly reduced the Yb_2O_3 and TiO_2 impurities, as determined from the observed decrease in relative intensity of reflections from these phases in the diffraction patterns, Figure 1.

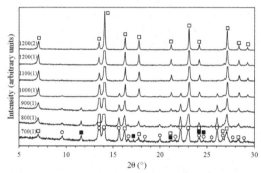

Figure 1. Mo K_α X-ray powder diffraction patterns obtained from $Yb_2Ti_2O_7$ samples synthesised by MSS for 1 h at different temperatures ($\square$ $Yb_2Ti_2O_7$, $\circ$ Yb_2O_3, $\blacksquare$ TiO_2 anatase). Text labels give temperatures in Celsius and reaction time in hours, e.g. 700(1) corresponds to 700 °C for 1 h.

Figure 2 shows powder XRD patterns of the products of 1 h solid state reactions (SSS) between Yb_2O_3 and TiO_2 (anatase), at between 700 and 1300 °C. Whereas near single phase $Yb_2Ti_2O_7$ was formed by MSS at 1100 °C for 1h, SSS under identical conditions produced only trace $Yb_2Ti_2O_7$. Reaction at 1200 °C for 1 h, produced $Yb_2Ti_2O_7$ as the major phase, but with significant quantities of unreacted Yb_2O_3 and TiO_2. Single phase $Yb_2Ti_2O_7$ was only produced by SSS after three 12 h reactions at 1350 °C, with intermittent grinding (Figure 2). It should be noted that reagent TiO_2 converts from the anatase to rutile polymorph above 900 °C in the products of SSS reaction shown in Figure 2.

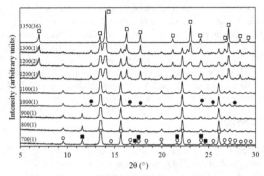

Figure 2. Mo K_α X-ray powder diffraction patterns obtained from $Yb_2Ti_2O_7$ samples synthesised by SSS after 1 h at different temperatures. ($\square$ $Yb_2Ti_2O_7$, $\circ$ Yb_2O_3, $\bullet$ TiO_2 rutile, $\blacksquare$ TiO_2 anatase). Text labels give temperatures in Celsius and reaction time in hours, e.g. 700(1) corresponds to 700 °C for 1 h.

Figure 3 shows the particle size distributions of Yb_2O_3, TiO_2, and single phase $Yb_2Ti_2O_7$ prepared by MSS after 2h at 1200 °C. The particle size distribution of $Yb_2Ti_2O_7$ prepared by MSS was found to be similar to that of TiO_2, with a maximum in distribution at 1.0 µm and 1.1 µm, respectively; in contrast Yb_2O_3 exhibited a maximum in distribution at 2.8 µm.

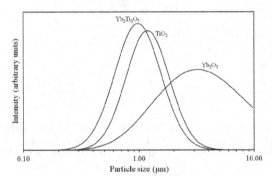

Figure 3. Particle size distribution of TiO_2, Yb_2O_3, and single phase $Yb_2Ti_2O_7$ produced by MSS (1200 °C, 2 h; 50:50% mol NaCl:KCl, 7:1 salt:reagent molar ratio), determined by laser scattering.

Scanning electron micrographs of Yb_2O_3, TiO_2 and $Yb_2Ti_2O_7$ prepared by both MSS and SSS routes, are shown in Figure 4. Inspection of the SEM images shows the average size of TiO_2 and MSS prepared $Yb_2Ti_2O_7$ particles was ~ 1 µm, significantly smaller than that of Yb_2O_3 particles which were ~ 4 µm in size. The globular particle morphology of the TiO_2 and the MSS prepared $Yb_2Ti_2O_7$ particles appear strikingly similar and clearly different from the acicular habit of Yb_2O_3. In contrast, the particles of $Yb_2Ti_2O_7$ prepared by SSS are ~ 3µm in size and exhibit a nodular appearance characteristic of sintered grains (note: this sample was ball milled for 5 minutes between each 12 h reaction at 1350 °C, which explains the particle size intermediate between that of the reagent materials).

Figure 4. SEM micrographs of reagent materials in secondary electron mode; (a) TiO_2, (b) Yb_2O_3, and (c) $Yb_2Ti_2O_7$ produced by MSS (1200 °C, 2 h; 50:50% mol NaCl:KCl, 7:1 salt:reagent molar ratio), (d) $Yb_2Ti_2O_7$ produced by SSS (1350 °C, 36 h total, intermittent grinding).

Molten salt mediated synthesis of lanthanide pyrochlores to immobilize lanthanides and actinides from the waste stream produced by molten salt reprocessing of nuclear fuel was investigated using $YbCl_3$ as a reagent (also acting as an analogue for $PuCl_3$).

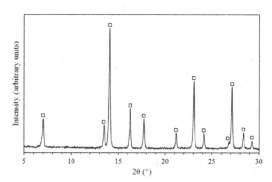

Figure 5. Mo K_α X-ray diffraction pattern obtained from MSS using $YbCl_3$ regent to form $Yb_2Ti_2O_7$ 4h at 1200 °C ($\square$ $Yb_2Ti_2O_7$).

Our preliminary experiments utilised a 50:50% mol NaCl:KCl salt composition and 7:1 salt:reagent molar ratio. In this case, the reagents comprised $YbCl_3 \cdot 6H_2O$ and TiO_2, with the

101

latter present in excess (15 mol%) to target complete formation of $Yb_2Ti_2O_7$ pyrochlore with a surplus of unreacted TiO_2 to mitigate against any batch non-stoichiometry. X-ray powder diffraction, Figure 5, revealed single phase $Yb_2Ti_2O_7$ pyrochlore.

CONCLUSIONS

It is clear from the X-ray diffraction patterns of Figures 1 and 2 that MSS is effective in achieving the formation of $Yb_2Ti_2O_7$ in shorter reaction time and at lower temperature (1200 °C, 1 h), relative to conventional SSS (1350 °C, 36 h). This is attributed to improved diffusion of reactive components mediated by the liquid alkali chloride salt medium. Although single phase $Yb_2Ti_2O_7$ was synthesised only after reaction at 1200 °C for 1 h, the formation of $Yb_2Ti_2O_7$, albeit with significant secondary phases, was observed at a temperature as low as 700 °C after reaction for only 1 h (Figure 1). MSS is therefore effective in reducing the formation temperature of $Yb_2Ti_2O_7$ by several hundred degrees, assisted by enhanced diffusion of reactive components through the molten salt medium. Our attempt to demonstrate the utility of MSS for immobilisation of lanthanides and actinides present in molten salt reprocessing wastes proved successful, since pyrochlore $Yb_2Ti_2O_7$ was formed as single phase from $YbCl_3 \cdot 6H_2O$ and TiO_2 precursors. This paper presents a new and effective method to decontaminate pyrochemical molten salts from nuclear fuel reprocessing. The reduced energy demand of such a low temperature MSS route is also clearly of environmental benefit, compared to the extended reaction schedule at high temperature required to form lanthanide pyrochlores by SSS.

ACKNOWLEDGMENTS

NCH is grateful to The Royal Academy of Engineering and Nuclear Decommissioning Authority for funding. NCH and MCS acknowledge part support from the Engineering and Physical Sciences Research Council, under grant numbers EP/I012214/1 and EP/G037140/1.

REFERENCES

1. D. Segal, J. Mater. Chem. **7**, 1297 (1983).
2. Z. Li, W. E. Lee, and S. Zhang, J. Am. Ceram. Soc. **90**, 364 (2007).
3. Z. Shixi, Q. Li, L. Wang, and Y Zhang, Mater. Lett. **60**, 425 (2006).
4. S. Zhang, D. D. Jayaseelan, G. Bhattacharya, and W. E. Lee, J. Am. Ceram. Soc. **89**, 1724 (2006).
5. S. Hashimoto, S. Zhang, W. E. Lee, and A. Yamaguchi, J. Am. Ceram. Soc. **86**, 1959 (2003).
6. R. C. Ewing, Prog. Nucl. Energy **49**, 635 (2007).
7. R. C. Ewing, W. J. Weber, J. Lian, J. Appl. Phys. **95**, 5949 (2004).
8. W. E. Lee, M. I. Ojovan, M. C. Stennett and N. C. Hyatt, Adv. Appl. Ceram. **105**, 3 (2006).
9. D. P. Reid, N. C. Hyatt, M. C. Stennett, and E. R. Maddrell, Mater. World **19**, 26 (2011).
10. M. A. Subramanian, G. Aravamudan, and G. V. Subba Rao, Prog. Solid State Chem. **15**, 55 (1983).
11. J. P. Ackerman, Ind. Eng. Chem. Res. **30**, 141 (1991).

Mater. Res. Soc. Symp. Proc. Vol. 1518 © 2013 Materials Research Society
DOI: 10.1557/opl.2013.72

Thermodynamic modeling and experimental tests of irradiated graphite molten salt decontamination

Olga Karlina, Michael Ojovan, Galina Pavlova and Vsevolod Klimov
Moscow SIA «Radon»,
119121, Moscow, 7-th Rostovsky per., 2/14, Russia

ABSTRACT

Molten salt flameless oxidation of graphite is one of the prospective methods of irradiated graphite waste processing. Molten salts are capable to retain a considerable part of radionuclides, to neutralize acidic off gases, moreover spent salts could be vitrified on completion of the process. We have used thermodynamic modelling to assess the efficiency of molten salt oxidation of graphite. Equilibrium compositions of both the melt and the off gas were calculated depending on graphite content and temperature. The feasibility of decontaminating the irradiated graphite of its near-surface layers using complete molten salt oxidation was investigated in a series of laboratory experiments. As the molten salt medium used to oxidize irradiated graphite we have investigated lithium, potassium and sodium carbonates. Sodium sulphate, boron oxide, barium and potassium chromates were also used as oxidizers. Tests were carried out at 870–1270 K. The efficiency of decontamination of graphite blocks has been assessed based on the activity of ^{137}Cs and ^{60}Co in the samples before and after molten salt oxidation. Data obtained demonstrated the feasibility of decontamination by molten salt removal of near surface layers on irradiated graphite blocks. Decontamination rate and efficiency depend on oxidizers used and temperature of process.

INTRODUCTION

The amount of accumulated irradiated graphite waste is continuing to grow worldwide. Decommissioning of uranium-graphite reactors generates the main part of irradiated graphite waste where the graphite has been used to moderate and reflect neutrons. More than a hundred of such reactors are located within UK, France, former USSR, USA and Spain.

Technical solutions and industrial technologies are not yet available to immobilize the irradiated graphite contaminated with nuclear fuel inclusions. Processing methods for irradiated graphite such as cementation, covering and impregnation with resins, melting with low-melt alloys, micro-capsulation, self-propagating high-temperature synthesis, do not provide any volume reduction of waste [1]. Considerable waste volume reduction (up to 1 – 3 % relative to initial volume) can be achieved using incineration. However incineration is typically used to treat low and intermediate level waste only. Besides that, radioactive ashes resulting from incineration, aerosols and acidic off gases are inherent to the combustion process and that makes the process complicated and expensive. Flameless molten salt oxidation (MSO) is one of the most promising methods to treat irradiated graphite waste [2,3]. MSO-based technology does not require fine crushing of waste graphite. Using molten mineral salts as the oxidizing medium provides a number of advantages such as universality, the capability to retain a considerable part of radionuclides due to formation of non-volatile compounds, acidic off gas neutralization and the possibility to vitrify spent salts on completion of the process.

THERMODYNAMIC ASSESSMENT

It is assumed that MSO occurs in a semi-closed system at atmospheric pressure. As the process duration is relatively long (hours) we assumed also that the distribution of components is homogeneous both in the gaseous and liquid phases and their composition is close to equilibrium. TERRA computer code has been used for the model calculations [4]. Equilibrium compositions of melt and off gas at various graphite contents were calculated as temperature functions. Thermodynamic properties of compounds formed in the system were taken from available sources [5, 6]. Based on calculations 2.67 and 1.33 kg of oxygen are required to oxidize 1 kg of carbon to CO_2 or CO respectively. Therefore only a small part of the graphite can be oxidized by alkali carbonates decomposition using oxidizing additives. The oxygen yield on lithium, sodium and potassium carbonates decomposition is about $0.1 - 0.2$ kg/kg.

Decomposition of sodium sulphate starts at ~1200 K. The gaseous phase is mainly composed of sodium vapours, sodium sulphates and sulphur dioxide, however significant amounts of oxygen appear only at T = 1300–1400 K. It is known from literature that at the same temperatures intensive oxidation of carbon starts. This perhaps explains the catalytic role of Na_2SO_4 oxidizing additive. Calculations of MSO of graphite with Na_2SO_4 oxidizing additives have been performed for temperatures within the range of 1100–1500 K in an inert argon atmosphere for component ratios (mass %): Na_2CO_3 (56), K_2CO_3 (44), Na_2SO_4 (25), C(0.1–10). They have shown that at these conditions the melt is composed of Na_2CO_3, K_2CO_3, K_2SO_4 and K_2S. It has been noted that complete replacement of sodium sulphate by potassium sulphate occurs in the melt due to the exchange reaction.

XRD phase analysis of solid samples containing 5 % of graphite retained at 1300 K and cooled down were performed using standard protocols. It has shown presence of K_2SO_4, $Na_2CO_3 \cdot H_2O$, as well as small amounts of Na_2SO_4, Na_2S, and K_2S. The appearance of $Na_2CO_3 \cdot H_2O$ instead of Na_2CO_3 is explained by the presence of water in the atmosphere. However it is not clear what caused the absence of crystalline K_2CO_3. Carbon dioxide and monoxide were found in the gas phase above the melt. Sulphur-containing compounds were always present with the highest concentrations of SO_2 and COS. Other compounds of sulphur (S, S_2, SO, S_2O, Na_2SO_4, K_2SO_4) were present at concentrations several orders of magnitude lower compared with SO_2 and COS. Calculations have shown that at carbon concentrations up to 5 mass %, the gaseous phase contains mainly CO_2 (90–95%), and that at higher content of carbon, CO is the main compound. All sulphur-containing compounds were present above 0.001 vol. % only at carbon content up to 7 mass %. The maximum volume content of SO_2 increases from 0.01% at 1200 K to 0.1% at 1700 K. The gaseous phase contains significant amounts of sodium and potassium vapours. On increase of the temperature from 1200 to 1700 K their content increases from 0.01 to 10–25%. Approximately 0.1% by volume is occupied by Na_2, K_2, NaK, K_2CO_3.

Thermodynamic modelling has shown that potassium chromate (K_2CrO_4) also has catalytic properties similar to sodium sulphate. Its melting temperature is almost a hundred degrees higher [5] than sodium sulphate. Hence its decomposition with oxygen release occurs at higher temperatures to assist carbon oxidation. Moreover decomposition products do not contain acidic products like sulphur oxides.

B_2O_3 is an important component of nuclear waste glasses [7, 8]. It has been therefore considered as a potential oxidiser added to the batch to replace completely or a part of sodium sulphate oxidiser. Both calculations and experiments confirmed the efficiency of additives of

B_2O_3 in combination with air blowing through the molten salts although the molten salt batch was more corrosive to alumina crucibles used in experiments.

EXPERIMENTAL

It is known that the depth of surface contamination of NPP graphite blocks by actinides and fission products does not exceed 2 mm [9, 10]. Hence the content of ^{137}Cs, ^{90}Sr, ^{238}Pu, ^{239}Pu, ^{241}Am, ^{243}Am, ^{244}Cm and other radionuclides is significantly lower in the irradiated graphite bulk compared to its near surface layers. This fact could be used to decontaminate large masses of irradiated graphite waste by removing the surface layers. MSO is an appropriate method to effectively achieve such a removal. Real reactor contaminated graphite sleeves have been used in experiments. These were taken from the AM research reactor at the Physical-Energy Institute named after A.I. Leypunsky. Fig. 1 shows a sleeve of mass 0.358 kg. The mass of fragments that have been used in experiments was not higher than 12 g.

Figure 1. An irradiated graphite sleeve of a research reactor.

The radionuclide content of the graphite sleeves was (in Bq/kg): ^{14}C – (0.9–1.4)$\cdot 10^6$, ^{3}H – 7.6$\cdot 10^5$, ^{60}Co – 2.88$\cdot 10^4$, ^{137}Cs – 2.27$\cdot 10^6$, ^{241}Am – 4.78$\cdot 10^4$, ^{154}Eu – 1.5$\cdot 10^3$, ^{144}Ce – 5.3$\cdot 10^3$, $\sum\alpha$ on ^{239}Pu – 3.7$\cdot 10^4$. The total activity of the irradiated graphite fragments used in the experiments was 1.1$\cdot 10^4$ Bq of ^{137}Cs and 1.4$\cdot 10^2$ Bq of ^{60}Co.

MSO Experiments

High temperature experiments on MSO decontamination of graphite were carried out in a laboratory chamber furnace. Working melts for experiments were compositions made of alkali metal carbonates (Table I) with oxidizing catalysts in form of sodium sulphate (Na_2SO_4), boron oxide (B_2O_3), barium and potassium chromates ($BaCrO_4$ and K_2CrO_4). The mass of the oxidizing catalysts was 25 % above the mass of the initial batch. Temperatures used for decontamination were dependent on the batch melting temperatures and were limited to 870–1270 K (Table I). The processing times for the irradiated graphite fragments at a given temperature were from one to three hours. The decontamination tests were carried out with two parallel samples. These were put and removed from the salts for analysis after processing using tongs. Remnant salts were not seen on sample surfaces although some small content of salts could be present in graphite pores. Processed samples have not been washed to avoid removal of loose graphite particles. The controlling parameters for irradiated graphite MSO decontamination were the activities of ^{137}Cs and ^{60}Co in samples before and after the decontamination process. Radiometric analysis of graphite sleeve fragments before and after MSO decontamination was done using the scintillating gamma-spectrometer with processing code «PROGRESS».

Table I. Molten salt batch compositions used in experiments.

Label	Batch composition, mass fraction	Process temperature, K	Melting temperature, K
L	Basic composition: Li_2CO_3 – 0.39; Na_2CO_3 – 0.28; K_2CO_3 – 0.33. Without oxidiser	870–1170	620
LS	Basic composition: Li_2CO_3 – 0.39; Na_2CO_3 – 0.28; K_2CO_3 – 0.33; Oxidizer: Na_2SO_4 – 25% above 100% of basic composition.	870–1170	740
S	Basic composition: Na_2CO_3 – 0.56; K_2CO_3 – 0.44; Oxidizer: Na_2SO_4 – 25% above 100% of basic composition.	1170, 1270	1120
B	Basic composition: Na_2CO_3 – 0.56; K_2CO_3 – 0.44; Oxidizer: B_2O_3 – 25% above 100% of basic composition.	1170, 1270	~1070
XBa	Basic composition: Na_2CO_3 – 0.56; K_2CO_3 – 0.44; Oxidizer: $BaCrO_4$ – 25% above 100% of basic composition.	1170, 1270	~1070
XK	Basic composition: Na_2CO_3 – 0.56; K_2CO_3 – 0.44; Oxidizer: K_2CrO_4 – 25% above 100% of basic composition.	1170, 1270	~1070

The potential to reuse the graphite after the contaminated outer layers have been oxidized has not been exploited in current study. Obviously the decontaminated graphite can be potentially incinerated without problems for radioactive components in off gas.

Experimental results

The decontamination efficiency (β_D) was estimated based on the removal of radionuclides from the irradiated graphite using the equation:

$$\beta_D = (A_i–A_e)/A_i*100\% ,$$

where A_i is the initial contamination, Bq/kg; A_e is the final contamination achieved, Bq/kg. The irradiated graphite MSO decontamination efficiencies are given in Table II and Fig. 2 as a function of batch composition, processing time and temperature.

Table II. Irradiated graphite MSO decontamination efficiencies.

Test	Sample label	T, K	Mean decontamination efficiency, %					
			For ^{137}Cs			For ^{60}Co		
			Decontamination process dwell time, h					
			1	2	3	1	2	3
1	L	870	14	25	28	11	16	18
2	L	970	22	23	30	6	9	18
3	L	1070	27	38	44	9	6	6
4	L	1170	65	78	86	0	0	0
5	LS	870	16	29	33	24	31	32
6	LS	970	24	38	49	18	21	29
7	LS	1070	37	57	63	8	14	18
8	LS	1170	65	88	96	45	60	73
9	S	1170	66	87	96	40	60	77
10	S	1270	95	-	-	85	-	-
11	B	1070	47	60	78	11	15	27
12	B	1170	54	68	82	12	11	17
13	XBa	1070	29	33	35	0	0	0
14	XBa	1170	50	56	62	10	5	10
15	XK	1070	35	37	33	7	10	4
16	XK	1170	43	57	77	12	14	15

Figure 2 (a) shows the decontamination efficiency of real reactor graphite waste for ^{137}Cs at various temperatures in experiments with processing time of three hours. A maximum decontamination efficiency of reactor graphite for ^{137}Cs (e.g. 96%) was obtained in the melt with oxidizer-catalyst in form of sodium sulphate (batch labels LS and S) at 1170 K.

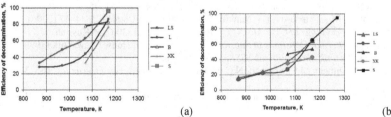

(a) (b)

Figure 2. Decontamination efficiency for ^{137}Cs at processing time – 3 h (a) and at processing time – 1 h (b).

Figure 2 (b) shows the efficiency of MSO decontamination for ^{137}Cs at various temperatures in experiments with a processing time of 1 hour. From the plot given in Fig. 2 (b), it can be seen that the highest decontamination efficiency for^{137}Cs (e.g. 95%) was achieved in 1 h at 1270 K in the molten salt batch labelled S, which contained sodium sulphate as the oxidizer-catalyst. The overall efficiency for ^{60}Co (as seen from Table II) is not high (< 32%) except for molten salt batch compositions which have sodium sulphate oxidizer-catalysts. Within the temperature interval 870–1070 K the decontamination efficiency for^{60}Co gradually decreases for all batches. In the temperature interval 1070–1170 K the temperature dependency of MSO decontamination efficiency for ^{60}Co has various trends. In the melt with boric oxide and without an oxidizer the efficiency decreases whereas in the melts with potassium chromate and sodium sulphate it increases. The efficiency of MSO decontamination for ^{60}Co at various temperatures in experiments with a processing time of 1 hour has reached maximum (e.g. 85 %) in the melt with oxidizer-catalyst in the form of sodium sulphate (batch label S) at 1270 K.

CONCLUSIONS

Thermodynamic modelling has shown high efficiency of sodium sulphate used as an oxidizer-catalyst due to its decomposition temperature close to the temperature of active carbon oxidation. For the same reasons potassium chromate can be used as an effective oxidizer-catalyst which does not form aggressive gases on decomposition. Experiments of MSO decontamination of real reactor graphite waste without air blow have demonstrated that: (a) MSO decontamination of real reactor graphite by removal of near-surface contamination is feasible; (b) The most effective media for MSO decontamination from ^{137}Cs and ^{60}Co are melts based on alkali metal carbonates with oxidizing-catalyst additives in the form of sodium sulphate, Na$_2$SO$_4$; (c) The optimal temperature range for MSO decontamination of irradiated graphite waste in molten salts with oxidizing-catalyst additives in the form of sodium sulphate is 1170–1270 K; (d) At 1270 K the rate of MSO decontamination of irradiated graphite waste in molten salts with oxidizing-catalyst additives in the form of sodium sulphate increases three times compared to 1170 K; (e) The efficiency of MSO decontamination from ^{60}Co is not high in melts without sodium sulphate oxidizer-catalysts (less than 32%).

REFERENCES

1. O.K. Karlina, V.L. Klimov, G.Yu. Pavlova, M.I. Ojovan. *Mater. Res. Soc. Symp. Proc.*, **1107**, 109-116 (2008).
2. R.L. Gay, Rockwell International Corporation, U.S. Patent 5 449 505, 1995.
3. A.A. Romenkov, M.A. Tuktarov, L.I. Minkin, V.P. Pyshkin, *Env. Safety*, **3**, 44–47 (2006).
4. B.G. Trusov, *Proceedings of GUP MosNPO "Radon"* **13**, 22-25 (2007, in Russian).
5. V.S. Yungman, *Thermal Constants of Substances*. V. 1–8, Begell House, New York (1999).
6. L.V. Gurvich, *Vestnik Akademiinauk SSSR*, **3**, 54-65 (1983).
7. W.E. Lee, M. I. Ojovan, M.C. Stennett, N.C. Hyatt. *Advances in Applied Ceramics*, **105**, 3 (2006).
8. M.I. Ojovan, W.E. Lee. *Metallurgical and Materials Transactions A*, **42** (4), 837-851 (2011).
9. A.V. Bushuev, Yu.M. Verzilov, V.N. Zubarev et al. *Atomic Energy*, **92**, 477-485 (2002).
10. A.V. Bushuev, Yu.M. Verzilov, V.N. Zubarev et al. *Atomic Energy*, **89**, 139-146 (2000).

Technetium Solutions

Mater. Res. Soc. Symp. Proc. Vol. 1518 © 2012 Materials Research Society
DOI: 10.1557/opl.2012.1567

Ceramic Immobilisation Options for Technetium

Martin C. Stennett*, Daniel J. Backhouse, Colin L. Freeman and Neil C. Hyatt.
Department of Materials Science and Engineering, The University of Sheffield, Sheffield,
S13JD, United Kingdom.

ABSTRACT

Technetium-99 (^{99}Tc) is a fission product produced during the burning of nuclear fuel and is particularly hazardous due to its long half life (210000 years), relatively high content in nuclear fuel (approx. 1 kg per ton of SNF), low sorption, and high mobility in aerobic environments. During spent nuclear fuel (SNF) reprocessing Tc is released either as a separate fraction or in complexes with actinides and zirconium. Although Tc has historically been discharged into the marine environment more stringent regulations mean that the preferred long term option is to immobilise Tc in a highly stable and durable matrix. This study investigated the feasibility of incorporating of Mo (as a Tc analogue) in a crystalline host matrix, synthesis by solid state synthesis under different atmospheres. Samples have been characterised with X-ray diffraction (XRD), scanning electron microscopy (SEM) and X-ray absorption spectroscopy (XAS).

INTRODUCTION

One of the main challenges facing the nuclear industry is the issue of waste management. All waste from the nuclear fuel cycle must be rendered passively safe so that it can either be placed in interim storage or sent for disposal. This is very challenging for some radionuclides, such as ^{99}Tc. ^{99}Tc is produced during nuclear fission of ^{235}U, with a yield of 6.06 %, and during fission of ^{239}Pu, with a yield of 5.9% [1]. This means that for every tonne of enriched ^{235}U that is used in a nuclear reactor, approximately 1 kg of ^{99}Tc is produced [2]. There are two main reasons for the difficulty in dealing with ^{99}Tc; firstly, it has a long half-life of approximately 210000 years, and secondly, it is present in spent fuel rods as the technetium (VII) oxide, Tc_2O_7. Tc_2O_7 is water-soluble and forms the anionic pertechnetate species (TcO_4^-) in solution which is extremely mobile in the environment.

An obvious end-route for ^{99}Tc is for it to be immobilised or encapsulated in a vitreous or ceramic host matrix. There are however difficulties in accomplishing this due to the high volatility of technetium species, particularly Tc_2O_7. One possible solution to this issue is immobilising the ^{99}Tc in its Tc^{4+} oxidation state. Technetium (IV) oxide is much less volatile than Tc_2O_7, and has a sublimation temperature of approximately 900°C, in contrast to the melting and boiling points of Tc_2O_7, which are 119.5°C and 311°C, respectively [3]. Tc^{4+} can potentially be immobilised by incorporation into the structure of a durable ceramic phase which has suitable sized crystallographic sites.

In this work $Gd_2Ti_2O_7$, which crystallises with the pyrochlore structure was explored as a suitable host for Tc [4]. The pyrochlore structure has an ideal formula $A_2B_2O_7$ and adopts cubic symmetry, with space group $Fd3$-m. The A-site can be occupied by tri- and tetravalent lanthanides and actinides, which adopt eight-fold coordination in the crystal lattice. The B-site can be occupied by tetravalent transition metals, such as Ti and Zr, which adopt six-fold coordination [4]. Pyrochlores exhibit excellent chemical durability, characterised by low rates of leaching [5], an ideal property for a radionuclide host. Mo was chosen as the analogue for Tc for several reasons; firstly, it exhibits similar oxidation states to Tc, specifically, it exhibits the +4 oxidation state and secondly, Mo^{4+} and Tc^{4+} have very similar atomic radii, 0.65 Å and 0.645 Å, respectively [6]. Although Mo does not exactly replicate the chemical behaviour of Tc, it is an adequate analogue for Tc for this synthesis work.

EXPERIMENTAL

Compositions in the series $Gd_2Ti_{2-x}Mo_xO_7$ were synthesised by a solid state reaction of Gd_2O_3, TiO_2 and MoO_2, where $0.0 \leq x \leq 0.6$ increasing in increments of 0.2 (argon) or 0.1 (air). The reagents (99.9% pure) were weighed out in the appropriate stoichiometric amounts to give a 2 gram batch of reacted product. The reagents were milled, using a Fritsch Pulverisette 23 mini-mill (3 minutes, 20 Hz), in a zirconia milling bowl containing zirconia milling media (3 x 10 mm diameter). Isopropanol was used as a carrier fluid. The resultant slurry was dried at 90 °C (1 hr), and then passed through a 500 μm sieve to break up any agglomerates. Samples were then split into two one gram fractions; one fraction underwent heat treatment in air, in an electric muffle furnace, and the other in flowing argon, in a horizontal electric tube furnace. All samples were heated in alumina boats at 1300 °C for a total of 24 hours with intermediate re-grinding. After each heat treatment, the phase-purity of the sample was determined using powder X-ray diffraction (XRD). The diffractometer (STOE Stadi P) was operated in transmission mode, using an imaging plate detector and Cu Kα radiation ($\lambda = 1.5418$ Å). Monolithic samples were prepared by pressing approx. 0.25 g of powder in a 3 mm diameter hardened tool steel die at 100 MPa for 1 min. The resulting powder compacts were sintered in air at 1600 °C (4 hrs) or in flowing argon at 1500 °C (4 hrs). The density of the sintered pellets was calculated using the geometric method. The pellets were then prepared for microstructural and chemical analysis by grinding and polishing the top surface using SiC laping paper (1200 grit finish) and diamond paste (1 micron finish), respectively. The pellets were then thermally annealed in air or flowing argon at 1450 °C and 1350 °C, respectively, to reveal the grain structure. Electron microscopy was conducted on a JEOL JSM6400 scanning electron microscope operating with a working voltage of 15 kV and a working distance of 15 mm. An Oxford Instruments INCA X-sight energy dispersive X-ray spectrometer was used for quantitative elemental analysis. Atomic number, absorption and fluorescence (ZAF) correction factors, for each element, were calculated by measuring standards of known composition. The reported standard deviations in composition represent the variation in ten individual point analyses conducted on separate grains. To minimize errors associated with the measurement procedure, the sample measurement position was fixed and the beam current was allowed to stabilise before being calibrated using a cobalt standard mounted flush with the sample in the holder as described elsewhere [7].

Mo K-edge X-ray absorption spectroscopy (XAS) data were acquired from $Gd_2Ti_{2-x}Mo_xO_7$ samples synthesised in air and argon. Mo standards were also measured to allow oxidation state determination. Data were collected on beamline X23A2 of the National Synchrotron Light Source (NSLS), Brookhaven National Laboratory (BNL), USA, in transmission mode using finely ground specimens dispersed in polyethylene glycol to achieve a thickness of one absorption length. Data reduction and analysis was performed using the programs Athena [8].

RESULTS AND DISCUSSION

Samples were initially characterised by X-ray diffraction (XRD) to investigate phase assemblage. Figure 1 shows the XRD patterns collected for $Gd_2Ti_{2-x}Mo_xO_7$ (x = 0.0 – 0.6) samples synthesized in flowing argon for a total of 24 hours with intermediate regrinding to improve homogeneity. All the reflections in the patterns could be indexed using the cubic pyrochlore structure with space group Fd-3m (No. 227). Reflections are labeled with the indices in the figure. The absence of any un-indexed reflections indicate that the samples are single phase within detection limits.

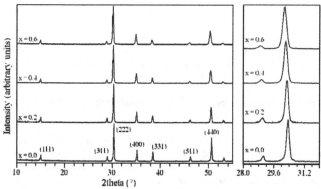

Figure 1. XRD patterns for $Gd_2Ti_{2-x}Mo_xO_7$ (x = 0.0 – 0.6) samples synthesized in flowing argon.

The right hand side of Figure 1 shows an enlargement of the region between 28.0 and 32.0 °2θ. A shift in the principle (222) reflection to lower angle as a function of increasing Mo content can clearly be seen. This indicates an increase in the unit cell size with increasing Mo content which is consistent with the substitution of Ti^{4+} cations (ionic radius of Ti^{4+} in VI-fold coordination = 0.605 Å) with larger Mo^{4+} cations (ionic radius of Mo^{4+} in Vi-fold coordination = 0.65 Å) on the B-site of the pyrochlore structure [6]. There is a corresponding shift in the other reflections to lower angle, in the XRD patterns, but only the (222) and (311) reflections are shown in the enlargement for the sake of clarity. This observed increase in unit cell parameter was confirmed by refinement of the unit cell parameter for each sample. Figure 2 shows the unit cell parameter as a function of Mo^{4+} content and shows a linear increase in the cubic unit cell parameter. These results indicate a solid solution limit of Mo^{4+} in $Gd_2Ti_2O_7$ of at least 0.6 formula units although synthesis of samples with increasing Mo content are underway.

The corresponding samples synthesized in air indicate a smaller solid solution limit (XRD patterns not shown). For Mo contents greater than 0.1 formula units (x = 0.1) a secondary phase was observed in the XRD patterns. The position of the extra reflections were consistent with the presence of Gd_6MoO_{12}. With increasing Mo content the relative intensity of the reflections, attributed to this phase, increased with respect to the pyrochlore reflections. Assuming this second phase to be fully oxygen stoichiometric its presence indicates oxidation of Mo^{4+} to Mo^{6+}. This was expected when processing these samples in air. These results indicate a very narrow solubility range for Mo^{6+} in $Gd_2Ti_2O_7$ with the charge balance proposed to be achieved by the incorporation of excess oxygen at the anion vacancy sites according to the following stoichiometry, $Gd_2Ti_{2-x}Mo_xO_{7+x}$.

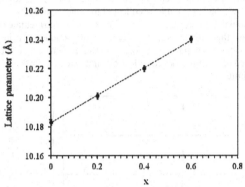

Figure 2. Pyrochlore unit cell parameter as a function of Mo content in $Gd_2Ti_{2-x}Mo_xO_7$ samples synthesized in argon.

Samples synthesized in argon (x = 0.0 – 0.6) and samples synthesized in air (x = 0.1 – 0.6) were pressed into pellets and sintered in argon and air respectively. These samples were analysed by scanning electron microscopy (SEM) and energy dispersive spectroscopy (EDX) to investigate the microstructure and phase assemblage. Typical backscattered electron micrographs are shown in Figure 3 along with the corresponding EDX spectra. Figure 3(a) shows the microstructure of $Gd_2Ti_{1.8}Mo^{4+}_{0.2}O_7$ sintered in argon at 1500 °C for 4 hours. The microstructure consists of equiaxed grains between 1 and 5 microns in diameter with isolated porosity consistent with a sintered density of approximately 90 - 95 % theoretical. Figure 3(b) shows the microstructure of $Gd_2Ti_{1.9}Mo^{6+}_{0.1}O_{7.1}$ sintered in air at 1600 °C for 4 hours; the microstructure also consists of equiaxed grains between 1 and 10 micron in diameter and very little porosity was observed. The increased densification was attributed to the higher sintering temperature. The uniform contrast indicates good chemical homogeneity which was confirmed by quantitative analysis of EDX spectra, Figures 3(c) and 3(d), acquired from multiple grains. Stoichimetries for the two phases were confirmed to be $Gd_{2.03(1)}Ti_{1.89(3)}Mo^{6+}_{0.08(3)}O_7$ and $Gd_{2.01(2)}Ti_{1.82(2)}Mo^{4+}_{0.18(1)}O_{7.1}$. In samples sintered in air (x > 0.1) a secondary phase lighter in contrast than the bulk was observed (not shown). This was consistent with the observations of a second phase by XRD and the relative volume of this phase increased with increasing x. EDX analysis (not shown) confirmed this phase was rich in Gd but also contained Mo consistent with the suggested composition of this phase (Gd_6MoO_{12}).

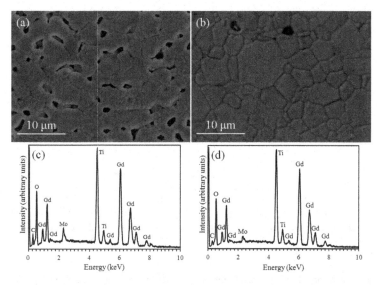

Figure 3. Backscattered electron micrographs and energy dispersive spectra (EDX) for $Gd_2Ti_{1.8}Mo^{4+}_{0.2}O_7$, 3(a) and 3(c), and $Gd_2Ti_{1.9}Mo^{6+}_{0.1}O_{7.1}$, 3(b) and 3(d), sintered in argon and air respectively.

To confirm the proposed Mo oxidation states the samples were analysed by X-ray absorption spectroscopy (XAS). The near edge regions (XANES) of the absorption spectra for $Gd_2Ti_{1.8}Mo_{0.2}O_7$ synthesised in argon and $Gd_2Ti_{1.9}Mo_{0.1}O_7$ synthesised in air are shown in Figure 4. The spectra for two oxidation state standards are also shown for comparison. XAS is element specific and a powerful technique for analysis of oxidation state and local structure in both amorphous and crystalline materials. The edge position and position of pre-edge feature can be used to identify oxidation state [9]. As can be seen from the figure the position of the edge and pre-edge feature for $Gd_2Ti_{1.8}Mo_{0.2}O_7$, synthesised in argon, are the same as for MoO_2. This indicates that the Mo in this sample is in the tetravalent oxidation state. The edge position and position of the pre-edge feature in $Gd_2Ti_{1.9}Mo_{0.1}O_7$ synthesised in air, are consistent with those observed for MoO_3 indicating that when fired under oxidizing conditions Mo^{4+} oxidises to Mo^{6+}.

CONCLUSIONS

Stabilisation and incorporation of Mo has been demonstrated in single phase pyrochlore ceramics, synthesized under inert atmosphere, in the system $Gd_2Ti_{2-x}Mo_xO_7$, where $0.0 < x \leq 0.6$. The tetravalent oxidation state of the Mo has been retained during synthesis, as confirmed by XAS, which indicates that this pyrochlore phase may be a promising host phase for Tc^{4+}. Modest solubility of Mo^{6+} has also been demonstrated in samples synthesized in air showing that control of processing atmosphere is of key importance to retaining reduced oxidation states.

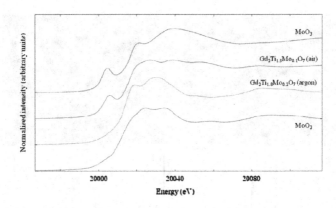

Figure 4. Mo K-edge X-ray absorption spectra for $Gd_2Ti_{1.8}Mo_{0.2}O_7$ synthesised in argon and $Gd_2Ti_{1.9}Mo_{0.1}O_7$ synthesised in air overlaid with MoO_2 and MoO_3 standards for comparison.

ACKNOWLEDGMENTS

Use of the National Synchrotron Light Source, Brookhaven National Laboratory, was supported by the US Department of Energy, Office of Science, Office of Basic Energy Sciences, under Contract no. DE-AC02-98CH10886. NCH is grateful to The Royal Academy of Engineering and Nuclear Decommissioning Authority for funding. NCH and MCS acknowledge part support from the Engineering and Physical Sciences Research Council, under grant numbers EP/I012214/1 and EP/G037140/1, and to the EPSRC Nuclear FiRST Doctoral Training Centre for funding DJB. CLF is grateful to funding from EPSRC grants EP/I001514/1, EP/I016589/1 and EP/G005001/1.

REFERENCES

1. J. E. Till, in *Technetium in the Environment*, edited by G. Desmet and C. Myttenaere (Springer, New York, 1986), p. 1.
2. F. Chen, P. C. Burns, and R. C. Ewing, J. Nucl. Mater. **278**, 225 (2000).
3. M. Y. Khalil and W. B. White, J. Am. Ceram. Soc. **66**, C197 (1983).
4. M. A. Subramanian, G. Aravamudan, and G. V. Subba Rao, Prog. Solid State Chem. **15**, 55 (1983).
5. S. S. Shoup, C. E. Bamberger, T. J. Haverlock, and J. R. Peterson, J. Nucl. Mater. **240**, 112 (1997).
6. R. D. Shannon, Acta Cryst. **A32**, 751 (1976).
7. E. Lifshin and R. Gauvin, Microsc. Microanal. **7**, 168 (2001).
8. B. Ravel and M. Newville, J. Synchrotron Radiat., **12** (2005) 537.
9. M. Wilke, F. Farges, P.E. Petit, G. E. Brown, and F. Martin, Am. Mineral. **86**, 714 (2001).

Mater. Res. Soc. Symp. Proc. Vol. 1518 © 2013 Materials Research Society
DOI: 10.1557/opl.2013.69

Technetium Incorporation into C14 and C15 Laves Intermetallic Phases

Edgar C. Buck, Alan L. Schemer-Kohrn, and Jonathan B. Wierschke

Energy and Environment Directorate, Pacific Northwest National Laboratory, 902 Battelle Blvd., Richland, Washington 99352, USA.

ABSTRACT

Laves-type intermetallic phases have been observed to be the dominant phases in a series of alloy compositions being designed for the immobilization of technetium in a metallic waste form. The dominant metals in the alloy compositions were Fe-Mo and Fe-Mo-Zr. The alloy composition, Fe-Mo-Zr, also contained Pd, Zr, Cr, and Ni. Both non-radioactive rhenium-containing and radioactive technetium-bearing alloy compositions were investigated. In the Fe-Mo series, the phases observed were Fe_2Mo (C14 Laves phase) and ferrite in agreement with predictions. Both Tc and Re resided predominantly in the Laves phases. In the Fe-Mo-Zr system, the phases included hexagonal C14 with the composition $(Fe,Cr)_2Mo$, cubic C15 phase with a $(Fe,Ni)_2Zr$ composition, and the hcp phase Pd_2Zr. The observation of these phases was in agreement with predictions. Re was found in the C14 intermetallic, $(Fe,Cr)_2Mo$. Technetium was also observed to be partitioned preferentially into the $(Fe,Cr)_2Mo$ phase; however, this phase exhibited a cubic structure consistent with the C15 structural type. The composition of Laves phases is influenced by both the atomic size and electro-negativity of the constituent elements. The long-term release behavior of technetium under nuclear waste disposal conditions may be more dependent on the corrosion characteristics of these individual Laves phases containing Tc than the other metallic phases.

INTRODUCTION

This paper discusses the microanalysis of a series of metallic waste forms designed for the incorporation of technetium. Microanalysis of alloy waste-form was performed with scanning electron microscopy (SEM) and transmission electron microscopy (TEM). Although the reprocessing of spent nuclear fuel may offer some significant advantages over direct disposal into a geologic repository, environmental issues associated with the long-term behavior of nuclear waste forms are unlikely to disappear. Advanced waste-forms, including metallic alloys, will occupy considerably smaller volumes than previously developed waste-forms; however, understanding long-term behavior remains important. The performance of the advanced waste forms need to be evaluated in terms of the full range of environmental conditions that are possible within a repository setting. Changes in environmental conditions are expected, and may originate from the waste forms themselves (e.g., heat, radiation damage, radiolysis, and corrosion), or may reflect the natural variability of the repository setting (e.g., water seepage rates, mechanical stability). In some cases, the environmental setting may directly reflect the durability of the waste-form itself. It is therefore important to have a good understanding of the microstructure of the nuclear waste-forms to develop effective long-term models for corrosion and to build scientific basis for the disposal strategy.

Iron-zirconium alloy waste forms were developed by Idaho National Laboratory (INL) for the Experimental Breeder Reactor-II wastes [1]. A minimum additive waste stabilization approach was adopted to produce a waste form that both addresses the performance acceptance requirements and minimizes the amount of additional materials that needs to be added to make a durable material. The alloy waste-form in this study was formulated using the nominal waste stream compositions for the undissolved solids (UDS), recovered Tc, and dissolved metals that might be recovered from the reprocessing of spent commercial light water nuclear fuel. The waste stream compositions were based on the expected composition of UO_2 fuel with a burn-up of 51 MWd/KgU that had been aged for 20 years. Zirconium was assumed to be from both nuclear fission and from Zircaloy cladding wastes [2].

Based on Fe-Mo phase system, the iron-rich Fe-Mo was expected to form hexagonal C14 Laves Fe_2Mo phase with the $MgZn_2$ structure type and bcc iron (ferrite) [3]. Whereas, the iron-rich Fe-Mo-Zr system was expected to form both ferrite, C14 Fe_2Mo, and cubic C15 Laves phase (Fe_2Zr)[4]. Abraham and co-workers [1] investigated the iron-rich Fe-Zr system and observed the formation of Fe_2Zr and ferrite (see Ref [1] and related papers). This metallic alloy was also alloyed with technetium which tended to reside in the ferrite structure rather than the intermetallic Laves phase. Poineau et al. [5] have investigated the metallurgy of Tc-Zr binary system and found the occurrence of Tc_2Zr and Tc_6Zr. Their results indicated that Tc_6Zr crystallized into the cubic R-Mn-type structure (I43m space group) with a variable stoichiometry of $Tc_{6.25-x}Zr$ ($0 < x < 1.45$); whereas, Tc_2Zr had a hexagonal crystal lattice consistent with the $MgZn_2$-type C14 Laves phase structure. These results suggest the importance of Laves phases.

METHODS AND MATERIALS

The metallic specimens were prepared and supplied by Dr. S. Frank at INL and heat treatments were conducted by Dr. W. Ebert at Argonne National Laboratory (ANL). The alloys for the mixed UDS and Tc waste streams were ~ 47 mass% Type 304L or Type 316L stainless steels. Using 47 mass% steel provides about 33 mass% iron. The composition was formulated to represent a combination of the Tc, other metals, and reprocessing waste alloyed with Type 316L stainless steel [2] (see Table 1). It was anticipated that more iron would be needed to form intermetallics with the wastes in this mixture, so a greater fraction of steel was used. The mixtures were heated to 1600°C over a 90 minute period then held at that temperature for 60 minutes. The furnace temperature was then reset to 1100°C and allowed to cool to that temperature over 30 minutes. The ingot was held at 1100°C for 90 minutes. The material was then allowed to cool slowly to room temperature [2].

Table 1. Composition of Alloys (wt%)

Materials	Fe	Cr	Ni	Zr	Mo	Ru	Pd	Tc	Re	Rh
RAW-1(Re)	38.4	11.0	7.4	12.2	12.4	7.6	5.1	-	4.7	1.3
RAW-1(Tc)	39.2	11.2	7.5	12.5	12.7	7.7	5.2	2.5	-	1.4
FeMo-Re	48.0	-	-	-	38.0	-	-	-	14.0	-
FeMo-Tc	52.0	-	-	-	41.0	-	-	7.0	-	-

Selected specimens were also heat treated at ANL to provide greater insight into the formation of the alloys. Some microstructural information on these heat-treated specimens is discussed in this paper. The samples were initially analyzed with an FEI (Hillsboro, OR, USA) Quanta 250 field emission gun (FEG) SEM equipped with an EDAX Instruments GENESIS x-ray energy dispersive spectroscopy (EDS) system. Samples for TEM analysis were cut into 3mm discs and polished and ion milled using a Gatan (Gatan Inc., Pleasanton, CA) Precision Ion Mill. TEM work was performed using a FEI Tecnai T30 at the Radiochemical Processing Laboratory (PNNL) instrument equipped with a Gatan (Gatan Inc., Pleasanton, CA) ORIUS™ digital camera, and an EDAX EDS system. Diffraction patterns and electron micrographs were analyzed with Gatan DigitalMicrograph 1.83.842 and aided with custom scripts from Mitchell [6]. Crystal models and simulated diffraction pattern were generated using CrystalMaker® 2.5, a crystal and molecular structures program for Mac and Windows, and SingleCrystal®2.0.1, an electron diffraction simulation program both distributed by CrystalMaker Software Ltd., Oxford, England (http://www.crystalmaker.com). Literature obtained structures were used to derive the electron diffraction patterns from the $MgCu_2$ (C15), $MgNi_2$ (C36), $MgZn_2$ (C14), and bcc iron structural types that were used to identify phases in the alloy samples.

RESULTS

Initial SEM analyses showed that the alloy materials were multi-phase alloys. An undissolved molybdenum phase was observed in the Re-containing Fe-Mo alloy; whereas a similar phase was not found in the Tc-bearing Fe-Mo alloy. This was considered to be due to undissolved solids in the alloy. Electron diffraction patterns from the two Fe-Mo alloys taken along a specific zone axis are shown in Figure 1. These were taken along the same zone axis in both metals and are a strong indication of the structural similarity of the two phases. Electron diffraction patterns were obtained from other zone axis and demonstrated the strong similarity in the phases. The rectangular patterns shown in Figure 1 were indexed to the $B[1\bar{2}10]$ direction consistent with the MgZn$_2$ structure of the C14 Laves intermetallic phase. Compositional analysis indicated that the Tc and Re was partitioned preferentially to the C14 intermetallic phase. There was little observable Tc or Re contained within the ferrite phase. The detection limit for EDS for this element was estimated to be 0.5 -1 wt% given the high energy of the x-ray line used to identify Tc in the alloy. The Fe content was consistent with an Fe$_2$Mo compositional region. These phases were extremely brittle and easily separated from the iron (bcc) phase. The concentration of Tc and Re in the intermetallic phases ranged from 2 to 11 wt%. The iron and molybdenum content was relatively stable and, generally, the Fe:Mo ratio was approximately 2:1. The major elements were quantified using the K-lines as there was no significant overlap in this energy region.

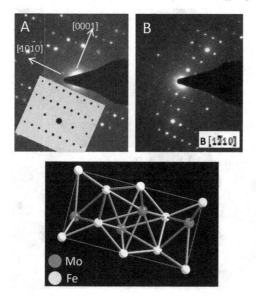

Figure 1. Zone axis pattern taken along $B[1\bar{2}10]$ from the (A) FeMo-Tc phase and the (B) FeMo-Re phase with simulated electron diffraction pattern (insert) based on the C14 Laves structure. Lower image shows a crystal model of Fe$_2$Mo.

The Fe-Mo-Zr alloy composition exhibited at least four different phases. As well as the major components of Fe, Mo, and Zr; this alloy also contained Pd, and lesser amounts of Ni, and Cr. The phases observed in the non-radioactive analog alloy containing Re were in agreement with the Fe-Mo-Zr phase diagram [4]. The identified phases included,; hexagonal C14 (Fe,Cr)$_2$Mo, cubic Laves phase (C15) with a composition of (Fe,Ni)$_2$Zr, and ferrite (*bcc* iron) together with a hexagonal PdZr$_2$ phase. Additionally, there were regions with a pure Zr phase. We used electron diffraction to identify the phases. Figure 2 has an example of a multiphase region where different Laves phases are present. This specimen had gone through a heat treatment to 850°C before being quenched. In this treated specimen, the (Fe,Ni)$_2$Zr composition was stabilized into the hexagonal C36 structure. As before, rhenium was found predominantly in the (Fe,Cr)$_2$Mo phase which possessed the C14 structure type. Abraham and co-workers demonstrated that Tc was not incorporated into C15 Fe$_2$Zr type phases in metallic alloys and this investigation indicated the same result. However, in the Abraham et al. [1] study, Tc entered the ferrite phase. In this study, even though ferrite was present both Tc and Re preferred the C14 phase. In the Tc-bearing Fe-Mo-Zr alloy composition, technetium was found preferentially in the (Fe,Cr)$_2$Mo phase; however electron diffraction indicated that the phase was not hexagonal but cubic (see Figure 3). Electron diffraction from the phase was not consistent with the hexagonal structure. Figure 3b shows the composition of the two major Laves phases in the Tc-bearing alloy and that the Tc was partitioned strongly to the Fe-Mo alloy composition.

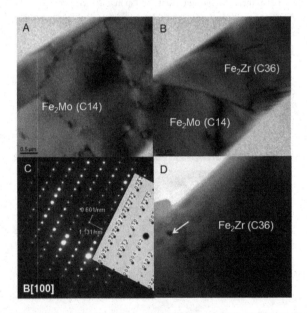

Figure 2. Images of a heat-treated Fe-Mo-Zr alloy showing the association of intermetallic phases and the occurrence of Laves intermetallic phases. The region contained the (A) (Fe,Cr)$_2$Mo intermetallic, (B) the boundary between the Fe$_2$Mo and Fe$_2$Zr phases, (C) electron diffraction analysis of the (Fe,Ni)$_2$Zr phase and computer simulated pattern (insert), and (D), the C36 intermetallic showing the occurrence of small iron particles in the matrix (arrow). The (Fe,Ni)$_2$Zr phase appeared to be C36 rather than C15 as was found in the non-heat treated specimens.

DISCUSSION

The C14 and C15 crystal structures are AB_2 type alloys belonging to a class of materials known as Laves phases. These phases are topologically highly close-packed intermetallic structures [1,7,9]. The C14 structure is hexagonal with the $MgZn_2$-type structure. The C15 possesses the $MgCu_2$-type structure and is cubic. During heat treatment, intercalation layers may be expected to form [7,8]. This layering can occur because the various Laves phases are so similar that they may easily shift from one polytype to another with only minor changes in composition. For example, the hexagonal layers in the C15 structure are along {111} planes, while similar stacking is along (0001) plane in C14.

The nearest neighbor distances in the various Laves phases are also very similar resulting in intergrowth structures. Evidence for layering or the occurrence of shear structures were seen in several micrographs during this examination. One of the $(Fe,Ni)_2Zr$ polytypes and the $(Fe,Cr)_2Mo$ Laves phase occur as four- and two- layered hexagonal structures, respectively, that are designated as C36 and C14.

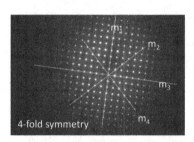 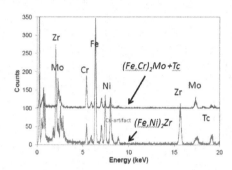

Figure 3. Electron diffraction pattern of $(Fe,Cr)_2Mo$ phase in Fe-Mo-Zr alloy composition showing 4-fold symmetry consistent with a cubic C15 laves phase and EDS analysis of the two major intermetallic phases showing that the Tc is incorporated into the $(Fe,Cr)_2Mo$ phase.

In contrast to the results of Abraham et al. [1], where the Fe-Zr intermetallic phase also contains Cr, the Fe-Zr intermetallic in these waste forms contained Ni rather than Cr as the tertiary species. The Cr tended to enter the Fe-Mo intermetallic. The C15 structure is also a layered phase but is described as a three-layer cubic structure. The preference for the C14 structure for chromium rather than the iron-zirconium system may be due its thermodynamic phase stability or rather the higher stability of the nickel-bearing iron-molybdenum intermetallic. This is further evidence that the $(Fe,Ni)_2Mo$ phase forms first in the melt. The formation energies of C36, C15, and C14 are no doubt very similar. Although compositionally different from the phases discussed elsewhere, we have shown that the C14 and C36 Laves intermetallics are more dominant in the higher temperature quenched materials. However, the excellent lattice match between the hexagonal C14 and the cubic C15 structures means that these phases may grow on one another. Although the Pd_2Zr phase is also a close-packed structure (possibly C36), it was not associated with either the $(Fe,Cr)_2Mo$ or $(Fe,Ni)_2Zr$ phases. This suggests that though it is structurally similar, it is forming last during cooling and is exsolving from the iron-rich solution.

The technetium may be stabilized into a cubic form of the $(Fe,Cr)_2Mo$ phase. This is consistent with the behavior of some solutes on Laves alloy structures where higher energy structures may be retained because of the presence of minor elements. Chu et al. [10] showed that Nb doping into HfV_2 stabilized the C15 structure.

121

CONCLUSION

Based on the analysis of Fe-Mo and Fe-Mo-Zr alloys together with the analysis of the heat-treated specimens, it was concluded that the hexagonal C14 Laves phase formed first with composition $(Fe,Cr)_2Mo$ in the Fe-Mo-Zr alloy system. As the material cooled, the L+C14 →C15 transformation occurred. The cubic C15 phase then formed with the composition $(Fe,Ni)_2Zr$. The C36 Pd_2Zr intermetallic segregated out within the bcc-iron phase that solidified interstitially last of all. This last formed phase tended to be highly deformed. Overall, the intermetallic phases that formed were extremely brittle. The Laves phase C36 was also identified for the $(Fe,Ni)_2Zr$ intermetallic composition. It is possible that both polytypes exist at all temperatures; however, this was readily identified in the highest heat-treated specimens investigated. The hexagonal Laves phases appeared to be more dominant in the heat-treated specimens. Knowledge of the specific phase that Tc is entering into may provide more accurate models for long-term behavior and enable computational studies to address the specific structures and compositions involved in the immobilization of Tc in the waste form.

ACKNOWLEDGEMENTS

This work was supported by the US Department of Energy, Office of Nuclear Energy, Fuel Cycle Research & Development Program under Contract DE-AC05-76RL01830. Alloy ingot specimens were produced by Dr. S. Frank (INL) and supplied by Dr. W. Ebert (ANL). We thank Dr. T. Todd (INL) and Dr. J. Vienna (PNNL) for programmatic guidance.

REFERENCES

1. D. P. Abraham and J.W. Richardson Jr., and S. M. McDeavitt, S.M. *Mater. Sci. Eng.* A239, 658-664 (1997).

2. W. Ebert, Advanced Fuel Cycle Initiative report, AFCI-SUI-WAST-WAST-MI-DV-2009-000001 (2009).

3. S. Ishikawa, T., Matsuo, and M., Takeyama, *Mater. Res. Soc. Symp. Proc.* 980 (2007).

4. M. Zinkevich and N. Mattern, N. *Acta Materialia,* 50, 3373-3383 (2002).

5. F. Poineau, T., Hartmann, P. F. Weck, E. Kim, G. W. Chinthaka Silva, G. D. Jarvinen, and K. R. Czerwinski, *Inorg. Chem.* 49, 1433-1438 (2010).

6. D. R. G. Mitchell, D. R. G., *Micro. Res. Tech.* 71, 588-593 (2008).

7. Z. Shi, S. Chumbley, F. C. Laabs, F. C. *J Alloys Compds,* 312, 41-52 (2000).

8. W. J. Boettinger, D. E. Newbury, K. Wang, L. A. Bendersky, C. Chiu, U. R. Kattner, K. Young, and B. Chao, *Metal. Mater. Trans.,* A41, 2030-2047 (2010).

9. F. Stein and M. Palm, G. Sauthoff, G. *Intermetallics* 13, 1056-1074 (2005).

10. F. Chu, D. J. Thoma, T. E. Mitchell, C. L. Lin, and M. Sob, *Phil. Mag.* B77, 121-136 (1998).

Mater. Res. Soc. Symp. Proc. Vol. 1518 © 2013 Materials Research Society
DOI: 10.1557/opl.2013.111

Technetium-99m Transport and Immobilisation in Porous Media: Development of a Novel Nuclear Imaging Technique

Claire L. Corkhill[1], Jonathan W. Bridge[2,3], Philip Hillel[4], Laura J. Gardner[1], Steven A. Banwart[3] and Neil C. Hyatt[1]

[1] The Immobilisation Science Laboratory, Department of Materials Science and Engineering, The University of Sheffield, UK.
[2] The Centre for Engineering Sustainability, School of Engineering, University of Liverpool, UK.
[3] Kroto Research Institute, Department of Civil and Structural Engineering, The University of Sheffield, UK.
[4] Department of Nuclear Medicine, Sheffield Teaching Hospitals NHS Foundation Trust, Sheffield, UK.

ABSTRACT

Technetium-99, a β-emitting radioactive fission product of ^{235}U, formed in nuclear reactors, presents a major challenge to nuclear waste disposal strategies. Its long half-life (2.1×10^5 years) and high solubility under oxic conditions as the pertechnetate anion $[Tc(VII)O_4]$ is particularly problematic for long-term disposal of radioactive waste in geological repositories. In this study, we demonstrate a novel technique for quantifying the transport and immobilisation of technetium-99m, a γ-emitting metastable isomer of technetium-99 commonly used in medical imaging. A standard medical gamma camera was used for non-invasive quantitative imaging of technetium-99m during co-advection through quartz sand and various cementitious materials commonly used in nuclear waste disposal strategies. Spatial moments analysis of the resulting ^{99m}Tc plume provided information about the relative changes in mass distribution of the radionuclide in the various test materials. ^{99m}Tc advected through quartz sand demonstrated typical conservative behaviour, while transport through the cementitious materials produced a significant reduction in radionuclide centre of mass transport velocity over time. Gamma camera imaging has proven an effective tool for helping to understand the factors which control the migration of radionuclides for surface, near-surface and deep geological disposal of nuclear waste.

INTRODUCTION

In Europe and the US, the preferred long-term disposal solution for nuclear waste is burial in a deep geological disposal facility (GDF) [1]. During the operational lifetime of such facilities ($\sim 10^6$ years) the influx of groundwater will occur, leading to the deterioration of the engineered barrier system (composed of a cement backfill or clay, and the canisters in which the waste are stored). This will lead to the release and transport of low concentrations of long-lived radionuclides from the GDF and into the near-field (the surrounding host geology). It is not yet fully understood how radionuclide species will be transported from these engineered facilities when the containment is breached. In a failure scenario, it is expected that the transport of most radionuclides will be inhibited by the backfill material, which provides alkaline conditions to suppress the solubility of cationic radionuclides and a high surface area to enhance radionuclide sorption [2]. However, it has been shown that anionic radionuclide species such as pertechnetate

(TcO_4^-) and iodide (I^-) are weakly sorbed by backfill cements [3 - 5]. ^{99}Tc is abundant in radioactive wastes due to its high fission yield of ca. 6% for thermal fission of ^{235}U. As a consequence of its long half-life (2.1 x 10^5 years) and high solubility of the pertechnetate species ($Tc^{VII}O_4^-$), ^{99}Tc is a key contributor to the potential radiological risk to future populations in post closure safety assessments.

In this study, we demonstrate a new technique for the non-invasive, quantitative imaging of the transport of technetium-99 as pertechnetate, using the γ-emitting metastable isomer, technetium-99m, which is commonly used in medical imaging. We quantify directly the transport of ^{99m}Tc in lab-scale flow cells through porous media relevant to geological disposal of nuclear waste. These include: quartz sand, as a simplified analogue for sandstone and cementitious materials commonly used in nuclear waste disposal strategies, including Ordinary Portland Cement (OPC) and OPC combined with pulverized fly ash (PFA).

METHODOLOGY

Perspex flow cells (10 x 5 x 0.5 cm) were filled with either: quartz sand (Fisher, particle size 500-700 μm); Ordinary Portland Cement; or Pulverised Fly Ash/OPC cement (5:4). Cements were prepared with a w/s ratio of 0.37, cured at room temperature for 28 days, and crushed and sieved to a 1-2 mm size fraction. Compositions are given in Table 1. Flow cells were saturated with pH 5.7 de-ionised water and continuous flow was maintained in the cells, from top to bottom, at 0.33 mL min^{-1} using an 8-channel peristaltic pump, beginning 30 minutes prior to injection of the radiotracer, and for the duration of the experiment.

Table 1. Chemical compositions of main cement formulations.

Component (wt %)	OPC	OPC/PFA
CaO	50.2	26.2
SiO_2	16.5	27.4
CO_2*	13.0	11.0
H_2O**	10.2	12.7
Al_2O_3	4.0	12.6
Fe_2O_3	2.0	4.8
MgO	1.4	1.4
SO_3	1.4	1.2
K_2O	0.6	1.7
P_2O_5	0.4	0.5
Na_2O	0.2	0.5

* determined by loss on ignition
** determined by loss of mass upon heating

^{99m}Tc as pertechnetate (TcO_4^-) was produced in the hospital radiopharmacy *via* saline-based elution of a GE Medical Systems Drytec ^{99m}Tc generator. Approximately 0.2 mL of diluted ^{99m}Tc was drawn into a syringe, corresponding to an activity of ~15-20 MBq. The activity in each syringe was accurately measured in a Capintec CRC-15R radionuclide calibrator before and after injection so that the exact activity injected into each cell could be determined. In all cases ^{99m}Tc activity readings were decay corrected to the time the gamma camera acquisition was started (^{99m}Tc half-life is 6.01 hrs). Imaging of the flow cells was performed on a dual-headed

GE Medical Systems Infinia gamma camera (GE Medical, Milwaukee, WI, USA). A dynamic acquisition of 330 frames taken at 30 s intervals was started just prior to injection of the ^{99m}Tc into the flow cells. Images were acquired with a matrix size of 256 x 256 resulting in a pixel size of 2.2 mm. Image counts were automatically decay corrected to the beginning of the acquisition. Raw image data were calibrated using a sensitivity factor [5] to give 2-D spatial arrays of tracer concentration data. Spatial moments analysis was applied to image data for the ^{99m}Tc to provide quantitative comparative data on radiotracer activity profile. Moments in the direction of travel were calculated using ImageJ software.

RESULTS

Transport data

First spatial moments analysis of the ^{99m}Tc plume gives activity profiles normal to the direction of travel, as a function of distance from the injection point, and time (Figure 1). For quartz sand, the activity profiles show a fairly high degree of symmetry and no tailing in the leading or trailing edges, indicating conservative transport, i.e. the sand does not impose a significant physical or chemical impediment to ^{99m}Tc transport. These results were reproducible across a number of flow cells with identical conditions, and also for sand that was saturated using a pH 10 buffer (NaOH, NaHCO$_3$) [5].

The characteristics of ^{99m}Tc transport through the saturated, crushed cement flow cells differed for each cement type. The end pH for all flow cells was between pH 12 and 13. Figure 1 shows that for all of the cements, the activity profiles are asymmetric with tails on the leading and trailing edges of the peaks. This suggests that transport of ^{99m}Tc is significantly affected physically and/or chemically by the cement, and is indicative of reactive transport. Figure 2 gives the ^{99m}Tc activity profiles for each material tested as a function of time, normalised to the input activity, and Table 2 shows the percentage of retained ^{99m}Tc in each material at the end of the experiment (160 minutes). These demonstrate that in the sand flow cells, all of the ^{99m}Tc exits the cell after approximately 40 minutes. For the OPC and OPC/PFA mix, some ^{99}Tc remains in the cell at 160 minutes. Calibrated concentration distribution data for ^{99m}Tc transport though the quartz sand, OPC and the OPC/PFA mix 40 minutes after injection are shown in Figure 3. In the saturated quartz sand, the ^{99m}Tc tracer passed through as well-defined plumes with peak concentrations in the centre of ~ 10 MBq mL^{-1}. In the saturated cement systems, the ^{99m}Tc tracer as transported as elongated plumes, with peak activities at the forefront of the plume of ~ 8 MBq mL^{-1} and 2 MBq mL^{-1} in the OPC and OPC/PFA, respectively, and activities in the tailing plumes of between 1 and 5 MBq mL^{-1}.

Table 2. Percentage ^{99m}Tc mass retention in porous media at 40, 80, and 160 minutes under flow conditions of 0.33 mL min^{-1}, as determined by γ-camera imaging.

Material	40 min	% mass retained after 80 min	160 min
Sand	0	0	0
Ordinary Portland Cement	55	19	10
Pulverised fly ash/OPC mix	52	8	4

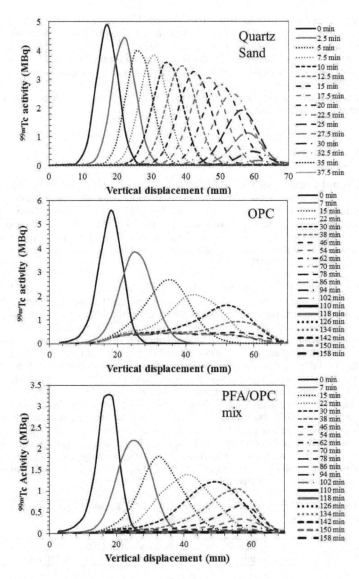

Figure 1. ^{99m}Tc activity profiles (derived from first spatial moment analysis) resulting from flow under water saturated conditions in quartz sand, Ordinary Portland Cement (OPC) and OPC mixed with pulverised fly ash (PFA). The input activity of ^{99m}Tc varied slightly for each flow cell.

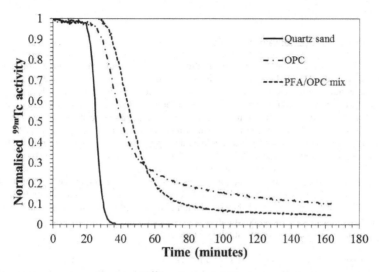

Figure 2. Total measured activity of ^{99m}Tc (normalised to input activity) as a function of time during transport through quartz sand, Ordinary Portland Cement (OPC) and OPC mixed with pulverised fly ash (PFA).

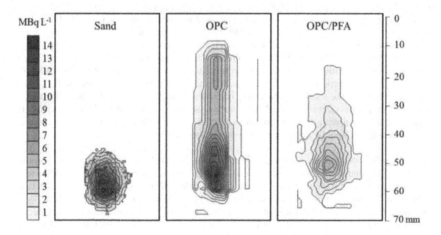

Figure 3. Calibrated concentration distributions derived from gamma camera images of ^{99m}Tc activity in quartz sand, Ordinary Portland Cement (OPC) and OPC mixed with pulverised fly ash (PFA) at 40 minutes after the injection of ^{99m}Tc.

DISCUSSION

Transport and immobilization of technetium

Quartz sand demonstrated poor retardation of ^{99m}Tc and conservative transport characteristics were displayed. This is unsurprising, due to the net negative surface charge of sand under the conditions in this study. Since the pertechnetate anion is also negatively charged, chemical sorption is prevented by repulsive electrostatic interactions [5]. As a simplified analogue for sandstone rock, results from this study suggest that should technetium be released from a GDF, it may be transported freely by groundwater. This result highlights the importance of the engineered components within the geological disposal facility, and in particular, the backfill material, to sorb and mitigate the transport of this radionuclide. Transport of the ^{99m}Tc tracer through the cement materials was slower and more dispersed than in the sand, as shown in Figure 2. As noted, the data indicate reactive transport of the ^{99m}Tc through the cement, with approximately 4% and 10% of the injected tracer remaining in the cement after 160 minutes in the OPC and OPC/PFA, respectively (Table 2). The normalised activity profiles shown in Figure 2 suggest that should the experiment continue for a longer period of time than allowed by the γ-camera, all of the ^{99m}Tc tracer would eventually exit the cell.

There was no sorption of ^{99m}Tc onto the OPC/PFA cement as it passed through the flow cell (indicated by the constant decrease of ^{99m}Tc tracer in the flow cells, Figure 2). This is in contrast to previous work that found that a PFA/OPC mix at a ratio of 1:1 was capable of sorbing almost all of the ^{99}Tc(VII) present in anaerobic, static batch experiments over a period of several months [4]. Previous studies have shown that Fe(II)-bearing minerals readily reduce Tc(VII) to Tc(IV) under reducing conditions [6, 7], therefore, it might be expected that the high Fe(II) content of PFA (~5 wt % in the current study, Table 1) may lead to a reduction of ^{99}Tc to insoluble, less-mobile Tc(IV). However, the redox conditions measured in the batch experiments described in [4] were found to be sufficiently positive for Tc(VII) to be the dominant oxidation state ((-110 to +60mV), implying that sorption to these cements was not *via* a reduction mechanism. These authors could not conclude a possible sorption mechanism. It is possible that the high sorption achieved in the batch experiments in [4] compared to the flow cell study described in the current study may be due to the smaller particle size of cement used (250 µm to 500 µm compared to 2 mm to 4 mm particles), leading to enhanced surface area for sorption in the batch experiments. However, in a study investigating the sorption of Tc(VII) to PFA alone (i.e. with no OPC added), particle size was not found to influence the sorption capacity [8].

It can be hypothesised that the high rates of transport used in the flow cells are such that sorption is reversible, compared to static batch methods. Further work is required to elucidate this, and to determine any possible sorption mechanisms. This work will aim to determine the effects of transport on ^{99}Tc sorption using a cement composition that is known to strongly sorb Tc(VII): an OPC Blast Furnace Slag (BFS) mixed cement (e.g. [9, 10]). This cement produces reducing conditions (Eh of ~-300 mV [11]) as it contains high concentrations of sulphide, and has been shown to form insoluble Tc(IV)-S phases [9, 10, 12].

Gamma camera imaging development

We have presented a new methodology that enables rapid measurement of the γ-emitting radionuclide ^{99m}Tc in porous media. Using this technique, we were able to quantitatively measure the movement of this radionuclide in flow through opaque natural and engineered materials relevant to nuclear waste geological disposal scenarios. Further work is directed towards interpretation of this data by simple conservative transport models, and in the case of cements that sorb ^{99m}Tc, reactive transport models. This novel quantitative methodology has the potential to rapidly and conveniently characterize design options for backfill materials, and to enhance the sorption capabilities of backfill materials towards anionic radionuclides such as ^{99}Tc.

ACKNOWLEDGMENTS

The research leading to these results has received funding from EPSRC Grant EP/F055412/1: Decommissioning, Immobilisation and Management of Nuclear Wastes for Disposal (DIAMOND). N. C. Hyatt is grateful to The Royal Academy of Engineering and the Nuclear Decommissioning Authority for funding.

REFERENCES

1. Geological Disposal: Steps toward implementation, Nuclear Decommissioning Authority, *Report number* NDA/RWMD/013, March 2010.
2. Development of the Nirex Reference Vault Backfill; Report on current status in 1994, United Kingdom Nirex Limited, *Report number* S/97/014, 1997.
3. S. Bayliss, R. McCrohon, P. Oliver, N. J. Pilkington and H. P. Thomason, Near-field sorption studies: January 1989 to June 1991. *AEA Report* NSS/R277, 1996.
4. S. Baker, A. Green and S. J. Williams, The removal of technetium(VII) from alkaline solution by NRVB, PFA/OPC and BFS/OPC. *Serco Assurance Report* SA/ENV-0606, 2004.
5. C. L. Corkhill, J. W. Bridge, X. C. Chen, P. Hillel, S. F. Thornton, N. C. Hyatt, S. A. Banwart. *Energy and Environ. Sci., Submitted* 2012.
6. G. Lear, J. M. McBeth, C. Boothman, D. J. Gunning, B. L. Ellis, R. S. Lawson, K. Morris, I. T. Burke, N. D. Bryan, A. P. Brown, F. R. Livens and J. R. Lloyd, *Environ. Sci. Technol.*, 2010, **44**, 156.
7. D. Cui and T. E. Eriksen, *Environ. Sci. Technol.*, 1996, **30**, 2259.
8. M. M. Cowper, A. Green and S. W. Swanton, The removal of Tc(VII) from alkaline solution by pulverised fly ash. *Serco Assurance Report* SA/ENV-0651, 2004.
9. P. G. Allen, G. S. Simering, D. K. Shuh, J. J. Bucher, N. M. Edelstein, C. A. Langton, S. B. Clark, T. Reich and M. A. Denecke, *Radiochim. Acta* 1997, **76**, 77.
10. W. W. Lukens, J. J. Bucher, D. K. Shuh and N. M. Edelstein, *Environ. Sci. Technol*, 2005, **39**, 8064.
11. F. P. Glasser and M. Atkins, *MRS Bull.*, 1994 **XIX**, 33.
12. Y. Liu and S. Jurisson. *Radiochim. Acta*, 2008, **96**, 823.

Spent Nuclear Fuel

Mater. Res. Soc. Symp. Proc. Vol. 1518 © 2013 Materials Research Society
DOI: 10.1557/opl.2013.76

Modelling the Activation of H_2 on Spent Fuel Surface and Inhibiting Effect of UO_2 Dissolution

L. Duro[1], O. Riba[1], A. Martínez-Esparza[2] and J. Bruno[1]
[1] Amphos 21 Consulting, S.L., P. Garcia Faria 49-51, 1-1, Barcelona, E-08019, Spain
[2] ENRESA C/ Emilio Vargas, 7 Madrid, E-28043 Spain.

ABSTRACT

The dissolution of spent nuclear fuel is defined in two different time steps, i) the Instant Release Fraction (IRF) occurring shortly after water contacts the solid spent fuel and responsible of the fast release of those radionuclides that have been accumulated in the zones of the spent fuel pellet with low confinement, such as gap and grain boundaries and ii) the long term release of radionuclides confined in the spent fuel matrix, much slower and dependent on the conditions of the water that contacts the spent fuel.

Several models have been developed to date to explain the dissolution behavior of spent nuclear fuel under disposal conditions. The Matrix Alteration Model (MAM) is one of the most evolved radiolytic models describing the dissolution mechanism in which an Alteration/Dissolution source term model is based on the oxidative dissolution of spent fuel. Under deep repository conditions and at the expected of water contacting time (after 1000 years of spent fuel storage), α radiation will be the main contributor to water radiolysis. In the current study, simulations evaluating the effect of surface area on the alteration/dissolution of spent fuel matrix are performed considering different particle sizes of spent fuel and simulations integrating the actinides dissolution have been performed considering the precipitation of secondary phases.

INTRODUCTION

Different models describing the dissolution mechanism of spent nuclear fuel under repository conditions have been developed in the last years. One of the most evolved is the Matrix Alteration Model (MAM). This source term model was developed within the frame of "Spent Fuel Stability under Repository Conditions" (SFS) project [1]. Since then, MAM has been improved under the umbrella of different European projects: "Understanding and Physical and Numerical Modelling of the Key Processes in the Near-Field and their Coupling for Different Host Rocks and Repository Strategies" (NF-PRO), "Model uncertainty for the mechanism of dissolution of spent fuel in nuclear waste repository" (MICADO) and also thanks to several projects framed in the ENRESA R&D programs.

MATRIX ALTERATION MODEL

MAM is a radiolytic source term model based on the oxidative dissolution of spent fuel under repository conditions. Radiolysis accounts for the generation of oxidising and reducing species in the vicinity of the spent fuel surface. The presence of oxidants produces an enhancement of the dissolution of the matrix, although the generation of molecular reductants, mainly hydrogen has been observed to inhibit importantly the dissolution of the matrix. In this work we present the latest updates incorporated to the MAM to account for the phenomenological observations made in the most recent experimental work. The experimentally observed inhibition of matrix dissolution by H_2 is rationalized by activation of hydrogen through two main pathways: a) the α, β and γ radiation through reactions with radiolytic radicals and b) by the metallic alloys contained in the fuel surface (ϵ particles). The activation of hydrogen by radiation was integrated into MAM by using the reaction: $OH\cdot + H_2 = H_2O + H\cdot$ and it has been validated with experimental data. The mechanism incorporated in the model consists of a first step controlled by diffusion of H_2 on the ϵ particle surface, which determines the kinetics of the process, and a fast electron transfer step from the ϵ particle to the matrix, in accordance with the literature [13, 14]. The main processes considered in MAM are: i) Generation of oxidants and reductants by water radiolysis [2], ii) Recombination reactions of radiolysis products [2], iii) Oxidation of spent fuel matrix surface by reaction with O_2 and H_2O_2 [3-6], iv) Dissolution of pre-oxidised spent fuel matrix [5, 7, 8], v) Inhibiting effect of H_2 activated by ϵ particles, vi) Non-oxidative matrix dissolution, vii) Decomposition of H_2O_2 by H_2 in absence of catalyst and viii) precipitation of secondary solid phases. The reaction mechanisms involved in processes i) to iv) have been described in [9]. Processes v) to viii) have also been recently incorporated in MAM and are described as follows:

- *Non-oxidative matrix dissolution*: The dissolution kinetics of $UO_2(s)$ under strong reducing conditions using $H_2(g)/Pd$ as a reducing agent was investigated at the pH range from 3 to 11 [10]. The resulting kinetic constant, once normalized with the UO_2 site density [11], was integrated in MAM. The backward kinetic constant was calculated from the thermodynamic dissolution constant for the $UO_2(am)$ selected in NEA database [12]. The non-oxidative matrix dissolution process was incorporated in MAM with the reactions (1) and (2). In order to account for the regeneration of a site beneath the dissolution of one surface site, the solid UO_2 was also defined as a product.

$$>UO_2 = U(OH)_4 \ + >UO_2 \qquad\qquad K = 6.9E\text{-}9 \ s^{-1} \qquad (1)$$
$$U(OH)_4 \ + >UO_2 = >UO_2 \qquad\qquad K = 2.19 \ M^{-1} \ s^{-1} \qquad (2)$$

- *Inhibiting effect of H_2 activated by ϵ particles*: the incorporation of this process in MAM is based on kinetic experiments reported in [13] with doped UO_2 pellets with Pd (0, 0.1, 1, 3%) in H_2O_2 solution and under N_2 o H_2 atmosphere. The experiments indicated the process is controlled by the diffusion of H_2 in >Pd with K $\sim10^{-6}$ m s^{-1}. The mechanism, reaction (3) and (4), agrees with SIMFUEL electrochemical studies reported in [14] which indicate that the catalytic reduction of oxidized UO_3 required galvanic coupling between matrix and ϵ particles resulting in a fast electron transfer once H_2 is activated.

$$H_2 + >Pd = H_2act + >Pd; \qquad\qquad K=3.65 \ M^{-1} \ s^{-1} \qquad (3)$$
$$>UO_3 + H_2act = >UO_2 + H_2O \qquad\qquad K=1E8 \ M^{-1} \ s^{-1} \qquad (4)$$

- *Incorporation of decomposition of H_2O_2 by H_2*: The reaction between H_2O_2 and H_2 has been studied at room temperature and in absence of catalytic products [15]. The kinetic experiments were performed using different hydrogen pressures, with a maximum of 48 bars, and H_2O_2 concentrations expected to be obtained by water radiolysis. The mechanism is presented by reaction (5):

$$H_2O_2 \ + \ H_2 = 2H_2O \qquad\qquad K = 0.029 \ M^{-1} \ s^{-1} \qquad (5)$$

DISCUSSION

Under repository conditions it is considered that spent fuel might enter in contact with water after $1,000 - 10,000$ years of storage, therefore, α radiation should represent the main contributor to the water radiolysis. The simulations were performed at constant α dose rate of 0.009 Gy s^{-1}, determined in this study, from radiological inventory calculated with SCALE 6.1 [16] and following the methodology described in [17], for spent fuel at BU 45 MWd/kgU and 10,000 years storage time. The simulations of MAM presented in this manuscript have been performed with the code ChemSimul [18]. The G-values from alpha radiolysis are selected from [2] and no initial pressure of H_2 was considered.

Modelling of the effect of H_2 activation by ε particles

The model considers that 1% of spent fuel is covered by ε particles in accordance with [19] and the effect of the ε particles is assumed to be similar to the effect of >Pd reported in [13]. In order to quantify the effect of hydrogen on the dissolution of spent fuel matrix, simulations of MAM were performed with and without activating H_2 with ε particles. The simulation results presented in Figure 1 are obtained considering >UO_2 surface area (SA) = 1.4E-3 m$^2 \cdot$g^{-1}. In Figure 1a is shown the evolution with time in contact with water of UO_3 oxidised sites ([>UO_3]) in the surface of the spent fuel matrix, the aqueous U(VI) resulting from the oxidative dissolution process of >UO_2 ([U(VI)aq]), with and without hydrogen being activated by ε particles. The representation (Figure 1a) indicates the [U(VI)aq] increase with the increase of [>UO_3], and a decrease of the latter implies [U(VI)aq] reaching a steady state, for both studied cases (H_2 non-activated and H_2 activated, with epsilon particles. In Figure 1a, [U(VI)aq] is allowed to evolve with time, and will be limited by secondary phase formation which, under the considered conditions of deionized water, is assumed to be metaschoepite ($UO_3 \cdot 2H_2O$). The equilibrium concentration of metaschoepite calculated with ThermoChimie database [20] is represented in the same graph (see Figure 1a) with a grey dotted line. The simulation of [H_2] and the main oxidizing species: [H_2O_2] and [O_2] of the two scenarios considered are presented in Figure 1b. As it is observed, during the first year of contact with water [H_2O_2], [H_2] and [O_2] concentrations are similar whether H_2 is activated by ε particles or not. At longer reaction times, the concentration of [H_2O_2] decreases due to the decomposition of H_2O_2 by H_2, reaction (5), the decrease of [O_2] is associated with the decrease of [H_2O_2] by recombination reactions of radiolysis products. The fact that this process is not observed in case of H_2 activated by ε particles seems to respond to the lower [H_2] under these conditions, not being enough for H_2O_2 decomposition.

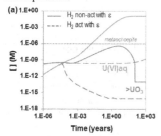

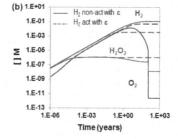

135

Figure 1. Evolution with time (in years) of [>UO₃], [U(VI)aq] (a) and [H₂O₂], [H₂] and [O₂] (b). Solid and dashed lines represent concentrations obtained when H_2 is non- activated and activated with ε particles, respectively.

In order to test the influence of >UO₂ reactive area, the simulation results obtained with SA = $1.4E^{-3}$ m²·g⁻¹ (SA2) were compared with those obtained considering SA1 = $1.4E^{-4}$ and SA3 = $2.1E^{-2}$ m²·g⁻¹. These values correspond to the surface area determined for fractured pellet (8, 8 and 6 radials, transversal and circumferential planes), non-fractured pellet and spherical particles with 150 µm size, respectively. In Figure 2 is presented the evolution with time of [>UO₃], [U(VI)], [U(IV)] (resulting from non-oxidative dissolution process) and residual [H₂act] (activated H_2 by reaction (3), which was not reacted in the reducing process of >UO₃, reaction (4)). It is shown in Figure 2a that the non-oxidative dissolution of the fuel, to form U(IV), increases with the reactive area until it reaches thermodynamic equilibrium for total dissolved uranium concentration in the order of $10^{-8.5}$ M.

The increase of [U(VI)aq] with the increase of SA is determined during the first 0.2 hours (2E-5 years) of water in contact with spent fuel, during this period, [H₂act] is not high enough to be able to reduce >UO₃. As it is observed in Figure 2b the [>UO₃] profile for the different SA is characterized by a decrease of concentration (reduction process of >UO₃) starting after ~0.2 hours until it reaches a steady state. The reaction time at which this steady state is reached follows the SA decreasing order and at the same reaction time, the activated H_2 concentration ([H₂act]), characteristic of each >UO₂ SA, reaches also a steady state. This mirroring behavior is indicative of activated H_2 being the species reducing >UO₃.

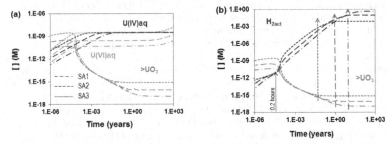

Figure 2. Evolution with time (in years) of [>UO₃], [U(VI)aq] and [U(IV)aq] (a) [>UO₃], [H₂act] (b), comparing concentrations resulting from the three different studied >UO₂ surface areas (SA1, SA2 and SA3).

Modelling precipitation of secondary phases

The behavior of the actinides Np, Pu, Am and Cm, present in the spent fuel matrix has been integrated in MAM considering that these fission products dissolve congruently with uranium [21 and 22]. The ratio [M]/[U] (M = Np, Pu, Am and Cm) has been calculated from the inventory simulated at 10,000 years with the code SCALE 6.1[16] for spent fuel with burn up 45 MWd/KgU (in g M inv. / g U inv.) and the FIAP values (fraction of inventory in the aqueous phase) for each element normalized with uranium and selected from [21]. As it shown in Figure 3, the concentration of the different considered elements increases until it reaches saturation with secondary phase limiting the concentration in solution (represented in grey colour together with

136

the phase name). The secondary phases for these elements have been determined from thermodynamic calculations using the code MEDUSA with ThermoChimie database [20] considering the solution composition of bicarbonate water (1mM [HCO_3^-], 19 mM NaCl, pH 7.2 and oxidizing conditions). The same solution composition was used in the static solubility experiments of spent fuel with burn up 48 MWd/KgU reported in [21] and [22], both studies also used the same starting material. The solubility data of U, reported in [22], and Np, Pu, Am and Cm, reported in [21], are also presented in Figure 3.

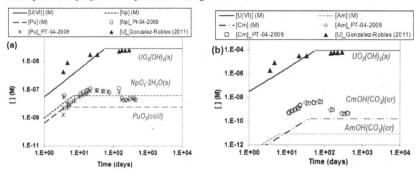

Figure 3. Evolution with time (in hours) of simulated [U(VI)aq], [Np(V)] and [Pu(IV)] (a) and [U(VI)aq], [Cm(III)] and [Am(III)] (b), together with spent fuel solubility data [21-22]. The determined secondary phases are represented.

Comparison of the simulations with the experimental data indicates that the simulated [U] evolution agrees with the experimental data and seems to reach a steady state after 200 days. The measured [Np(V)] and [Pu(IV)] after 200 days reaction time is in accordance with $NpO_2\cdot2H_2O$ and PuO_2(coll), respectively, limiting their solubility. The measured [Cm] and [Am] shows a significant decrease after 200 days reaction time reaching concentrations that tend to agree with the expected [Cm(III)] concentration limited by $CmOH(CO_3)$(cr) solubility, however, the measured [Am(III)] remains one order of magnitude over the expected $AmOH(CO_3)$(cr) solubility.

CONCLUSIONS

The integration in MAM of the inhibiting effect of H_2 by ε particles has been quantified by comparing simulations over time with H_2 activated and non-activated by ε particles. The activation of H_2 induces a decrease of [U(VI)aq] of about three orders of magnitude with respect [U(VI)] limited by metaschoepite. A sensitivity analysis of the effect of $>UO_2$ reactive area indicates that [U(VI)ac] reaches a steady state from 0.2 hours with a ratio [U(VI)aq)] / SA ($>UO_2$) = 2E-7. The incorporation of secondary phases in MAM have been performed by using the [M] / [U] ratio (M = Np, Pu, Am and Cm) calculated from inventory and the selected FIAP values [21]. The simulation results considering bicarbonate water and oxidizing conditions are validated with solubility experimental data. The results indicate that the formation of $UO_2(OH)_2$(s), $NpO_2\cdot2H_2O$, PuO_2(coll), $CmOH(CO_3)$(cr) and $AmOH(CO_3)$(cr) as secondary phases limit the concentration of U, Np, Pu, Cm and Am, respectively, under these conditions.

REFERENCES

1. (a) A. Martínez-Esparza, E. Cera (eds.). Z. Andriambololona, J. Bruno, J. Cáceres, C. Cachoir, P. Carbol, I. Casas, J. M. Cavedon, E. Cera, F. Clarens, J. Cobos, J. de Pablo, J. A. Gago, J. Giménez, J. P. Glatz,. "Deliverable D13 (WP4), SFS Project (2004), contract n° FIKW-CT-2001-00192 . (b) A. Martínez Esparza, M. A. Cuñado, J. A Gago, J. Quiñones, E. Iglesias, J. Cobos, A. González de la Huebra, E. Cera, J. Merino, J. Bruno, J. de Pablo, I. Casas, F. Clarens, J. Giménez. ENRESA PT-01-2005.
2. M. Kelm, E. Bohnert. "Deliverable D4", SFS project (2004). FZKA 6977.
3. W. E. Schortmann, M.A. DeSesa. "Kinetics of the dissolution of uranium dioxide in carbonate-bicarbonate solutions". 2nd United Nations International Conference on the Peaceful Uses of Atomic Energy, Geneve. Proceedings vol. 3, 333-341(1958).
4. J. de Pablo, I. Casas, J. Giménez, F. Clarens, L. Duro, J. Bruno. *Mat. Res. Soc. Symp. Proc.* 807, 83-88 (2004).
5. J. de Pablo, I. Casas, F. Clarens, J. Giménez, M. Rovira. ENRESA PT-01-2003.
6. E. Ekeroth, M. Jonsson. J. *Nucl. Mat.*, 322, 242-248 (2003).
7. J. de Pablo, I. Casas, J. Giménez, M. Molera, M. Rovira, L. Duro, J. Bruno. *Geochim. Cosmochim. Acta* 63, 3097-3103 (1999).
8. J. Merino, E. Cera, J. Bruno, J. Quiñones, I. Casas, F. Clarens, J. Giménez, J. de Pablo, M. Rovira, A. Martínez-Esparza. *J. Nucl. Mater.*, 346, 40-47 (2005).
9. L. Duro, A. Tamayo, J. Bruno, A. Martinez-Esparza. Integration of the H_2 Inhibition effect of UO_2 Matrix Dissolution Into Radiolytic Models. Proceedings of the 12th International Conference on Environmental Remediation and Radioactive Waste Management October 1115, 2009, Liverpool. ICEM-2009-16239.
10. J. Bruno, I. Casas, I. Puigdomènech. *Geochim. Cosmochim. Acta* 55, 647-658 (1991).
11. F. Clarens, J. de Pablo, I. Casas, J. Giménez, M. Rovira. *Mat. Res. Soc. Symp. Proc.*, Vol 807, 71-76 (2003)
12. R. Guillaumont, J. Fanghänel, V. Neck, J. Fuger, D.A. Palmer, I. Grenthe, M.H. Rand. "Chemical Thermodynamics 5. Update on the Chemical Thermodynamics of Uranium, Neptunium, Plutonium, Americium and Technetium". NEA OECD, Elsevier (2003).
13. M. Trummer, S. Nilsson, M. Jonsson. *J. Nucl. Mater.*, 378, 55 (2008).
14. M.E. Broczkowski, J.J. Noël, D.W. Shoesmith, *J. Nucl. Mater.*, 346, 16-23(2005).
15. J. Giménez, I. Casas, R. Sureda, J. de Pablo. *Radiochim. Acta* 100, 1–4(2012).
16. SCALE: A Modulas Code System for Performing Standardized Computer Analyses for Licensing Evaluation (2009). OAK Ridge National Laboratory.
17. A. Poulesquen, C. Jégou, S. Peuget, *Mat. Res. Soc. Symp. Proc.*, 932 (2006).
18. P. Kirkegaard, E. Bjergbakke, CHEMSIMUL: A simulator for chemical kinetics. Risø-R-1085(EN). http://www.risoe.dk/ita/chemsimul (2002)
19. Håkansson, R. (1999). SKB Technical Report, R-99-74.
20. D.L. Parkhurst, C.A.J Appelo PHREEQC v. 2.17.5: User's guide to Phreeqc (version 2) - A computer program for speciation, batch-reaction, one-dimensional transport, and inverse geochemical calculations. Water-Resources Investigations Report 99-4259 (1999).
21. A. Martínez-Esparza, F. Clarens, E. Gonzalez-Robles, F.J. Giménez, I. Casas, J. De Pablo, D. Serrano, D.Wegen, J.P. Glatz. ENRESA PT- 04-2009
22. E. González-Robles Corrales (2011). Study of radionuclide release in commercial UO_2 spent nuclear fuels. Effect of burn-up and high burn-up structure. Ph. D. Thesis, UPC.

Mater. Res. Soc. Symp. Proc. Vol. 1518 © 2013 Materials Research Society
DOI: 10.1557/opl.2013.77

Corrosion Study of SIMFUEL in Aerated Carbonate Solution Containing Calcium and Silicate

Hundal Jung,[1*] Tae Ahn,[2] Roberto Pabalan,[1] and David Pickett[1]
[1]Center for Nuclear Waste Regulatory Analyses (CNWRA), 6220 Culebra Rd, San Antonio, TX 78238, U.S.A.
*Currently at: Xodus Group Inc., 10111 Richmond Avenue, Ste 150, Houston, TX 77042, U.S.A.
[2]U.S. Nuclear Regulatory Commission (NRC), MS E2–B2, Washington, DC 20555-0001, U.S.A.

ABSTRACT

The corrosion behavior of simulated spent nuclear fuel (SIMFUEL) was investigated using electrochemical impedance spectroscopy and solution chemistry analyses. The SIMFUEL was exposed to aerated solutions of NaCl+NaHCO$_3$ with and without calcium (Ca) and silicate. Two SIMFUEL compositions were studied, representing spent nuclear fuel (SNF) corresponding to 3 or 6 at % burnup in terms of fission product equivalents of surrogate elements. For all tested cases, the polarization resistance increased with increased immersion time, indicating possible blocking effects due to accumulation of corrosion products on the SIMFUEL surface. The potential-pH diagram suggests formation of schoepite that may cause the increase in the polarization resistance. The addition of Ca and silicate produced no measureable change in the polarization resistance measured at the corrosion potential. The dissolution rate ranged from 1 to 3 mg/m^2-day, which is similar to the range of dissolution rates for SIMFUEL and SNF reported in the literature for comparable conditions. SIMFUEL burnup did not have a major effect on the dissolution rate. Analysis of the solution chemistry shows that uranium is the dominant element dissolved in the posttest solutions, and the dissolution rates calculated from uranium (U) concentrations are consistent with the dissolution rates obtained from impedance measurements. Simulated-fission product elements (i.e., barium, molybdenum, strontium, and zirconium) dissolved from the SIMFUEL electrode at a relatively high rate. Sorption test results indicated significant sorption of U onto the oxide formed on stainless steel. Electrochemical methods were found to be effective for measuring the uranium dissolution rate in real time.

INTRODUCTION

The SNF dissolution rate in groundwater is generally represented by the rate of uranium mass loss per unit area of the uranium dioxide (UO$_2$) matrix [1]. One of the important parameters influencing the SNF dissolution rate is groundwater chemistry, including carbonate concentration, pH, the presence of cations dissolved in the groundwater (e.g., calcium ions), dissolved silica, dissolved oxygen, as well as oxidants (e.g., radiolysis products) and reducing species (e.g., iron). The SNF dissolution rate usually increases with increases in concentrations of carbonate and oxygen, and decreases with an increase in calcium, silicate, and reducing species, and as pH becomes more neutral.

As an analogue for the behavior of SNF, the corrosion behavior of SIMFUEL was investigated using electrochemical and solution chemistry analyses. SIMFUEL is an un-irradiated analogue of irradiated SNF, produced by doping a natural uranium dioxide matrix with a series of nonradioactive elements in appropriate proportions to replicate the chemical and microstructural effects of irradiation on UO$_2$ fuel [2]. Potential sorption of radionuclides onto stainless steel was also investigated by immersing stainless steel coupons in the posttest solutions. This paper presents the results of dissolution rate measurements of SIMFUEL using both electrochemical methods, solution chemistry analyses, and the sorption behavior of the radionuclides onto stainless steel under oxidizing conditions. More details on the experiments and results are available in Jung, et al. [3].

EXPERIMENT

The working electrodes used for the corrosion tests were disk-shaped SIMFUEL pieces, 2 to 3 mm thick and 12 mm in diameter, simulating SNF with a 3% or 6% burnup. To enable rotation of the working electrode, the disk was inserted into a threaded Teflon cylinder and the inner face of the disk was bonded to a stainless steel wire with a highly conductive silver epoxy. The sorption test used a stainless steel Type 316L disk, 6.35 mm thick and 20.32 mm in diameter. Two different concentrations of carbonate solutions (i.e., simulated groundwater and hypothetical in-package chemistry water) were used. The simulated groundwater (pH = 8) contained 0.24, 0.61, 0.61, and 0.13 mM of NaCl, NaHCO$_3$, Na$_2$SiO$_4$•5H$_2$O, and CaCl$_2$, respectively. For the in-package chemistry water (pH = 7), 0.2, 0.2, 0.1, and 0.05 mM of NaCl, NaHCO$_3$, Na$_2$SiO$_4$•5H$_2$O, and CaCl$_2$ were added based on literature data, as summarized in Jung, et al. [3]. The pH of the test solution after the corrosion tests was close to the pH before the corrosion tests. The test solution was aerated by keeping the test vessel open to air.

The electrochemical tests were conducted at room temperature (~22 °C) in a 200-mL glass cell. A saturated calomel reference electrode and a platinum foil counter electrode were used. The average exposed surface area of the SIMFUEL working electrodes was 1.13 cm^2. In the tests using simulated groundwater, test samples were rotated at a speed of 1,000 rpm in order to avoid oxygen diffusion effects in the solution; tests showed that the results were independent of sample rotation. In the tests using in-package chemistry water, test samples were not rotated in order to better simulate stagnant in-package conditions. Impedance measurements were carried out in the frequency range of 100 kHz to 0.1 mHz with an alternating current voltage amplitude of 20 mV at the corrosion potential. A sorption test subsequently was conducted by transferring 40 mL of each post-test solution into a set of glass cells and immersing a stainless steel disk in each solution. The chemistry of the post-test solution samples and two blank solutions with and without calcium (Ca) and silicon (Si) were analyzed using ICP-AES for Ca and Si concentrations, and ICP-MS for barium (Ba), chromium (Cr), iron (Fe), molybdenum (Mo), nickel (Ni), strontium (Sr), uranium (U), and zirconium (Zr) concentrations.

RESULTS AND DISCUSSION

SIMFUEL characterization

Using optical micrographs of the SIMFUEL surfaces, grain size was determined to be ~15 μm for the 3% burnup and <10 μm for the 6% burnup SIMFUEL. Both SIMFUEL specimens have a compact surface and include many small-sized precipitates, present mainly along grain boundaries (Figure 1). Lucuta, et al. [2] reported a similar microstructure of SIMFUEL, showing spherical, intergranular metallic precipitates (0.5–1.5 μm diameter) along the grain boundary. In the same reference, results indicated the precipitates consisted mainly of ε-phase enriched with Mo, Ru, Rh, and Pd. The microstructures of SIMFUEL samples in this study were consistent with those produced by prior research [2,4].

The chemical composition of the general surface area (~200 μm by 200 μm) and precipitate (arrows in Figure 1) was analyzed by energy dispersive x-ray (EDX) analysis. For the general surface area consisting of SIMFUEL grains and grain boundaries, the matrix is predominantly

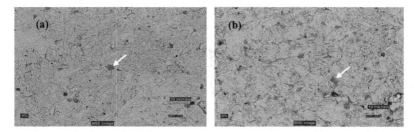

Figure 1. Backscattered electron micrographs of SIMFUEL surfaces: (a) 3% and (b) 6%

increases with burnup (i.e., 2.54 and 4.98 wt.% for 3 and 6% burnups, respectively). The 6% SIMFUEL has more precipitates along the grain boundaries, indicating a more heterogeneous nature. The precipitates consist predominantly of Mo, Ru, Pd, and Rh, in agreement with results by Lucuta, et al. [2]. In the general surface area, other elements such as Sr, Zr, La, Ce, Nd, and Ba were detected but were not detected in the grain boundary precipitates. Previous results of SIMFUEL analyses [2] indicate that Sr, Zr, and rare earth elements (e.g., Ce and La) were fully or partially dissolved in the matrix grains and Ba was a major element for the very small-sized oxide precipitates (about 0.1 μm) present at grain boundaries.

Electrochemical impedance spectroscopy

Figure 2 shows typical impedance spectra for 3% SIMFUEL after 24 and 77 hours immersion in a simulated groundwater (without the addition of Ca and silicate) measured at the corrosion potential. With an increase of immersion time from 24 to 77 hours, two changes were observed: (i) increase in the impedance modulus |z|, indicating higher polarization with time, and (ii) a tendency to broaden the phase angle curve in the low frequency range, indicating another time constant most likely due to formation of a secondary phase (e.g., schoepite), as often reported in the literature. For the 3% SIMFUEL with the addition of Ca and silicate, the spectra were very similar to the no-addition case, indicating similar polarization resistance and the same or similar dissolution mechanism. Even if the shape of the impedance spectrum for the 6% SIMFUEL was very similar to that of the 3% SIMFUEL (indicating no changes in the corrosion mechanisms), the 6% SIMFUEL exhibited a lower polarization resistance compared to that of the 3% SIMFUEL for both cases.

The dissolution (corrosion) rate of SIMFUEL was estimated by fitting the impedance spectra with a simple Randle circuit. The dissolution rate is calculated from the polarization resistance using the Stern-Geary equation and Faraday's law, as shown in equations 1 and 2, respectively.

$$i_{corr} = B / R_p \tag{1}$$
$$B = b_a \times b_c / [2.303(b_a + b_c)]$$

where i_{corr} is corrosion current density, R_p is polarization resistance, and b_a and b_c are anodic and cathodic Tafel slopes, respectively. The value of the composite Tafel parameter B for uranium dissolution used in this study is selected to be 25 mV based on literature data [5].

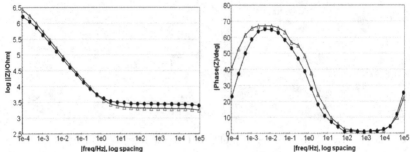

Figure 2. Electrochemical impedance spectra for 3% SIMFUEL measured after 24 (•) and 77 hours (Δ) immersion in a simulated groundwater without Ca and silicate addition

$$Dissolution\ Rate = K_2 \times i_{corr} \times EW \qquad (2)$$

where K_2 is a constant (0.0895 mg cm^2/μA dm^2 day) and EW is the equivalent weight for UO$_2$ (33.75 assuming +6 and -2 valences for U and O, respectively).

Table I lists the calculated dissolution rates of 3 and 6% SIMFUEL after 77 hours immersion in two different solutions. SIMFUEL exhibited very high polarization resistance on the order of 10^6 ohms, indicating high corrosion resistance, and the calculated dissolution rates of SIMFUEL range from 1 to 3 mg/m^2-day. For the simulated groundwater tests, the burnup effect on SIMFUEL dissolution rates was consistent. The dissolution rate of 6% SIMFUEL is relatively higher than that of 3% SIMFUEL for both cases with and without Ca and silicate by a factor of two to three. Forsyth [6] reported that the cumulative fractional release of radionuclides (i.e., Cs-137 and Sr-90) from SNF increased with burnup almost linearly up to values of 40–45 MWd/kg U, but the rates decreased as burnup further increased up to 49 MWd/kg U. The effect of the addition of Ca and silicate on the dissolution rate of SIMFUEL was not significant under the conditions tested in this study. This result could be due to very low concentrations of Ca and silicate added and/or the masking effect of schoepite formation. In the in-package chemistry water, the dissolution rate of 6% SIMFUEL was slightly lower compared to the simulated groundwater. This result could be due to a lower carbonate concentration in the in-package chemistry water.

Table I. Calculated dissolution rates of SIMFUEL obtained from impedance measurements

	Simulated groundwater				In-package chemistry	
Burnup (at.%)	3%		6%		6%	
Ca and silicate	Without	With	Without	With	Without	With
R_p (Ohms)	5.76×10^6	6.66×10^6	2.95×10^6	2.20×10^6	3.48×10^6	3.04×10^6
i_{corr} (A/cm^2)	3.1×10^{-9}	2.7×10^{-9}	6.0×10^{-9}	8.1×10^{-9}	5.1×10^{-9}	5.8×10^{-9}
Dissolution rate (mg/m^2-day)	1.2	1.0	2.3	3.0	1.9	2.2

Solution chemistry analyses

The measured solution concentrations of U, Ba, Mo, Sr, and Zr are shown in Figure 3(a). The concentrations in the figure were corrected for the amounts present in the blank solutions. The U concentration, which is the dominant element on the SIMFUEL surface, ranges from

7.5 × 10⁻⁹ to 1.9 × 10⁻⁸ mol/L. U concentration for the 6% SIMFUEL is almost twice the U concentration for the 3% SIMFUEL. Mo concentration, which is a dominant element in the grain boundary precipitate, is similar to that of U, ranging from 6.6 × 10⁻⁹ to 1.2 × 10⁻⁸ mol/L. Considering that a relatively small amount of Mo is present on the SIMFUEL surface compared to U, it is likely that Mo was preferentially dissolved from the precipitates at the grain boundary. The other fission-product-simulating elements, such as Ba and Zr, have a similar behavior.

Figure 3(b) shows normalized dissolution rates, which decrease in the order Ba > Mo ≈ Zr > Sr > U. The higher release rates for Ba, Mo, and Zr could be due to the preferential dissolution of the precipitate content present in the matrix (e.g., Zr) and at the grain boundary (e.g., Mo, Ba, and Zr). The relatively low U release rate could be a result of secondary phase formation. As mentioned previously, there was an indication of the possible formation of secondary phases from the impedance spectra after 77 hours immersion. In the literature, the formation of schoepite has been reported in similar solutions under oxidizing conditions [7]. In Jung, et al. [3], the calculated potential-pH for a U-simulated groundwater system at 22 °C showed a stable region of schoepite (UO₃·2H₂O) from pH of 6 to 8. Therefore, during the corrosion test conducted in this study, formation of schoepite was very likely. The calculated U dissolution rates range from 1.0 to 2.6 mg/m²-day, which are very close to the rates of 1 to 3 mg/m² day determined by impedance measurements after a 77-hour immersion time (see table I). The calculated U dissolution rates suggest that SIMFUEL dissolution is controlled by the U dissolution process. In particular, the dependence of the dissolution rate on burnup and solution chemistry observed from the solution analysis has the same trend as that determined from the electrochemical impedance measurements. The dissolution rate of Sr is higher than U, which is consistent with literature data [8,9] and has been attributed to fission product migration to the grain boundaries during irradiation.

The sorption test results showed that U, Ba, and Mo concentrations decreased by more than 25% after the disks were immersed for 21 days in the posttest solutions. The decrease in the concentrations is likely due to sorption onto the stainless steel disk. Sr concentration in the solution increased, but the reason for this increase is unknown. The change in Zr solution concentration showed an opposite and inconsistent trend. The tests that used a 6% SIMFUEL showed a decrease in Zr concentration, whereas the tests that used a 3% SIMFUEL showed a slight increase in solution concentration. The reason for this inconsistency is not known.

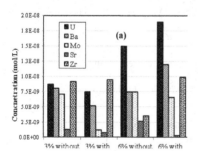 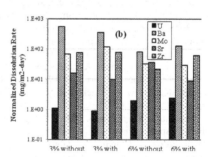

Figure 3. Concentrations of dissolved species (a) and normalized dissolution rates (b) of U, Ba, and Mo at different test conditions, including 3 and 6% burnups with and without the addition of Ca and silicate in a simulated groundwater.

CONCLUSIONS

Under test conditions in this study, measured dissolution rates of 3 and 6% burnup SIMFUEL ranged from 1 to 3 mg/m^2 day and the 6% burnup SIMFUEL had a slightly higher dissolution rate than the 3% SIMFUEL. The addition of Ca and silicate produced no measureable change in dissolution rate. U was the dominant element in the posttest solutions and its dissolution rate was very close to the rates measured by electrochemical methods. The dissolution of other fission product surrogates, such as Ba, Mo, Sr, and Zr, were not congruent with U. The uranium oxide dissolution appeared to be protected by secondary phases, such as schoepite, that formed on the SIMFUEL surface. The sorption test results suggested significant sorption of U onto oxide formed on stainless steel. The electrochemical impedance technique was found to be an effective tool for measuring the U dissolution rate in real time and for improving the understanding of U dissolution processes. SIMFUEL dissolution reasonably represents SNF dissolution rates expected after a long-term containment period when gamma and beta radiation decreases significantly.

DISCLAIMER

This paper is an independent product of CNWRA and does not necessarily reflect the view or regulatory position of NRC. The NRC staff views expressed herein are preliminary and do not constitute a final judgment or determination of the matters addressed or of any licensing action that may be under consideration at NRC.

ACKNOWLEDGMENTS

The authors thank English Pearcy for his review and Arturo Ramos for his administrative support. Special thanks to Dirk van Oostendorp and Lawrence Lambon at Xodus Group Inc. for their support in preparing the manuscript.

REFERENCES

1. T. Ahn and S. Mohanty, NUREG–1914, Washington, DC: U.S. Nuclear Regulatory Commission (NRC) (2008).
2. P.G. Lucuta, R.A. Verrall, H.J. Matzke, and B.J. Palmer, *J. Nucl. Mater.* **178**, 48 (1991).
3. H. Jung, T. Ahn, K. Axler, R. Pabalan, and D. Pickett, NRC ADAMS Accession Number ML112520488. Washington, DC: NRC (2011).
4. V. Rondinella and Hj. Matzke, *J. Nucl. Mater.* **238**, 44 (1996).
5. B. Grambow, A. Loida, A. Martinez-Esparza, P. Diaz-Arocas, J. De Pablo, J.L. Paul, G. Marx, J.P. Glatz, K. Lemmens, K. Ollila, and H. Christensen, European Commission, Nuclear Science and Technology, Report EUR 19140 EN (2000).
6. R. Forsyth, SKB TR 97-25. Stockholm, Sweden: Swedish Nuclear Fuel and Waste Management Company (1997).
7. D.W. Shoesmith, *J. Nucl. Mater.* **282**, 1, (2000).
8. S. Röllin, K. Spahiu, and U.-B. Eklund, *J. Nucl. Mater.* **297**, 231 (2001).
9. D. Serrano-Purroy, F. Clarens, J.-P. Glatz, D. Wegen, B. Christiansen, J. de Pablo, J. Gimennez, I. Casa, and A. Martinez-Esparza, *Radiochim. Acta* **97**, 491 (2009).

Mater. Res. Soc. Symp. Proc. Vol. 1518 © 2013 Materials Research Society
DOI: 10.1557/opl.2013.78

Effects of matrix composition on instant release fractions from high burn-up nuclear fuel

Olivia Roth[1], Jeanett Low[1], Michael Granfors[1], Kastriot Spahiu[2]
[1]Studsvik Nuclear AB, Hot Cell Laboratory, SE-611 82 Nyköping, Sweden
[2]SKB, Box 250, SE-101 24, Stockholm, Sweden.

ABSTRACT

The release of radionuclides from spent nuclear fuel in contact with water is controlled by two processes – the dissolution of the UO_2 grains and the rapid release of fission products segregated either to the gap between the fuel and the cladding or to the UO_2 grain boundaries. The rapid release is often referred to as the Instant Release Fraction (IRF) and is of interest for the safety assessment of geological repositories for spent fuel due to the potential dose contribution.

Previous studies have shown that the instant release fraction can be correlated to the fission gas release (FGR) from the spent fuel. Studies comparing results from samples in the form of pellets, fragments, powders and a fuel rodlet have shown that the sample preparation has a significant impact on the instant release, indicating that the differentiation between gap release and grain boundary release should be further explored.

Today, there are trends towards power uprates, longer fuel cycles and increasing burn-up putting additional requirements on the nuclear fuel. These requirements are met by the development of new fuel types, such as UO_2 fuels containing dopants or additives. The additives and dopants affect fuel properties such as grain size and fission gas release. In the present study we have performed experimental leaching studies using two high burnup fuels with and without additives/dopants and compared the fuel types with respect to their instant release behavior. The results of the leaching of the samples for the 3 initial contact periods; 1, 7 and 23 days are reported here.

INTRODUCTION

The release of toxic and radioactive species from spent fuel in contact with water is expected to depend mainly on the dissolution rate of the UO_2 matrix, since the majority of the inventory is uniformly distributed in the UO_2 fuel matrix. However, a fraction of certain volatile and segregated fission products will be leached from the gap between the fuel and the cladding or from the UO_2 grain boundaries. Matrix dissolution is considered to be a slow process, whereas the release from grain boundaries and gap is rapid upon contact with water. For the performance assessment of direct disposal of spent fuel in a geological repository, both release processes need to be quantified and the chemical reactions understood.

The rapidly released part of the inventory is designated as Instant Release Fraction (IRF) and consists of fission products (e.g. I and Cs) that have migrated to grain boundaries and radially from the hot center to cooler regions close to the pellet periphery and the gap between the fuel and the cladding.

The fraction of non-gaseous nuclides present in the spent fuel-cladding gap and in fuel cracks is considered to be comparable to the released fission gases as measured in gas release testing of fuel rods [1-4]. This relationship can be explained be the fact that under reactor operation temperatures volatile and segregated fission products are usually in gaseous state which facilitates migration to the grain boundaries and the fuel/cladding gap.

In addition to the release of gaseous fission products, other elements that are incompatible with the structure of the uranium dioxide are also segregated to form separate phases. Light water reactor fuel contains inclusions of metal alloys of Mo-Tc-Ru-Rh-Pd (4d metals), known as ε-phases.

The Fission Gas Release (FGR) depend partly on the burnup of the fuel and studies investigating instant release fractions at varying fuel burnup have been performed [5-7].However, the Fission Gas Release (FGR) is more strongly correlated to the linear heat rating than to the burnup of the fuel [8] and previous studies have also showed that the instant release fraction depends on the power history of the fuel [5].

Today, there are trends towards power uprates, longer fuel cycles and increasing burn-up putting additional requirements on the nuclear fuel. These requirements are met by the development of new fuel types, such as UO_2 fuels containing dopants or additives. The aim of the present study is to investigate the effect of fuel additives on the instant release behavior of the fuel. In this work we have tested an ADOPT (Advanced DOped Pellet Technology) fuel, produced by Westinghouse, containing additions of Cr_2O_3 and Al_2O_3 and compared the instant release from the ADOPT fuel to the release from a standard UO_2 fuel with similar burnup and power history.

Addition of Cr_2O_3 and Al_2O_3 facilitates densification and diffusion during sintering resulting in approximately 0.5% higher density within a shorter sintering time. The grain size is increased about five times compared to standard UO_2. At high burnups the doped fuel also shows a significant decrease in fission gas release (FGR) [9], better resistance to pellet-cladding interaction and to fuel washout during reactor operation.

EXPERIMENTAL DETAILS

Sample preparation:

Sample preparation (cutting, cladding detachment, fragment collection) and leaching tests were carried out in hot cells; for the leaching tests, a hot cell dedicated to leaching experiments was used. The experiments are performed in air. More details on experimental setup can be found in [5] where similar experiments are described. In Table I, the fuel rod data are listed including sample positions for the samples used in the leaching experiments.

Table I. Fuel rod data.

Sample designation (Studsvik)		5A2	C1
Irradiation time		7 years	
Reactor		Oskarshamn 3	
Fuel type		Standard UO_2	Al/Cr doped UO_2
Initial enrichment [wt% U^{235}]		3.5	4.1
Rod average burnup [MWd/kgU]		57.1	59.1
Fission gas release [%]	Kr	2.5	1.3
	Xe	2.4	1.5
Sample position (from rod bottom) [mm]		2648-2668	2672-2692

Leaching experiments:
The samples were leached as fuel fragments + separated cladding. The samples were prepared by cutting ~20 mm long pieces of each rod. The cladding of the rod segments were afterwards cut longitudinally by sawing on opposite sides of the segment. Force was applied to the halves until the fuel broke away from the cladding.

The two cladding halves, together with the detached fuel fragments were weighed, collected in a glass vessel with glass filter bottom (100-160 µm pores) and immersed into 200 ml leaching solution (10 mM NaCl + 2 mM NaHCO$_3$). The results of the leaching of the samples for the 3 initial contact periods; 1, 7 and 23 days are reported here.

Radionuclide analyses:
After each contact period, samples were collected for ICP-MS (Inductively Coupled Plasma-Mass Spectrometry) isotopic and γ-spectrometric analyses, as well as for pH and carbonate determination (pH and carbonate results were fairly constant during all the tests). ^{129}I and ^{127}I were analyzed by ICP-MS equipped with a Dynamic Reaction Cell (DRC) [10]. Oxygen was used as reaction gas to remove the interfering isotope ^{129}Xe on mass 129. Details on analyses and corrections (e.g. isobaric interferences) that are applied to the raw ICP-MS data can be found in Zwicky et al. [6].

Release fractions are calculated by dividing the total amount of a nuclide of concern in the analyzed solution by the total amount in the corroding fuel sample. The total amount of each nuclide present in the fuel sample was determined by general model calculations for fuel with burnup ~59 MWd/kgU. Cumulative release fractions are the sum of release fractions up to a certain cumulative contact time.

RESULTS AND DISCUSSION
The cumulative release fractions of ^{137}Cs and ^{129}I for the standard UO$_2$ sample and the Cr$_2$O$_3$ and Al$_2$O$_3$ doped sample are shown in Figure 1 and 2 respectively. The measured fission gas release (FGR) for each of the corresponding fuel rods from which the samples were cut is also indicated in the figures by a horizontal line.

As seen in Figures 1 and 2, the cumulative release of ^{129}I and ^{137}Cs levels out with time, but is still much lower than the FGR for the corresponding rod. Thus it is expected that the release will continue with increasing contact time. The maximum cumulated contact time shown in the figures is 31 days and it is still too early to judge the total IRF from these fuel samples. In any case, it is clearly seen that both fission gas release and the release of ^{137}Cs is much lower for the doped fuel sample than for the standard fuel sample. This is as expected due to the larger size of grains in the doped fuel. The release if ^{129}I is similar for the both fuel types.

Data from previous experiments [5,7] generally show that ^{129}I and ^{137}Cs release fractions are less than the FGR with ^{137}Cs releases being smaller than ^{129}I. The generally lower release of ^{137}Cs can (partly) be attributed to its lower diffusion coefficient, assuming the amount of Cs available for release during leaching is determined by diffusion under reactor operating conditions. However, previous data have shown that even taking the lower diffusion coefficient into account the lower release cannot fully be explained [5].

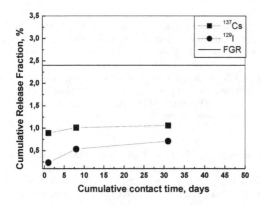

Figure 1. Cumulative release fractions for ^{137}Cs and ^{129}I for standard UO_2 sample.

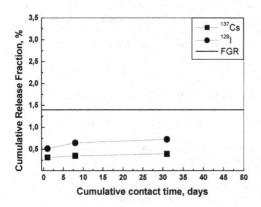

Figure 2. Cumulative release fractions for ^{137}Cs and ^{129}I for Cr_2O_3 and Al_2O_3 doped sample.

The results presented in Figures 1 and 2 show that for the fuel types used in this study, the doped fuel has higher ^{129}I release than ^{137}Cs release, as expected from the diffusion coefficients and as noted in previous studies [7]. For the non-doped fuel investigated in this study, the release of ^{129}I is lower than that of ^{137}Cs and significantly lower than the fission gas release, at least at this relatively short leaching period. This further supports the statement that other factors than diffusion under reactor operating conditions influence the ^{137}Cs and ^{129}I release rates.

In Figure 3 and 4 the cumulative release rates of potentially segregated fission products as compared to uranium (matrix) releases are shown. The release of these nuclides is generally lower for the doped sample than for the standard UO_2.

148

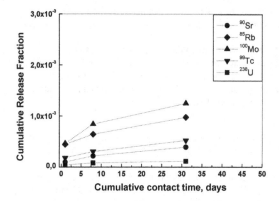

Figure 3. Cumulative release fractions for potentially segregated fission products as compared to uranium (matrix) releases for Cr_2O_3 and Al_2O_3 doped sample.

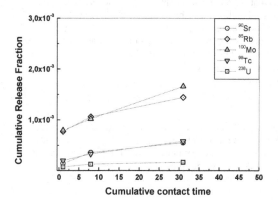

Figure 4. Cumulative release fractions for potentially segregated fission products as compared to uranium (matrix) releases for standard UO_2 sample.

The data in Figure 3 and 4 show that [100]Mo and [85]Rb are released to similar extent whereas ^{90}Sr and ^{99}Tc show lower release. As discussed in ref [6] no satisfying corrections procedure of the ICP-MS data for mass 85 have been established. Therefore, the [85]Rb data should be considered as preliminary.

Comparing the standard UO_2 fuel used in the present study (Figure 4) with previous data [6] obtained for pellet samples (i.e. with the cladding attached) the cumulative release fraction of [85]Rb, ^{137}Cs, ^{90}Sr and ^{99}Tc is similar in both cases, whereas the release of [100]Mo and [238]U is about one order of magnitude higher in the present study. The difference between the data series can probably be explained by the different sample types (cladded samples vs. cladding + separated fragments).

The difference in release of ^{129}I and ^{137}Cs between the two sample types has been explored in ref. [7] and although the release rates vary with sample type the results were found to be within the same order of magnitude. The increased release of ^{100}Mo and ^{238}U for the samples consisting of cladding + separated fragments suggests that the samples have been pre-oxidized, possibly during the cutting and separation of the cladding from the fuel.

SUMMARY

In this study we have investigated the effect of the addition of additives on the instant release behavior of the fuel by comparing the instant release from a UO_2 fuel doped with Cr_2O_3 and Al_2O_3 to the release from a standard UO_2 fuel with similar burnup and power history.

The results presented are evaluated from an ongoing experiment where a cumulative leaching time of 31 days has been reached. These initial data show that the release from the doped fuel sample is generally lower than the release from the standard sample, which is expected given the lower FGR and larger grain size of the doped fuel.
The cumulative release of ^{129}I and ^{137}Cs from the both samples is still well below the FGR for the corresponding rod. Thus it is expected that the release will continue with increasing contact time.

For the non-doped fuel investigated in this study, the release of ^{129}I is lower than that of ^{137}Cs which contradicts findings in previous studies. The release of ^{85}Rb, ^{137}Cs, ^{90}Sr and ^{99}Tc is similar to previous data whereas the release of ^{100}Mo and ^{238}U is about one order of magnitude higher in the present study, indicating potential pre-oxidation of the fuel.

The results presented here represent a work in progress the data should be interpreted with care until the final data has been analyzed.

ACKNOWLEDGEMENTS

The fuel samples, fabrication and irradiation data were made available by Westinghouse Electric Sweden and OKG Aktiebolag. The project has received funding from the European Union's European Atomic Energy Community's (Euratom) Seventh Framework Programme FP7/2007-2011 under grant agreement n° 295722 (FIRST-Nuclides project).

REFERENCES

1. L. H. Johnson and D. F. McGinnes, Nagra Technical Report **NTB 02-07**, (2002).
2. L. H. Johnson, C. Poinssot, C. Ferry and P. Lovera, Nagra Technical Report **NTB 04-08**, (2004).
3. L. H. Johnson and J. C. Tait, **SKB TR 97-18**, (1997).
4. C. Ferry, J-P. Piron, A. Poulesquen and C. Poinssot, Mater. Res. Soc. Symp. Proc. **1107**, 2008, pp. 447–454.
5. L. Johnson et. al., *J. Nucl. Mater.*, **420**, 54-62 (2012).
6. H-U. Zwicky, J. Low, E. Ekeroth, **SKB TR 11-03**, (2011).
7. E. Ekeroth et al. Mater. Res. Soc. Symp. Proc. **1475**, paper O36 (2012).
8. K. Kamimura. IAEA-TECDOC-697, International Atomic Energy Agency, pp 82–88 (1992).
9. K. Backman et. al., in: **IAEA-TECDOC-1654**, pp. 117-126 (2010).
10. A. Izmer, S. F. Boulyga and J. S. Becker, *J. Anal. At. Spectrom.* **18** 1339-1345 (2003).

Mater. Res. Soc. Symp. Proc. Vol. 1518 © 2013 Materials Research Society
DOI: 10.1557/opl.2013.90

Reducing the uncertainty of nuclear fuel dissolution: an investigation of UO$_2$ analogue CeO$_2$

Claire L. Corkhill[1], Daniel J. Bailey[1], Stephanie M. Thornber[1], Martin C. Stennett[1] and Neil C. Hyatt[1]

[1]The Immobilisation Science Laboratory, Department of Materials Science and Engineering, The University of Sheffield, UK, S1 3JD.

ABSTRACT

In this investigation, CeO$_2$ analogues, which approximate as closely as possible the characteristics of fuel-grade UO$_2$, were characterised after dissolution under a wide range of conditions. Powdered samples were subject to a range of aggressive and environmentally relevant alteration media with different solubility controls, and reacted at 70 °C and 90 °C. Dissolution kinetics were monitored through analysis of the coexisting aqueous solution. Monolith samples were monitored for development of surface defects such as pores and dissolution pits, in addition to morphological changes at grain boundaries and surface pores upon dissolution under aggressive conditions. The surfaces were analysed using confocal profilometry, vertical scanning interferometry and scanning electron microscopy. Dissolution rates were found to be greatest in low pH solutions and at higher temperatures. Preferential dissolution appears to occur at grain boundaries and on particular grains, suggesting a crystallographic control on dissolution.

INTRODUCTION

In the safety case for the geological disposal of nuclear waste, the release of radioactivity from the repository is controlled by the dissolution of the spent fuel in groundwater [1]. Therefore, to assess the performance of the repository after infiltration of groundwater and contact with spent fuel, the dissolution characteristics must be determined. The use of spent nuclear fuel, or its main component UO$_2$ is problematic due to safety and redox sensitivity issues. As such, in experiments using UO$_2$, special care must be taken to avoid oxidation of the sample surfaces [2, 3]. It is useful to use a non-redox sensitive analogue for UO$_2$ with the same fluorite-type structure for investigations of surface alteration, such as CeO$_2$ [4].

In spent nuclear fuel, high energy sites occur at grain boundaries and within the material as naturally occurring surface defects. Current studies of spent nuclear fuel dissolution have not considered the effect of high energy surface sites within the material structure. In this preliminary investigation, CeO$_2$ analogues were powdered and subjected to a range of aggressive and environmentally relevant alteration media with different solubility controls. Dissolution was monitored through analysis of the aqueous solution. To complement these experiments, the dissolution of monoliths of CeO$_2$ is described. These samples were monitored for the evolution of high energy sites, including pores, steps and dissolution pits. Preliminary results are discussed, which will form the basis for future investigation.

METHODOLOGY

CeO$_2$, prepared as described in [4] was ground to a powder (75-150 µm) according to the product consistency test ASTM standard (ASTM C 1285-02). Prior to use, the powder was inspected by SEM to ensure no fine particles <1 µm in size remained. 0.1 g of powder was placed in 50 ml PTFE vessels and filled with 40 ml of alteration solution, either: 0.01 M HNO$_3$ (pH 2); a simplified groundwater of dilute NaCl (10mM) buffered to pH 8.5 by NaHCO$_3$ (2mM); or 0.005M NaOH (pH 11.7), giving a surface area to volume ratio of 200 m^{-1}. Experiments were performed in triplicate at 70 °C and 90 °C. Sampling was conducted at 0, 1, 3, 7, 14, 21, 28, 35, 42, 56 and 70 days. An aliquot (1.2 ml) of each sample was removed prior to analysis by ICP-MS. The removed volume was replaced after each sampling. Leaching is expressed as the normalised elemental leaching N$_L$(Ce) (g m^{-2}) according to:

$$N_L(Ce) = \frac{m_{Ce}}{S/V} \tag{1}$$

where m$_{Ce}$ is the total amount of Ce released in the solution and S/V is the surface area to volume ratio. The normalised element leaching rate R$_L$(Ce) (g m^{-2} d^{-1}) is determined by:

$$R_L(Ce) = \frac{m_{Ce}}{\frac{S}{V} \times \Delta t} \tag{2}$$

where Δt is the leaching time in days.

Monolith experiments were conducted using pressed and polished (0.05 µm, diamond paste) CeO$_2$ pellets [4]. The pellets were placed within pre-cleaned (ASTM C 1285-02) cold-sealed Teflon reactors (3 cm^3) on a spacer to ensure fluid contact all around the pellet. 2 ml 2M HCl was added to the vessel prior to closure. Reaction vessels were placed in an aluminium heating block, and heated to either 90 °C or 150 °C. Samples were removed at 1, 3, 7, 14 and 21 days for imaging by Scanning Electron Microscopy (SEM) and Optical Profilometry. The reaction fluid was also removed at these time periods and analysed for Ce concentration by ICP-MS. This was replaced with fresh reaction solution prior to re-starting the experiment for the next sampling point.

RESULTS

CeO$_2$ powder dissolution

The normalised leaching of CeO$_2$ at 90 °C as function of pH is shown in Figure 1. Fast initial kinetic leaching (R$_L$0) is observed for pH 2, pH 8.5 and pH 11.7 batch experiments, with rates of (1.98 ± 0.4) x 10^{-4} g m^{-2} d^{-1}, (0.6 ± 0.01) x 10^{-4} g m^{-2} d^{-1} and (0.50 ± 0.09) x 10^{-4}, respectively (Table 1). Under the redox and pH conditions of these experiments, Ce(IV) is the dominant Ce species, as determined by geochemical modelling.

Table 1. Normalised leaching rates for CeO_2 dissolution. R_L0 indicates the fast initial kinetic dissolution and R_Lt indicates the slower solution saturation limited dissolution rate.

Solution	pH	Alteration temperature (°C)	R_L0 $(g\,m^{-2}\,d^{-1})$	R_Lt $(g\,m^{-2}\,d^{-1})$
HNO_3	2	90	$(1.98 \pm 0.4) \times 10^{-4}$	$(0.10 \pm 0.02) \times 10^{-4}$
HNO_3	2	70	$(0.90 \pm 0.4) \times 10^{-4}$	$(0.15 \pm 0.02) \times 10^{-4}$
$NaCl/NaHCO_3$	8.5	90	$(0.60 \pm 0.1) \times 10^{-4}$	$(0.02 \pm 0.003) \times 10^{-4}$
$NaCl/NaHCO_3$	8.5	70	$(0.26 \pm 0.06) \times 10^{-4}$	$(0.008 \pm 0.001) \times 10^{-4}$
$NaOH$	11.7	90	$(0.50 \pm 0.09) \times 10^{-4}$	$(0.01 \pm 0.001) \times 10^{-4}$
$NaOH$	11.7	70	$(0.28 \pm 0.01) \times 10^{-4}$	$(0.005 \pm 0.001) \times 10^{-4}$

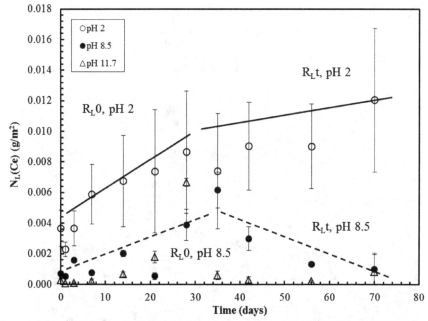

Figure 1. Normalised leaching $N_L(Ce)$ obtained during leaching tests (T = 90 °C, pH 2 = 0.01M HNO_3, pH 8.5 = NaCl (10mM) / $NaHCO_3$ (2mM) and pH 11.7 = 0.005M NaOH) of crushed CeO_2 (75-150 μm). R_L0 indicates the fast initial kinetic dissolution, and R_Lt indicates the slower solution saturation limited dissolution rate. Rate data are given in Table 1.

At pH 2, as the solution becomes saturated, the concentration of Ce in solution continues to increase, but the leaching rate (R_Lt) decreases (Table 1). However, for powders reacted at pH 8.5 and pH 11.7, the leach rate due to saturation (R_Lt) is also influenced by removal of Ce from solution, most likely through precipitation of $Ce(OH)_4$. The evolution of normalised leaching of

CeO$_2$ at 70 °C and 90 °C is also shown in Table 1. As expected, for powders reacted at pH 2, the initial kinetic normalised leaching rates (R$_L$0) are greater than those for 70 °C. The same trend is observed for powders leached at pH 8.5 and pH 11.7 (Fig, 2b), although the normalised leaching rates are considerably lower (Table 1).

This is consistent with results obtained for CeO$_2$ powders leached at 60 °C in nitric acid (3.2 ± 0.2) x 10^{-6} g m^{-2} d^{-1} [5] and slightly lower, but consistent with results obtained for Ce leaching in multi-component Ce-containing ceramic oxides (e.g. [6]), which gave rates of approximately 3 x 10^{-3} g m^{-2} d^{-1}.

CeO$_2$ monolith dissolution

Figure 2 shows the extent of CeO$_2$ dissolution at 150 °C after 9 and 21 days. After 9 days (Fig. 2b), it is clear that the grain boundaries have undergone extensive dissolution compared to the pristine sample (Fig. 2a), etch pits have formed in some of the grain surfaces (mostly triangular in shape), and some grains appear to have dissolved more than others, as shown by the height contrast between grains. After 21 days (Fig. 2c and d), the surface of the sample has been severely transformed, with no observable grain boundaries. It is also clear that triangular facets are present on the surface of the least dissolved grains (Fig. 2d, insert).

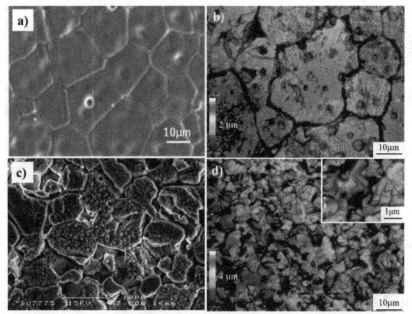

Figure 2. SEM and confocal profilometer images of CeO$_2$ monolith samples a) pristine, b) after 9 days reaction in 2M HCl at 150 °C , and c) and d) after 21 days reaction in 2M HCl at 150 °C. Insert in d) shows x 4 magnified area of image in d).

For CeO_2 reacted at 90 °C under identical conditions, relatively little change was observed at the sample surface. This is evident in the leached Ce concentration data for CeO_2 leached at 90 °C compared to 150 °C, as shown in Figure 3. Dissolution at both temperatures show a linear trend, but for dissolution at 150 °C, the concentration of Ce is an order of magnitude higher than that for 90 °C, giving concentrations after 14 days reaction of 83.3 ppm and 8.3 ppm, respectively.

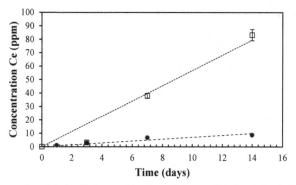

Figure 3. Concentration of Ce in solution resulting from CeO_2 monolith dissolution in 2M HCl at 150 °C (□) and 90 °C (●). Instrument error of 5% is shown. Lines are to guide the eye.

CONCLUSIONS

It is clear that several factors control the dissolution of the UO_2 spent nuclear fuel analogue, CeO_2. The results presented herein demonstrate the expected dissolution kinetic dependence on pH and temperature for powdered samples. Further work will elucidate the effect of particle size, and hence high surface energy site density under the same experimental conditions. Preliminary evidence is shown for preferential leaching of some grains and the presence of triangular facets, which suggest a crystallographic control on leaching. This is consistent with investigations on fluorite, which has the same crystal structure as CeO_2 [7, 8]. Further work is directed towards determining the relative contribution of grain boundary dissolution to the reaction kinetics.

ACKNOWLEDGMENTS

The research leading to these results has received funding from the European Atomic Energy Community's Seventh Framework Programme (FP7) under Grant Agreement No. 269903 (REDUPP). N. C. Hyatt is grateful to The Royal Academy of Engineering and the Nuclear Decommissioning Authority for funding. We also thank the EPSRC Nuclear FiRST Doctoral Training Centre for project funding for S. M. Thornber and D. J. Bailey.

REFERENCES

1. D. W. Shoesmith, *Journal of Nuclear Materials* **282**, 1 (2000).
2. K. Ollila and V Oversby, *SKB Technical Report* TR-05-07 (2005).
3. K. Ollila, *Posiva Work Report* 2008-75 (2008).

4. M. C. Stennett, C. L. Corkhill, L. A. Marshall and N. C. Hyatt *Journal of Nuclear Materials* **432**, 182 (2013).

5. L. Claparede, N. Clavier, N. Dacheux, P. Moisy, R. Podor and J. Ravaux, *Inorganic Chemistry* **50**, 9059 (2011).

6. J. P. Icenhower, D. M. Strachan, M. M. Lindberg, E. A. Rodriguez and J. L Steele. *PNNL Report* 14252 (2003).

7. J. R. A. Godinho, S. Piazolo, M. C. Stennett and N. C Hyatt. *Journal of Nuclear Materials* **419**, 46 (2011).

8. J. R. A. Godinho, S. Piazolo and L. Z. Evins. *Geochimica et Cosmochimica Acta* **86**, 392 (2012).

Mater. Res. Soc. Symp. Proc. Vol. 1518 © 2013 Materials Research Society
DOI: 10.1557/opl.2013.96

Research and Development on Cementation of Liquid Radioactive Waste (LRW) Resulting from Spent Nuclear Fuel Reprocessing in the Experimental Demonstration Center (EDC) of Mountain Chemical Combine

Lebedeva A.V.[1], Sukhanov L.P.[1]., Ustinov O.A[1]

[1]Bochvar High-Technology Research Institute of Inorganic Materials, 5a Rogova street, Moscow, 123098, Russian Federation

ABSTRACT

The report presents the results of research and development on cementation technology of liquid radioactive waste (LRW) generated in spent nuclear fuel reprocessing in Experimental Demonstration Center (EDC) at the Mountain Chemical Combine, Russian Federation.

INTRODUCTION

Currently Experimental Demonstration Center for testing of innovative spent nuclear fuel reprocessing technologies is being established at the Mountain Chemical Combine in Russia. One of the main requirements to the technologies being tested is to minimize the radioactive waste generation.

Cementation was chosen as a method of low and intermediate-level LRW conditioning. The reasons for the choice of cementing technology were its following advantages over other LRW conditioning technologies:

- The cement matrix ability to hold U and Pu due to the formation of complex compounds;

- The possibility of the most mobile nuclide (Cs-137) fixing in the cement matrix by means of sorption additives;

- The relative equipment simplicity as cementation is carried out at low temperatures;

- The availability and low cost of the matrix material;

- No secondary waste.

EDC LRWs can be conditionally divided into non-technological and technological.

The non-technological EDC LRWs includes the so-called "drain waters" – low-salt and low-active solutions consisting of water from the decontamination of facilities, equipment leaks, water from sanitary locks, sanitary inspection rooms, showers, etc.

The technological EDC LRWs are solutions containing tritium and acetates.

The quality of the cement compound resulting from pulps solidification must meet the regulatory requirements [1-2] on the mechanical strength, radionuclide leachability, radiation, water and frost resistance.

In addition to the standard criteria of cement compound quality, some other parameters should be considered. These include, in particular, the compound flowability at the stage of preparation (flowability value must be sufficient to discharge the compound from a mixer) and heat generation during the cement mixture hardening (increase in temperature above 90^0C may lead to a deterioration of the resulting compound properties).

EXPERIMENTAL DETAILS

Major experiments in VNIINM have been conducted with the use of simulators of radioactive waste. Also, to determine the radionuclide's leaching rate, samples were marked with Cs-137 and tritium.

For *the non-technological waste* the bottom residue simulator was used, estimated composition of which is shown in table I.

Table I. Estimated composition of simulated bottom residue

LRW components		Content in the solution g/l	Content in the dry residue, % by weight
Solution	Sediment		
NaNO$_3$	-	256	78.0
NaCl	-	4.2	1.3
Na$_2$CO$_3$	-	9.3	2.8
Na$_2$C$_2$O$_4$	-	1.4	0.4
Na$_2$SO$_4$	-	6.0	1.8
S.A.S.*	-	2.0	0.6
-	CaCO$_3$	13.1	4.0
-	Mg(OH)$_2$	5.7	1.7
-	Na$_2$C$_2$O$_4$	24.9	7.6
-	Suspensions	5.5	1.7
* surface-active substance			

The total content of dry residue in the simulator of non-technological LRW to be cemented is about 330 g/l.

For technological waste:
- *for tritium waste* a 5% solution of sodium nitrate was used as a tritium water simulator
- *for acetate waste* the simulator of CH$_3$COONa×3H$_2$O – 329.3 g/l, NaOH – 5.2 g/l composition was used.

Portland cement (PC) and Slag Portland cement (SPC) of various brands, and metallurgical slag were used as the matrix material, and bentonite clay M4T1K and clinoptilolite (Kholinsk deposit, Siberia) were used as sorption additives. In investigations with tritium water the additional waterproofing covering was used to keep it in a cement compound. Bitumen was chosen as the most available material for the covering. In the present work bitumen BND 90/130 was used.

A binder type, a sorption additive type, their content in the compound, a cement-compound loading capacity, and a water-to-binder ratio (W/B) were varied in the course of the research. The research of the cement compound properties were carried out on samples 2x2x2 cm in size, which were placed in air-humid conditions for hardening, then subjected to the necessary measurements.

DISCUSSION

For all types of LRW mechanical strength (after 7, 14, 28, 56 and 90 testing days), water and frost resistance, radionuclides leaching were determined. For non-technological and acetate waste the flowability and specific heat generation were also tested.
The experimental results are shown below.

For non-technological waste:

The results of experiments to determine the mechanical strength, water and frost resistance of cement compounds, radionuclides leaching, and flowability of cement paste are shown in Tables II and III.

To determine the radionuclides leaching rate samples number 1 and number 2 (Table II) were selected, marked with Cs-137. The choice of cesium as the marking agent was made due to the highest solubility of its compounds. The specific activity of compounds was $4,67 \cdot 10^3$ Bq/g. The samples surface - 24 cm^2. Samples 1 and 2 differ in the type of sorption additive. In the first case, bentonite was used as sorbent, and in the second - clinoptilolite. The sorbent content in the matrix material, in both cases was 10% by weight. Following the results, the tested samples have similar radionuclides leaching rate, which after 28 days reached a value not exceeding the standard value of $- 1 \cdot 10^{-3}$ g/(cm$^2 \cdot$day).

Comparative evaluation of the various compounds specific heat generation was carried out upon the results of calorimetric measurements. Example of the experiments results is shown in Figure 1 as a dependence diagram of the change in the temperature difference (between the ambient temperature and the temperature of the cement compound sample) on time.

The test results showed the following:

- Replacement of Portland cement with slag Portland cement and bentonite – with clinoptilolite reduces the heat generation in the process of cementation;

- Increase of clinoptilolite content in the matrix material from 10 to 30% leads to a slight increase of the specific heat generation;

- Minimum heat generation was characteristic for sample number 5 (Table II), consisting of slag Portland cement as a binder and clinoptilolite as sorbent additive.

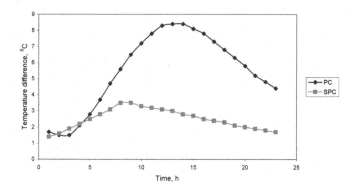

Figure 1. Influence of Cement Compound Binder Type on Heat Generation

Table II. Composition and quality of the cement compounds samples with the non-technological EDC LRW simulator

Sample №	Compound composition, % be weight				Flowability, mm	Increase in mechanical strength in MPa for time, days			Mechanical strength, MPa			
									Frost resistance		Water resistance	
	PC	SPC	Bento-nite	Clinopt ilolite		28	42	56	after 30 thermal cycles	reference samples	after 90 days	reference samples
1	90	-	10	-	95	16,2	15,0	14,2	20,0 25,0	17,7	30,0 30,0	27,5 30,0
2	90	-	-	10	122	28,7	23,2	25,0	30,0 32,5	30,0	30,0 35,0	22,5
3	18 (20)	72 (80)	10	-	140	14,2	15,5	19,0	19,0 20,0	18,5 20,3	25,0 21,3	23,8
4	18 (20)	72 (80)	-	10	180	7,3	10,8	9,5	11,0 9,5	11,0 11,3	11,8 10,8	10,8 14,5
5	70	-	-	30	107	35,0	35,0	35,0	40,0 40,0	40,0	35,0 35,0	35,0

Table III. Leachability of Cs-137 from cement compounds samples with non-technological EDC LRW simulator. Leaching rate and the total leached activity at different exposure time

Sample №	Sample weight, g	Sample activity, Bq	1 day g/(cm²·day)	%	2 days g/(cm²·day)	%	3 days g/(cm²·day)	%	7 days g/(cm²·day)	%	14 days g/(cm²·day)	%	28 days g/(cm²·day)	%
1	14,79	$6,9 \cdot 10^{4}$	$7,0 \cdot 10^{-3}$	1,5	$2,0 \cdot 10^{-2}$	4,4	$9,0 \cdot 10^{-3}$	5,9	$2,3 \cdot 10^{-3}$	7,3	$1,3 \cdot 10^{-3}$	8,8	$7,3 \cdot 10^{-4}$	10,5
	14,65	$6,8 \cdot 10^{4}$	$9,0 \cdot 10^{-3}$	1,1	$1,2 \cdot 10^{-2}$	3,4	$5,8 \cdot 10^{-3}$	4,4	$2,1 \cdot 10^{-3}$	5,8	$1,4 \cdot 10^{-3}$	7,4	$6,5 \cdot 10^{-4}$	8,9
2	16,29	$7,6 \cdot 10^{4}$	$1,6 \cdot 10^{-2}$	2,4	$9,6 \cdot 10^{-3}$	3,8	$5,0 \cdot 10^{-3}$	4,6	$2,5 \cdot 10^{-3}$	6,0	$1,4 \cdot 10^{-3}$	7,5	$8,7 \cdot 10^{-4}$	9,3
	16,30	$7,6 \cdot 10^{4}$	$1,6 \cdot 10^{-2}$	1,5	$7,2 \cdot 10^{-3}$	3,0	$4,8 \cdot 10^{-3}$	3,6	$2,1 \cdot 10^{-3}$	4,9	$1,5 \cdot 10^{-3}$	6,4	$8,2 \cdot 10^{-4}$	8,1

For tritium waste an important factor is reduction of the tritium water release from the cement compound. This effect was achieved by creating a protective hydrolytic covering. The experiments results with model solutions and real tritium water are given in Tables IV and V.

Table IV. Water resistance of tritium waste simulator samples with bitumen covering

Cementing composition, %		Bitumen covering thickness	Leaching rate (g/cm^2 day) and total leached activity (%) at different time, days:			
PC	SPC		7	14	21	28
100	-	-	$8 \cdot 10^{-4}$ (20,0 %)	$3 \cdot 10^{-4}$ (23,7 %)	$2 \cdot 10^{-4}$ (26,3 %)	$8 \cdot 10^{-5}$ (27,2 %)
100	-	~3	$3 \cdot 10^{-5}$ (0,5 %)	$2 \cdot 10^{-5}$ (0,5 %)	$2 \cdot 10^{-5}$ (0,8 %)	$3 \cdot 10^{-5}$ (1,0 %)
-	100	-	$4 \cdot 10^{-4}$ (15,1 %)	$2 \cdot 10^{-4}$ (17,2 %)	$1 \cdot 10^{-4}$ (18,5 %)	$8 \cdot 10^{-5}$ (19,4 %)
-	100	~5	$7 \cdot 10^{-6}$ (0,38 %)	-	$3 \cdot 10^{-6}$ (0,4 %)	-

Table V. Tritium leaching rate from cement compound samples with real tritium water with bitumen covering

Sample composition	Sample activity, Bq	Leaching rate (g/cm^2 day) and total leached activity (%) at different time, days:				
		1	2	3	7	14
PC – 54 g Bentonite – 6 g Tritium water – 29 g	$4,2 \cdot 10^9$	$8 \cdot 10^{-4}$ ($2,3 \cdot 10^{-2}$ %)	$2 \cdot 10^{-4}$ ($2,9 \cdot 10^{-2}$ %)	$1 \cdot 10^{-4}$ ($3,3 \cdot 10^{-2}$ %)	$1 \cdot 10^{-4}$ ($4,8 \cdot 10^{-2}$ %)	$1 \cdot 10^{-4}$ ($7,7 \cdot 10^{-2}$ %)
SPC – 93,3 g Tritium water – 46,7 g	$4,1 \cdot 10^9$	$8 \cdot 10^{-5}$ ($2,4 \cdot 10^{-3}$ %)	$4 \cdot 10^{-5}$ ($3,8 \cdot 10^{-3}$ %)	$3 \cdot 10^{-5}$ ($4,6 \cdot 10^{-3}$ %)	$2 \cdot 10^{-5}$ ($6,7 \cdot 10^{-3}$ %)	$1 \cdot 10^{-5}$ ($9,8 \cdot 10^{-3}$ %)

The results of experiments conducted with acetate waste simulators are listed in Table VI.

Table VI. Composition and quality of the cement compounds samples with the acetate waste simulator

Sample №	Matrix material		Flowability, mm	Mechanical strength		Frost resistance, MPa		Water resistance, MPa	
	Sorbent	Binder		age, days	MPa	after 30 thermal cycles	reference samples	after 90 days	reference samples
1	Bentonite 10%	PC 500-D0 (CEM I 42.5B) 90%	130	28 42 56	20,0 18,0 20,8 21,0 25,0	30,0 30,0	36,3	32,5 35,0	21,2
2	Bentonite 10%	80% SPC 400 (CEM III 32.,5) and 20% of PC 500-D0 (CEM I 42.5B)90%	109	28 42 56	3,5 7,5 9,0	8,5 12,5	13,0 15,0	18,7 18,5	15,0 16,3
4	Clinoptilolite 10%	40% SPC 400 (CEM III 32.,5) and 60% PC 500-D0 (CEM I 42.5B) 90%	135	28 42 56	25,0 45,0 45,0	50,0 50,0	35,0 42,5	50,0 52,5	50,0 45,0
5	Clinoptilolite 10%	PC 500-D0 (CEM I 42.5B)90%	147	28 42 56	35,0 40,0	43,8 45,0	40,0	50,0 51,3	50,0
6	Clinoptilolite 30%	PC 500-D0 (CEM I 42.5B) 70%	130	28 42 56	47,5 52,5	60,0 40,0	50,0	45,0 65,0	60,0

Preliminary experiments to determine the compounds heat generation showed that compounds containing LRW simulator are characterized by slow-curing and, therefore, low amount of heat is generated in the process of hydration. This may be explained by the high content of sodium acetate in the compound. Therefore, further studies of the heat generation is inexpedient.

Researches of radionuclides leaching rate were carried out the same way as for non-technological waste. The specific activity of the samples was $4,67 \cdot 10^3$ Bq/g.

CONCLUSIONS

The cement compounds based on Portland cement and Slag Portland cement with clinoptilolite and bentonite additives, including solutions simulating the non-technological and technological EDC LRWs, satisfy the regulatory requirements for mechanical strength, water and frost resistance.

To fix the most easily leached radionuclide - Cs-137 - in the compound both bentonite and clinoptilolite can be used as a sorption additive. With the same content (10% by weight of the matrix material), the cesium leaching rate is almost the same and after 28 days it reaches the regulatory value.

The flowability value of cement paste isn't reduced below 95-100 mm, which is sufficient to discharge compound from the mixer. Increase of bentonite content over 10% leads to poor cement paste flowability and reduced mechanical strength.

Recommended dry residue content in the resulting compounds for non-technological and technological waste is 12% and 15% by weight, respectively. Limitations are the maximum permissible LRW concentration set forth in the design documentation and impossibility to increase the water-to-binder ratio without the compound quality degradation.

The study of heat generation during cement compound hardening showed that the hydration heat depends on the binder type, sorption additive type and can vary widely. Compounds based on slag Portland cement with clinoptilolite as sorbent additive are characterized by the lowest heat generation.

For tritium water localization the compound must be cover with the bitumen film and put in plastic and steel containers.

The results obtained in laboratory were used to develop hardware-technological diagram of EDC LRW conditioning.

REFERENCES

1. Maslennikov I.A., Fedorov Ju.S., Shadrin A.Ju., Zilberman B.Ya., Saprykin V.F., Beznosjuk V.I., Ryabkov D.A. Experimental Demonstration Center: objectives, technologies, prospects// Environmental safety, №1- 2010, P. 30-33.

2. NP-019-2000. Collection, Processing, Storage and Conditioning of Liquid Radioactive Waste. Safety Requirements.

3. GOST R 51883-2002. Cemented Radioactive Waste. General Technical Requirements.

4. Lokken R.O. et al. Results of deviations' studies in the cement compound recipe of Hanford tank 241-AN-106, PNL-8226, 1993.

5. Lebedeva A.V. The first stage of complex researches of LRW cementation process resulting from the spent nuclear fuel reprocessed at the Experimental Demonstration Center (EDC), created at Mountain Chemical Combine. Proceedings of the International Workshop on nuclear technology "Cheremshanskie reading-2012", part 2, p. 215-221, Dimitrovgrad , 2012.

6. Akhmad I.K., Mikhalchenko A.G., Doilnicin V.A., Motornaya V.N. Research ways to reduce the tritium leaching from the cement matrix. Radiation safety, 2009, № 2, p. 21-25.

Waste Repositories

Mater. Res. Soc. Symp. Proc. Vol. 1518 © 2013 Materials Research Society
DOI: 10.1557/opl.2013.93

NMR Study of Interlayer and Non-interlayer Porewater in Water-saturated Bentonite and Montmorillonite

Torbjörn Carlsson[1], Arto Muurinen[1], and Andrew Root[2]

[1]VTT Technical Research Centre of Finland, Espoo, P.O. Box 1000, FI-02044 VTT, Finland.
[2]Magsol, Tuhkanummenkuja 2, 00970 Helsinki, Finland.

ABSTRACT

Bentonite is planned to be used in many countries as an important barrier in high-level waste repositories. Assessment of the barrier with regard to, *inter alia*, its ability to hinder transport of dissolved radionuclides leaking from a damaged canister containing spent nuclear fuel, requires quantitative data about the pore structure inside bentonite. The present NMR study was made in order to determine the number of distinguishable porewater phases in compacted water-saturated samples of MX-80 bentonite and Na-montmorillonite. The samples were compacted to dry densities in the interval 0.7-1.6 g/cm^3 and subsequently saturated with Milli-Q water or 0.1 M NaCl solution in equilibrium cells. The NMR measurements were performed with a high-field 270 MHz NMR spectrometer using a short inter-pulse CPMG method to study proton $T_{1\rho}$ relaxation. The measured relaxation curves were found to consist of one faster and one slower proton relaxation. Subsequent analysis of the data indicated that the faster relaxation was associated with interlayer (IL) water between montmorillonite unit layers, while the slower one was associated with non-interlayer (non-IL) water located outside the interlayer spaces. The results indicate for compacted samples with a dry density of $\geq$ 1.0 g/cm^3, that Na montmorillonite contains a larger relative volume of non-IL water than the corresponding MX-80 bentonite. This in turn, suggests that the stacking number in Na-montmorillonite is smaller than in MX-80 bentonite. Changing the porewater chemistry seemed to have some effect on the non-IL water content in the Na montmorillonite but not in the MX-80 bentonite.

INTRODUCTION

Bentonite is planned to be used as a barrier material in, e.g. Finnish and Swedish repositories for spent nuclear fuel [1, 2]. To ensure that the barrier will function properly, it is necessary to have detailed knowledge about the behaviour of water-saturated bentonite, with regard to *inter alia* the distribution of water between different pores in the bentonite. The materials used in this study were Wyoming MX-80 Na/Ca bentonite and Na montmorillonite prepared by purifying the above MX-80 bentonite. Water-saturated samples with dry densities between 0.7 and 1.6 g/cm^3 were analysed with ^{1}H NMR and the results were used to distinguish between two major types of porewater; which in the following are referred to as "interlayer water" (IL water) and "non-interlayer water (non-IL water). Briefly, IL water refers to porewater between montmorillonite unit layers, while non-IL water refers to the rest of the water in the sample. This division may not be perfectly unambiguous, since the structure of real samples may exhibit, e.g., turbostratic stacking [e.g. 3], and other deviations from simple structural models. Nevertheless, the distinction between IL and non-IL water seems appropriate for the present discussion. The bimodal porewater model is discussed in detail in [4-6].

EXPERIMENT

The clay materials used in this study were: i) Wyoming MX-80 Na/Ca bentonite (in the following referred to as MX-80) in the state 'as delivered' and ii) Na montmorillonite prepared by purifying Wyoming MX-80 and thereafter saturating it with Na ions. This material is hereafter referred to as PNa. Details on the preparation of the Na montmorillonite are found in [7]. The main component in MX-80 is montmorillonite, a swelling clay that comprises more than 80 weight per cent of the bentonite, while other components are mainly inorganic non-swelling minerals. Recent studies state montmorillonite contents of 81-83 % [8] and 88 % [9] by weight. The exchangeable cations in MX-80 bentonite are Na (62%), Ca (27%), Mg (9%), and K (2%), where the percentages show the relative saturation of exchangeable sites for the respective ion [9].

The samples were prepared by uniaxial compaction of bentonite and montmorillonite powder to dry densities in the range 0.7 to 1.6 g/cm^3 and subsequent saturation with either Milli-Q water or a 0.1 M NaCl solution in an equilibrium cell. Details concerning the sample preparation technique are found in [10]. Subsamples of the saturated material were finally transferred to NMR glass tubes, which were sealed with PTFE stoppers to prevent drying.

^{1}H NMR measurements of water-saturated MX-80 and Na montmorillonite were carried out with a high-field Chemagnetics CMX Infinity 270 MHz spectrometer. The sample temperature was kept at 23˚C by gas flushing. Spin-locking CPMG technique [11] was used to measure the $T_{1\rho}$ relaxation times. The pulse-spacing in the CPMG sequence was the same (22μs) in all measurements, which is necessary in order to allow meaningful comparison between results from different samples.

The relative amounts of water phases were determined by peak-o-mat, a data analysis and curve fitting program for fitting discrete exponential decays [12]. The analysis was very similar to those applied by others on similar materials [e.g. 6, 13]. Briefly, the discrete analysis aims at determining the number of water phases and their relative amounts from and exponential equation in the form:

$$M_t = \sum_{i=1}^{n} M_i \exp\left[\frac{-t}{T_{1\rho,i}}\right]$$

(1)

where t is time, M_t is the spin locked macroscopic magnetization vector at time t, $M_{0,i}$ is the magnetization vector of water phase i at time zero, n is the number of water phases, and $T_{1,\rho}$ is the spin lattice relaxation time in the rotating-frame associated with decay i. Further details about the NMR measurements are found in [14] and references therein.

RESULTS AND DISCUSSION

The peak-o-mat analysis yielded for most of the samples one minor and two major $T_{1\rho}$ values, see table 1. Since water in smaller pores interacts more with the paramagnetic Fe^{3+} in the montmorillonite structure than that in larger pores, the decay with the shorter relaxation time was associated with the IL water and the decay with the longer relaxation time with the non-IL water. In addition to the above phases, traces of a third phase were indicated, see table I, but their presence was considered insignificant for the discussion here. Data interpretation and discussion

Table I. Data for water-saturated samples and associated NMR results.

SAMPLE DATA				NMR RESULTS				
Solid	Ext. soln.	Time[1] d	ρ_{dry} g/cm³	$M_{0,1}{}^2$, $T_{1\rho}$	$M_{0,2}{}^2$, $T_{1\rho}$	$M_{0,3}{}^2$, $T_{1\rho}$	Short $T_{1\rho}$ %	Long $T_{1\rho}$ %
MX-80	Milli-Q	7	0.67	**38.8**, *538*	**60.1**, *1391*	**1.1**, *17642*	**39**	**61**
MX-80	Milli-Q	14	0.97	**50.2**, *364*	**48.9**, *818*	**0.9**, *20935*	**50**	**49**
MX-80	Milli-Q	26	1.27	**61.6**, *287*	**37.3**, *586*	**1.1**, *19919*	**62**	**38**
MX-80	Milli-Q	33	1.55	**70.1**, *210*	**29.1**, *441*	**0.8**, *22876*	**70**	**30**
MX-80	0.1M NaCl	7	0.71	**35.8**, *467*	**63.4**, *1213*	**0.8**, *16210*	**36**	**64**
MX-80	0.1 M NaCl	15	0.96	**51.2**, *341*	**47.6**, *744*	**0.9**, *20106*	**52**	**48**
MX-80	0.1 M NaCl	24	1.27	**63.1**, *286*	**35.7**, *590*	**1.2**, *21667*	**63**	**36**
MX-80	0.1 M NaCl	33	1.57	**66.7**, *189*	**32.5**, *390*	**0.8**, *24497*	**67**	**33**
PNa	Milli-Q	6	0.66	**25.8**, *793*	**74.2**, *1386*	**0**, -	**26**	**74**
PNa	Milli-Q	13	0.93	**44.0**, *528*	**56.0**, *974*	**0**, -	**44**	**56**
PNa	Milli-Q	6	1.20	**42.6**, *278*	**57.4**, *571*	**0**, -	**43**	**57**
PNa	Milli-Q	21	1.26	**41.2**, *287*	**58.7**, *527*	**0.1**, *69432*	**41**	**59**
PNa	Milli-Q	31	1.59	**46.2**, *208*	**53.7**, *391*	**0.1**, *22655*	**46**	**54**
PNa	0.1 M NaCl	6	0.69	**34.8**, *700*	**65.0**, *1251*	**0.1**, *3353*	**35**	**65**
PNa	0.1 M NaCl	14	1.02	**45.1**, *355*	**54.8**, *672*	**0.1**, *20592*	**45**	**55**
PNa	0.1 M NaCl	21	1.28	**37.2**, *241*	**62.7**, *478*	**0.1**, *24338*	**37**	**63**
PNa	0.1 M NaCl	32	1.57	**42.2**, *190*	**57.7**, *365*	**0.1**, *15690*	**42**	**58**

[1] Time for saturation in an equilibrium cell. [2] $M_{0,1}$, $M_{0,2}$ and $M_{0,3}$: components in the magnetization decay in per cent of total magnetisation, M_0, followed by the corresponding $T_{1\rho}$ in μs.

is made in terms of the well-known microstructural model by Suzuki et al. [15]. This idealized model describes the structure of water-saturated swelling clay in terms of a unit cell containing water and a stack of montmorillonite unit layers. The unit layers are separated by interlayer water with a given thickness. The unit cell contains in addition non-interlayer water associated with external basal surfaces and, optionally, with edge surfaces. The corresponding macroscopic sample is achieved by close-packing copies of the unit cell in three dimensions. Suzuki et al. [15] expressed the thickness, L, of the external porewater on the basal surface in the unit cell as:

$$L = n \cdot \delta \left(\frac{1}{(1+\lambda)^2} \frac{\rho_{grain}}{\rho_{dry}} - 1 \right) - (d_{001} - \delta)(n-1) \qquad (2)$$

where n is the stacking number (i.e. number of unit layers in the stack), δ is the thickness of the montmorillonite unit layer, ρ_{dry} is the dry density, ρ_{grain} is the grain density of the montmorillonite, d_{001} is the basal spacing, and λ is a 'structural parameter'. The unit layer thickness is about 0.92 - 1.01 nm [16], while the lateral unit layer dimension is about 50-200 nm [17]. The amount of water associated with edge surfaces is thus negligible in comparison to the amount of water associated with basal surfaces, which justifies our approximation $\lambda=0$.

Equation (2) can be used in a slightly modified form to express the ratio of the relative volumes of the interlayer and non-interlayer water, V_{non-IL} and V_{IL}, respectively, in the macroscopic sample as a function of the inverse dry density. Thus, for $\lambda=0$:

$$\frac{V_{non-IL}}{V_{IL}} = A \frac{1}{\rho_{dry}} + B \qquad (3)$$

where $A = n\delta\rho_{grain}/(d_{IL}(n-1))$ and $B = -1 - n\delta/(d_{IL}(n-1))$, d_{IL} is the interlayer spacing defined as $d_{IL} = d_{001} - \delta$ and other symbols have the same meaning as before. The value of V_{non-IL}/V_{IL} is thus coupled to three structural variables (n, d_{IL} and ρ_{dry}) and two constant parameters (ρ_{grain} and δ, both of which are well-known from the literature), which means that equation (3) is a potential tool for getting microstructural information about the clay sample. The appearance of a plot of V_{non-IL}/V_{IL} vs. $1/\rho_{dry}$ depends, for example, on the values of n and d_{IL} and on whether these values are constant or not. In case n and d_{IL} are constant, the graph becomes linear and gets a value of -1 for ρ_{dry} equal to ρ_{grain}. (This is mathematically correct, but, needless to say, V_{non-IL}/V_{IL} does not take negative values). In case n and d_{IL} vary, the graph might still be linear but with a slope that is not given by equation (3), see figure 1, but it may also depart from linearity (see below). Another example of information that is inherent in the model by Suzuki et al. is the stacking number n, which can be calculated by simple re-arrangement of equation (3) and by using the relation $\delta = 2/(\rho_{grain} \cdot S_{sp})$, where δ is the unit layer thickness (cm), S_{sp} is the specific surface area (cm^2/g) and ρ_{grain} (g/cm^3) is the density of the solid. The stacking number is now given by:

$$n = \left[1 + \frac{2 \cdot (1 - \rho_{grain}/\rho_{dry})}{\rho_{grain} S_{sp} d_{IL} (1 + V_{non-IL}/V_{IL})} \right]^{-1} \qquad (4)$$

Equation (4) thus expresses the stacking number in terms of parameters, all of which are measurable with various experimental methods. Equation (4) was used for estimating stacking numbers for MX-80 and PNa (see below) by inserting the corresponding montmorillonite surface areas, $610 \cdot 10^4$ and $750 \cdot 10^4$ cm^2/g, respectively.

The NMR measurements indicated the presence of two relaxation components, presumably due to water in domains with different sizes, which agrees with previous results [14]. In case when the water-saturated compacted PNa or MX-80 samples had dry densities in the range of about 1.0–1.6 g/cm^3, i.e. inverse dry densities of 0.625–1.0 cm^3/g, the estimated volumes of water in smaller and larger domains were considered to represent the respective volumes of interlayer and non-interlayer water volumes in the above structure model. However, at lower dry densities (i.e. higher water contents), the separation of the montmorillonite unit layers increased to the extent that pore coupling between IL and non-IL water becomes a problem.

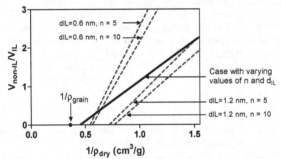

Figure 1. V_{non-IL}/V_{IL} vs. $1/\rho_{dry}$ calculated from equation (2) for different constant values of d_{IL} and n (dotted lines). The solid line describes a situation where d_{IL} and n vary with the inverse dry density.

Briefly, the effect of coupling in this case is that part of the IL water is seen by the NMR instrument as non-IL water, leading to overestimation of the V_{non-IL}/V_{IL} ratio. Figure 2 shows plots of the ratio V_{non-IL}/V_{IL} based on NMR measurements as a function of the inverse dry density. The lower limit for the inverse dry density corresponds to the case were the sample contains no water at all and $1/\rho_{dry} = 1/\rho_{grain}$ (≈ 0.36 cm^3/g). As was indicated in figure 1, increasing the water content first leads to a situation where all porewater is in the form of IL water, until at some point the water content is high enough to allow of the presence of non-IL water, and the increase of the V_{non-IL}/V_{IL} ratio to positive values. The diagram in figure 2 shows V_{non-IL}/V_{IL} ratios vs. $1/\rho_{dry}$ for MX-80 and PNa. The graphs for the MX-80 samples look very similar irrespective of whether the saturating solution was Milli-Q water or a 0.1 M NaCl solution. One reason for this may be that the equilibrium solutions actually did not produce porewaters with physicochemical differences large enough to be detectable by NMR. However, it is also possible that the equilibration times used were too short (see table I), so that the samples were only saturated with water but not with NaCl.

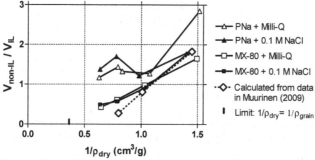

Figure 2. The ratio V_{non-IL}/V_{IL} vs. $1/\rho_{dry}$ for water-saturated samples of PNa and MX-80. Full lines combine dots based on NMR measurements. The dotted line was calculated by using data from SAXS measurements on water-saturated MX-80 [10].

The graphs for MX-80 based on NMR measurements agree roughly with the one showing calculated V_{non-IL}/V_{IL} ratios based on data from previous SAXS measurements on MX-80/water [10], which supports the idea that NMR and SAXS provide similar information about the distribution of water between different phases in the type of materials considered here. Satisfactorily agreement between NMR and SAXS measurements on water-saturated MX-80 has also been reported previously [14]. The PNa samples exhibits for each inverse dry density a higher V_{non-IL}/V_{IL} ratio than the corresponding MX-80 sample. This might be explained in terms of the stacking number. The diagram in figure 3 shows V_{non-IL}/V_{IL} ratios according to equation (4)

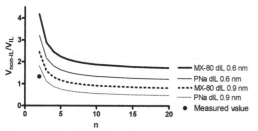

Figure 3. Calculated V_{non-IL}/V_{IL} ratios vs. the stacking number n for PNa and MX-80 samples with ρ_{dry} = 1.2 g/cm³. The point represents PNa with $d_{IL} \approx 1.1$ nm determined by SAXS.

vs. n in water-saturated MX-80 and PNa with a dry density of 1.2 g/cm³ and interlayer distances of 0.6 and 0.9 nm. Other values used on the calculations: ρ_{grain} = 2.78 g/cm³ [9], S_{sp} = 749 m²/g for montmorillonite [17], S_{sp} = 610 m²/g for MX-80, δ = 0.96 nm [18]. The effect of surface area is seen by comparing MX-80 and PNa for the same d_{IL} and n; Due to the larger surface area for PNa, it exhibits a significantly lower V_{non-IL}/V_{IL} ratio than the corresponding MX-80 sample. It is also obvious that the only case where PNa has a higher V_{non-IL}/V_{IL} ratio than the corresponding MX-80 sample is the one where n is low for PNa and high for MX-80, see figure 3.

Measurements of porewater distributions and knowledge about the physico-chemical factors that govern such distributions, contribute to the understanding of how bentonite functions as, for example, a barrier in a KBS-3 repository. This understanding may later result in the development of geochemical transport models that offer the possibility of having porewater distributed between different phases and thereby provide a more realistic picture of the porewater system than is presently available. The NMR study is ongoing, and further data are hoped to shed more light on the microstructure of bentonite and montmorillonite.

CONCLUSIONS

The interpretation of the NMR data was made in terms of the microstructural model by Suzuki et al. [15]. It was concluded for corresponding samples of MX-80 bentonite and montmorillonite (PNa) that i) the presence of 0.1 M NaCl increased significantly the V_{non-IL}/V_{IL} ratio in PNa but not in MX-80 bentonite, ii) purified montmorillonite (PNa) exhibited higher V_{non-IL}/V_{IL} ratios than MX-80 bentonite, and iii) the stacking number in the montmorillonite sample is smaller than in the MX-80 bentonite. Further experiments are needed however, to determine the extent of which the results can be generalised to other clay-porewater systems.

ACKNOWLEDGMENTS
The work was funded by Posiva Oy.

REFERENCES

1. Posiva. TKS-2009, Eurajoki, Finland (2010).
2. SKB. SKB TR-10-12, Stockholm, Sweden (2010).
3. A. Meunier, Clays Clay Miner., 41, 551-566 (2006).
4. L. Van Loon, M. Glaus, and W. Muller, Appl. Geochem., 22(11), 2536-2552 (2007).
5. R. Pusch, O. Karnland, and H. Hökmark, SKB TR-90-43 (1990).
6. T. Ohkubo, H. Kikuchi, M., and Yamaguchi, Phys. Chem. Earth 33, S169-S176 (2008).
7. A. Muurinen and T. Carlsson, Posiva Report. (To be published.)
8. O. Karnland, S. Olsson, and U. Nilsson, SKB TR-06-30 (2006).
9. L. Kiviranta and S. Kumpulainen, Posiva Working Report 2011-84 (2011).
10. A. Muurinen, Posiva Working Report 2009-42 (2009).
11. G.E. Santyr, R.M. Henkelman, and M.J. Bronskill, J. Magn. Reson., 79, 28-44 (1988).
12. C. Kristukat, Peak-o-mat. https://sourceforge.net/project/showfiles.php?group_id=67624. (2008).
13. G. Montavon, Z. Guo, C. Tournassat, B. Grambow, and D. Le Botlan, Geochim. Cosmochim. Acta, 73, 7290-7302. (2009).
14. T. Carlsson, A. Muurinen, M. Matusewicz, and A. Root. Mater. Res. Soc. Symp. Proc. 1475, Materials Research Society, 397-402 (2012).
15. S. Suzuki, H. Sato, T. Ishidera, and N. Fujii, J. Contam. Hydrol, 68(1-2), 23-37 (2004).
16. M. Holmboe, S. Wold, and M. Jonsson, J. Contam. Hydrol., 128, 19-32 (2012).
17. Tournassat, C. and C.A.J. Appelo, Geochim. Cosmochim. Acta 75, 3698-3710 (2011).
18. A. Muurinen, Posiva Working Report 2010-11 (2010).

Mater. Res. Soc. Symp. Proc. Vol. 1518 © 2013 Materials Research Society
DOI: 10.1557/opl.2013.66

Long-term Corrosion of Zircaloy-4 and Zircaloy-2 by Continuous Hydrogen Measurement under Repository Condition

Tomofumi Sakuragi, Hideaki Miyakawa, Tsutomu Nishimura[1] and Tsuyoshi Tateishi[2]
Repository Engineering and EBS Technology Research Project, Radioactive Waste Management
Funding and Research Center, 1-15-7 Tsukishima, Chuo City, Tokyo 104-0052, Japan
[1]Kobe Steel, Ltd., 4-7-2 Iwaya-Nakamachi, Nada-ku, Kobe 657-0845, Japan
[2]Kobelco Research Institute, Inc., 1-5-5 Takatsukadai, Nishi-ku, Kobe 657-2271, Japan

ABSTRACT

Corrosion behavior is a key issue for the waste disposal of irradiated metals, such as hulls
and endpieces, and is considered to be a leaching source of radionuclides including C-14.
However, little information about Zircaloy corrosion in anticorrosive conditions has been
provided.

In the present study, long-term corrosion tests of Zircaloy-4 and Zircaloy-2 were
performed in assumed disposal conditions (dilute NaOH solution, pH 12.5, 303 K) by using the
gas flow system for 1500 days. The corrosion rate, which was determined by measuring gaseous
hydrogen and the hydrogen absorbed in Zircaloy, decreased with immersion time and was lower
than the value of 2×10^{-2} μm/y used in performance assessment (1500-day values: 5.84×10^{-3} and
5.66×10^{-3} μm/y for Zircaloy-4, 1000-day values: 8.81×10^{-3} μm/y for Zircaloy-2). The difference
in corrosion behavior between Zircaloy 4 and Zircaloy-2 was negligible. The average values of
the hydrogen absorption ratios for Zircaloy-4 and Zircaloy-2 during corrosion were 91% and
94%, respectively.

The hydrogen generation kinetics of both gas evolution and absorption into metal can be
shown by a parabolic curve. This result indicates that the diffusion process controls the Zircaloy
corrosion in the early corrosion stage of the present study, and that the thickness of the oxide
film in this stage is limited to approximately 25 nm and may therefore be in the form of dense
tetragonal zirconia.

INTRODUCTION

Spent fuel claddings after reprocessing are expected to be disposed in a deep
underground repository. Corrosion of the irradiated metals raises a concern in the safety
assessment of gas generation and the source term of activated radionuclides (e.g. C-14) [1].
However, the corrosion rate of Zircaloy under the anticorrosive repository conditions (low
oxygen, high alkaline, and low temperature) is extremely slow. Therefore, the available data on
Zircaloy corrosion are very limited.

On the basis of numerous out-pile studies, it is widely accepted that the corrosion kinetics
of Zircaloy follows a cubic rate law before transition ("breakaway") in the temperature range of
approximately 561 K to 673 K [2]. Although some disagreement remains, this empirical kinetic
behavior is considered not only a simple diffusion process through the zirconia as a corrosion
product, but also one of cracks and stress effects of the oxide on mass transport. One study has
extrapolated the cubic law data to the corrosion rate at low temperatures [3], but the sufficiency
of the extrapolation to corrosion behavior in a disposal environment has not been investigated.

In our previous work, a sensitivity analysis of hydrogen measurements for both gaseous hydrogen and hydrogen absorbed in metal was demonstrated to allow quantitative evaluation for up to 90 days of the corrosion rate of Zircaloy-4 under repository conditions [4]. The purpose of the present study is to show the long-term corrosion behavior of Zircaloy-4 and Zircaloy-2. Gaseous hydrogen evolved by the corrosion reaction was measured continuously by using the gas flow system. The hydrogen generation kinetics of both gaseous hydrogen and hydrogen absorption into metal are discussed.

EXPERIMENTAL

Zircaloy-4 and Zircaloy-2 obtained from CEZUS Co., Ltd. were pretreated to produce a sample of appropriate thickness by undergoing two sets of cold-rolling, followed by vacuum annealing after each set to remove the hydrogen absorbed during cold-rolling. The accumulated annealing parameter before purchase is not known. Final polishing was done with 0.02 mm alumina powder. The initial hydrogen content was lower than 10 ppm, as shown together with other Zircaloy element in Table 1.

Zircaloy foils (100 sheets: 100 mm×100 mm×0.1 mm, surface area of 2.0 m^2) were immersed in a dilute NaOH solution (pH 12.5) kept at 303 K. Argon carrier gas with oxygen concentration below 1×10^{-3} ppm was passed through the glass flask at a flow rate of 0.9 dm^3/min, and the hydrogen concentration in the carrier gas was measured periodically by using atmospheric pressure ionization mass spectrometry (API-MS). After immersion, Zircaloy coupons (20 mm×30 mm×0.05 mm) were collected in a separate flask, and the absorbed hydrogen content was determined by an inert gas melting system together with gas chromatography (LECO RH-404).

The sample pretreatment, experimental setup, and other details for the gas flow experiment are described in a separate report [4].

Table I. Composition of Zircaloys-2 and Zircaly-4 used for corrosion tests (wt%).

	Sn	Fe	Cr	Ni	O	H
Zircaloy-2	1.23	0.13	0.1	0.04	0.13	0.0006
	(1.20 - 1.70)	(0.07 - 0.20)	(0.05 - 0.15)	(0.03 - 0.08)	(-)	(< 0.0025)
Zircaloy-4	1.24	0.18	0.1	< 0.006	0.15	0.0009
	(1.20 - 1.70)	(0.18 - 0.24)	(0.07 - 0.13)	(< 0.007)	(-)	(< 0.0025)

Note: Measured after pretreatment. Values in parenthesis represent the specification values (JIS H 4751).

RESULTS AND DISCUSSION

Figure 1 shows the change in gaseous hydrogen concentration (C_{ti}) in the argon carrier gas. The C_{ti} value decreased over time from 100 to 1 ppb. This amount is adequate for the blank run, which shows a nearly constant value of 0.61 ± 0.11 ppb.

The following corrosion reaction was assumed for hydrogen absorption ratio x:

$$(2x + 1)Zr + 2H_2O \rightarrow ZrO_2 + 2x \cdot ZrH_2 + 2(1 - x)H_2 \qquad (1)$$

The cumulative atomic molar amount of gaseous hydrogen per unit surface area, A_{gas} (mol/m^2), can be obtained from

$$A_{gas} = \frac{2}{S_g} \sum \frac{C_{t_i} + C_{t_{i-1}}}{2} \cdot \frac{v}{V^\circ} \cdot (t_i - t_{i-1}), \quad (2)$$

where C_{ti} is the concentration of hydrogen gas (H_2) at time t_i, t_{i-1} is one time increment before t_i, v is the gas flow rate (0.9 dm³/min.), V° is the molar volume of a perfect gas (22.4 dm³/mol), and S_g is the surface area (2.0 m²). The molar amount of hydrogen absorbed in Zircaloy metal per unit surface area A_{abs} (mol/m²) is

$$A_{abs} = \frac{H_{abs} \cdot W \cdot 10^{-6}}{M_H \cdot S_a}, \quad (3)$$

where H_{abs} is absorbed hydrogen in ppm, W is the sample weight, M_H is the molecular weight of hydrogen (1 g/mol), and S_a is the surface area (1.2×10^{-3} m²). The results for A_{gas} and A_{abs} as a function of time and the least square fit in accordance with the following allometric equation are shown in Figure 2.

$$A_{gas,abs} = a \times t^b \quad (4)$$

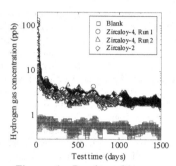

Figure 1. Results of gaseous hydrogen concentration evolved in the gas flow experiment.

The corrosion rate, R_c (μm/y), which represents the amount of zirconium consumed by oxidation and hydrogenation under stoichiometry in equation (1), can be obtained from

$$R_c = \frac{(2x+1) \cdot A_{gas} \cdot M_{Zr} \cdot 10^6}{4(1-x) \cdot \rho \cdot t}. \quad (5)$$

where M_{Zr} is the molecular weight of zirconium (91.22 g/mol), ρ is the zirconium density (6.44×10^6 g/m³), t is test time, and x represents the hydrogen absorption ratio $x = A_{abs}/(A_{gas}+A_{abs})$. Here, hydrogen absorption can be assumed constant during experiment with values of 91% for Zircaloy-4 and 94% for Zircaloy-2 as shown in Table II.

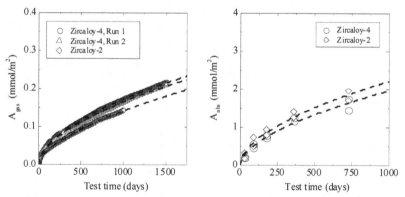

Figure 2. Gaseous hydrogen (left) and absorbed hydrogen (right) generation by the gas flow system. Dashed lines represent the least squares fit under the allometric equation of $A_{gas,\ abs} = a \times t^b$.

Figure 3 shows the corrosion rate of Zircaloy-4 and Zircaloy-2 with the corrosion rate of 2×10^{-2} μm/y estimated in the TRU-2 report [1]. The corrosion rate of Zircaloy-2 is slight larger than that of Zircaloy-4. The corrosion rates decrease with time, and are smaller than the assumed corrosion rate in the TRU-2 report [1]. As reported previously, the ratio of hydrogen absorbed into metal is over 90% [4] and remains constant in the long term.

It is generally accepted that the corrosion kinetics of zirconium alloys before transition can be represented by the cubic rate law, based on numerous out-pile studies at high temperatures of 561 K to 673 K [2]. This suggests that corrosion is not a simple diffusion process. As shown in Figure 2, however, the low temperature corrosion kinetics of both gaseous and absorbed hydrogen generation in the present study

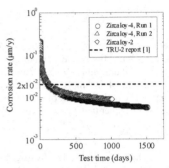

Figure 3. Corrosion rates of Zircaloy-4 and Zircaloy-2 obtained from gas flow test together with the value used in the TRU-2 report [1].

follows a parabolic curve, where the approximate values of the exponent b summarized in Table III are in the range of 0.55 to 0.60. This indicates that the diffusion process controls the low temperature corrosion. It is difficult to discuss the corrosion behavior because the gap in corrosion rate between high temperature and low temperature is extremely large; the cubic rate constant is 1.77×10^{-2} $(g/m^2)^3$/day at 561 K and extrapolated value at 303 K is 1.81×10^{-11} $(g/m^2)^3$/day [2]. However, above discrepancy may be explained by oxide characteristics.

Table II. Corrosion behavior of Zircaloy-4 and Zircaloy-2 obtained from the gas flow system.

Material	Day	A_{gas} (mol/m²)		A_{abs} (mol/m²)		x (%)	R_c (μm/y)		ΔW (g/m²)	
		Run 1	Run 2	Coupon 1	Coupon 2		Run 1	Run 2	Run 1	Run 2
Zircaloy -4	1	6.18×10^{-6}	3.69×10^{-6}	-	-	-	2.15×10^{-1}	1.59×10^{-1}	5.51×10^{-5}	3.29×10^{-5}
	10	1.89×10^{-5}	1.94×10^{-5}	-	-	-	7.51×10^{-2}	7.84×10^{-2}	1.68×10^{-4}	1.72×10^{-4}
	30	2.95×10^{-5}	3.08×10^{-5}	1.85×10^{-4}	1.92×10^{-4}	86	4.01×10^{-2}	4.11×10^{-2}	2.63×10^{-4}	2.74×10^{-4}
	90	-	4.96×10^{-5}	5.53×10^{-4}	4.64×10^{-4}	91	-	2.21×10^{-2}	-	4.42×10^{-4}
	180	7.11×10^{-5}	6.69×10^{-5}	7.86×10^{-4}	7.04×10^{-4}	92	1.60×10^{-2}	1.51×10^{-2}	6.33×10^{-4}	5.96×10^{-4}
	365	9.71×10^{-5}	9.01×10^{-5}	1.14×10^{-3}	1.26×10^{-3}	93	1.08×10^{-2}	1.03×10^{-2}	8.65×10^{-4}	8.03×10^{-4}
	730	1.45×10^{-4}	1.36×10^{-4}	1.44×10^{-3}	1.74×10^{-3}	92	8.04×10^{-3}	7.51×10^{-3}	1.29×10^{-3}	1.21×10^{-3}
	1000	1.71×10^{-4}	1.63×10^{-4}	(Average 91%)			6.94×10^{-3}	6.60×10^{-3}	1.53×10^{-3}	1.45×10^{-3}
	1500	2.16×10^{-4}	2.10×10^{-4}				5.84×10^{-3}	5.66×10^{-3}	1.93×10^{-3}	1.87×10^{-3}
Zircaloy -2	1	3.39×10^{-6}	-	-	-	-	2.13×10^{-1}	-	3.02×10^{-5}	-
	10	1.39×10^{-5}	-	-	-	-	8.36×10^{-2}	-	1.24×10^{-4}	-
	30	2.26×10^{-5}	-	1.65×10^{-4}	3.04×10^{-4}	92	4.60×10^{-2}	-	2.01×10^{-4}	-
	90	3.66×10^{-5}	-	4.94×10^{-4}	7.23×10^{-4}	94	2.51×10^{-2}	-	3.26×10^{-4}	-
	180	5.07×10^{-5}	-	9.34×10^{-4}	7.90×10^{-4}	94	1.75×10^{-2}	-	4.52×10^{-4}	-
	365	7.60×10^{-5}	-	1.26×10^{-3}	1.41×10^{-3}	95	1.29×10^{-2}	-	6.77×10^{-4}	-
	730	1.18×10^{-4}	-	1.68×10^{-3}	1.95×10^{-3}	94	1.00×10^{-2}	-	1.05×10^{-3}	-
	1000	1.42×10^{-4}	-	(Average 94%)			8.81×10^{-3}	-	1.27×10^{-3}	-

Table III. Results of the fitting parameters for the gaseous hydrogen evolution (A_{gas}) and hydrogen absorption (A_{abs}) using an allometric equation of $A_{gas, abs} (mol/m^2) = a \times t^b$.

		a	b
A_{gas}	Zircaloy-4, Run 1	$3.87 \times 10^{-6} \pm 0.02 \times 10^{-6}$	0.55 ± 0.0006
	Zircaloy-4, Run 2	$3.62 \times 10^{-6} \pm 0.03 \times 10^{-6}$	0.55 ± 0.0011
	Zircaloy-2	$2.26 \times 10^{-6} \pm 0.02 \times 10^{-6}$	0.60 ± 0.0011
A_{abs}	Zircaloy-4	$4.03 \times 10^{-5} \pm 1.1 \times 10^{-5}$	0.56 ± 0.050
	Zircaloy-2	$4.96 \times 10^{-5} \pm 1.5 \times 10^{-5}$	0.55 ± 0.049

Godlewski et al. reported on a Raman spectroscopic study of oxide film characteristics for oxidized Zircaloy-4 at 673 K in an oxide thickness region of approximately 0.5 to 10 µm [5]. As the oxide film become thinner, the 15% tetragonal ZrO_2 (t-ZrO_2) layer decreases, whereas a phase with constant thickness of 200 nm and containing 40% t-ZrO_2 is observed at the metal interface. Based on the above observation, the oxide in the present study, which ranges from 5 nm to 50 nm as observed previously by TEM [4], can be indicated a tetragonal rich phase. The t-ZrO_2 which is denser than monoclinic ZrO_2 could therefore play a diffusion role. The presence of parabolic kinetics in the early stage of out-pile corrosion has been conjectured [6] but has never observed because the high temperature corrosion occurs too fast to find early corrosion behavior, as stated above. The discussion above is based on the hypothesis that the oxide generation process and characteristics at low temperature is considering as high temperature one.

Assuming a constant hydrogen absorption ratio and no loss of corrosion products, the oxidation and hydrogenation weight gain, ΔW (g/m^2), can be obtained from

$$\Delta W = \frac{M_O + 2x \cdot M_H}{2} \cdot A_{gas} \quad (6)$$

where M_O is the molecular weight of oxygen (16 g/mol). The discussion above leads to the establishment of the parabolic rate constants k_c ($g^2/m^4/day$) as

$$\Delta W^2 = k_c \cdot t. \quad (7)$$

The weight gain changes are shown in Figure 4 and the approximate results for k_c are summarized in Table IV. Note that a part of ZrO_2 must be dissolved in the dilute NaOH (pH 12.5). The present ΔW, therefore, is not practical weigh gain but is theoretical one based on the corrosion reaction of eq. (1) and the hydrogen measurement.

Further discussion is needed for the Zircaloy corrosion kinetics and its long-term behavior under repository conditions. It should be noted that the

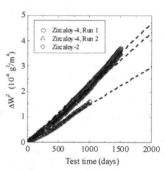

Figure 4. Weight gain changes for Zircaloy-4 and Zircaloy-2 as a function of time. Lines are least square fits.

Table IV. Results of the corrosion rate constant, k_c ($g^2/m^4/day$), assuming parabolic behavior.

	$K_c (g^2/m^4/day)$
Zircaloy-4, Run 1	$2.36 \times 10^{-9} \pm 2.1 \times 10^{-12}$
Zircaloy-4, Run 2	$2.17 \times 10^{-9} \pm 2.8 \times 10^{-12}$
Zircaloy-2	$1.49 \times 10^{-9} \pm 2.9 \times 10^{-12}$

general out-pile experiment was performed in pure water but that the present study used dilute NaOH solution. One issue is whether or not the parabolic corrosion at low temperature will approach out-pile cubic kinetics. Future research should be done at medium-range temperature from activation energy point of view and combined with studies of oxide characteristics such as the crystallographic approach [7], which are a challenge for extremely thin oxide film.

CONCLUSION

This paper presents the corrosion behavior of Zircaloy-2 and Zircaloy-4 in simulated deep repository water by continuous measurement of evolved gaseous hydrogen gas together with absorbed hydrogen in Zircaloy metal by use of the gas flow experimental system. Corrosion rates are smaller than previously reported. Both gaseous and absorbed hydrogen generation kinetics follow a parabolic curve, indicating that the diffusion process controls the Zircaloy corrosion. The oxide film which is approximately 25 nm thick may be in a form of dense tetragonal ZrO_2.

ACKNOWLEDGMENT

This research is a part of "Research and development of processing and disposal technique for TRU waste containing I-129 and C-14 (FY2011)" program funded by the Agency of Natural Resources and Energy, the Ministry of Economy Trade and Industry of Japan.

REFERENCES

1. Federation of Electric Power Companies (FEPC) and Japan Atomic Energy Agency (JAEA), Second Progress Report on Research and Development for TRU Waste Disposal in Japan (2007).
2. E. Hillner, Corrosion of Zirconium-Base Alloys–An Overview, Zirconium in the Nuclear Industry, 3rd Int. Symp., ASTM STP **633**, 211–235 (1977).
3. E. Hillner et al., Long-term Corrosion of Zircaloy Before and After Irradiation, J. Nucl. Mater. **278**, 334 (2000).
4. T. Sakuragi et al., Corrosion Rates of Zircaloy 4 by Hydrogen Measurement under High pH, Low Oxygen and Low Temperature Conditions, Mater. Res. Soc. Symp. Proc. **1475**, 311 (2012).
5. J. Godlewski et al., Raman Spectroscopy Study of the Tetragonal-to-Monoclinic Transition in Zirconium Oxide Scales and Determination of Overall Oxygen Diffusion by Nuclear Microanalysis of O18, Zirconium in the Nuclear Industry, 9th Int. Symp., ASTM STP **1132**, 416–436 (1991).
6. IAEA, Waterside Corrosion of Zirconium Alloys in Nuclear Power Plants, IAEA-TECDOC-996 (1998).
7. T. Sawabe et al., Microstructure of Oxide Layers Formed on Zirconium Alloy by Air Oxidation, Uniform Corrosion and Fresh-Green Surface Modification, J. Nucl. Mater. **419**, 310–319 (2011).

Mater. Res. Soc. Symp. Proc. Vol. 1518 © 2013 Materials Research Society
DOI: 10.1557/opl.2013.39

Radioelement Solubilities in SR-Site, the Influence of Variability and Uncertainty

Christina Greis Dahlberg[1], Patrik Sellin[1], Mireia Grivé[2], Lara Duro[2] and Kastriot Spahiu[1]
[1]Svensk Kärnbränslehantering AB, Swedish Nuclear Fuel and Waste Management Co, Box 250, SE-101 24 Stockholm, Sweden
[2]Amphos 21,Consulting S.L. Passeig de García i Faria, 49-51, E08019 Barcelona, Spain

ABSTRACT

If groundwater enters a damaged canister and comes in contact with the spent fuel, radionuclides are released into the water in the void inside the canister when fuel dissolves. Solubility limits restrict the amount of radioelements that may migrate with the water flowing from the canister. In this study the impact of variability in groundwater chemistry compositions and the impact of uncertainties in thermodynamic data on solubility limits for Np, Pb, Pu, Ra, Se, Th, U and Zr were looked into. The solubility limits for all the studied radioelements seemed to be more sensitive to uncertainties in thermodynamic data than to differences in groundwater chemistry. The sole exception was radium, where variability in water composition has a somewhat larger impact. Radium is also the most safety critical element in the safety assessment SR-Site and groundwater compositions are expected to vary during the assessment period of one million years.

INTRODUCTION

The safety assessment project SR-Site is undertaken by Swedish Nuclear Fuel and Waste Management Co (SKB) to assess the safety of a potential geologic repository for spent nuclear fuel [1]. The assessment supports SKB's license application for a final repository of KBS-3 type at the Forsmark site in Sweden.

If a disposed canister containing spent fuel is damaged, groundwater may enter the canister and come in contact with the spent fuel inside the canister. The result may be that radionuclides are released into the water in the void inside the canister. If the aqueous concentration of an element reaches saturation with respect to the solid phase, then its solubility limit is attained and the element will precipitate, provided that this is kinetically favored. As a result, only the aqueous fraction of the element may migrate with the water flowing from the canister while the fraction that has precipitated remains in the canister.

The usual approach taken in safety assessments, and also in SR-Site, is to work with scenarios and variants that are designed to capture the broad features of a number of representative possible future evolutions. Furthermore, each variant, represented by a specific calculation case, may be evaluated probabilistically in order to determine the mean exposure given the data uncertainties for the particular variant. This means that probability density functions are needed for all data that are used for radionuclide transport. In SR-Site the radionuclide transportation codes COMP23 and FARF31 were used [2] and the @risk software [3], which is an Excel add-in application that performs quantitative risk analysis using Monte Carlo simulations, was used to derive distributions of data.

To produce probability density functions for elemental solubilities, the Simple Functions tool was developed [4]. Simple Functions performs geochemical equilibrium calculations, but contains only a limited subset of data and reactions that is needed to calculate solubilities for the conditions that can be expected at the Forsmark site. In SR-Site, groundwater data was supplied to Simple Functions and the output produced was the solubility limit for the radiolelements. Simple Functions was used in combination with @risk to calculate distributions of solubility limits to be used in the radionuclide transport calculations.

The key factors that affect the elemental solubility limits were identified as:
- The assumed solubility limiting phase
- The geochemical conditions inside the damaged canister
- The thermodynamic database used

In SR-Site, solubility limiting phases were selected by "expert judgment", favoring phases that would be likely to precipitate without any kinetic restrictions. The geochemical conditions inside the damaged canister were assumed to be identical to the conditions in the groundwater with the exception that the redox conditions were controlled by the magnetite/goethite equilibrium. The thermodynamic database used was based on the Nagra/PSI Chemical Thermodynamic Data [5]. This database was updated with more recent data for certain elements as Zr, Th and U and the database was also completed with thermodynamic data for additional elements such as Pb [6-8].

The assessment in SR-Site covers just over 6 000 canister positions and the assessment period is one million years. This means that there will be a natural spatial and temporal variability in the composition of the groundwater that interacts with each canister. To calculate solubility limits for the safety assessment that cover the entire time period, a data set with mixtures of groundwater compositions representing all different expected climates were used [2]. This data set of mixtures of groundwater compositions consisted of 25% of groundwater compositions representing the temperate climate, 25% representing the permafrost climate, 25% representing the glacial climate and 25% representing the submerged climate. Data were randomly chosen from four large data sets representing each climate period. For the uncertainties in thermodynamic data a normal distribution was applied to the equilibrium constants ($\mu = log_{10} K^0$ and $\sigma = \frac{\Delta log_{10} K^0}{2}$).

THEORY

To evaluate the above described method to handle the spatial and temporal variability in the groundwater compositions for solubility calculations, the relative importance of variability in groundwater composition was compared to uncertainties in underlying thermodynamic data.

Impact of variability in groundwater compositions

The impact of variability in groundwater compositions on radionuclide solubilities was evaluated by calculating four sets of solubility limits representing the climate periods: temperate, permafrost, glacial and submerged. The combined tool of Simple Functions and @risk used Latin hypercube sampling to pick a set of 6,916 groundwater compositions from the large set representing a climate period. A normal distribution was applied to the equilibrium constants and the tool was run for 6,916 iterations to achieve a set of solubility limits. The procedure was repeated for each of the four groundwater compositions to achieve four sets of solubility limits representing the climate periods. Solubility limits for the four alternatives were calculated for the elements: Np, Pb, Pu, Ra, Se, Th, U and Zr.

Impact of uncertainty in thermodynamic data

The impact of uncertainty in the thermodynamic data on radionuclide solubility limits was evaluated by calculating two sets of solubility limits for a temperate groundwater composition.
The first set was calculated by choosing a representative chemical groundwater composition from the temperate climate period, see Table I. This groundwater composition was kept constant while a normal distribution was applied to the equilibrium constants and Simple functions in combination with @risk was run for 6,916 iterations to achieve a set of solubility limits.
The second set of solubility limits was calculated by using the combined tool of Simple Functions and @risk and Latin hypercube sampling to pick a set of 6,916 groundwater compositions from the large set representing the temperate climate period. Equilibrium constants were kept invariable (without any distribution) while the tool was run for 6,916 iterations to achieve a set of solubility limits. The solubility limits for the two alternatives were calculated for the elements: Np, Pb, Pu, Ra, Se, Th, U and Zr.

Table I. Representative chemical groundwater composition from the temperate climate period.

Input parameters	
pH	7.43
Eh (mV)	-437
IS (mol/kg)	$2.25 \cdot 10^{-1}$
$[HCO_3^{-}]$ (m)	$3.20 \cdot 10^{-4}$
$[SO_4^{2-}]$ (m)	$1.58 \cdot 10^{-3}$
$[Cl]_{tot}$ (m)	$1.66 \cdot 10^{-1}$
$[Ca]_{tot}$ (m)	$5.12 \cdot 10^{-2}$
$[Na]_{tot}$ (m)	$6.47 \cdot 10^{-2}$
$[Si]_{tot}$ (m)	$1.28 \cdot 10^{-4}$

DISCUSSION

Impact of variability in groundwater compositions

The relative importance of variability in groundwater compositions within the limits of the above mentioned climate scenarios on solubility limits was insignificant for almost every radioelement in this study (Np, Pb, Pu, Ra, Se, Th, U and Zr). The sole exception was radium, whose solubility limits are about three orders of magnitude lower for temperate conditions than for submerged conditions. In Figure 1 the solubility limits for radium are compared for groundwater compositions representing temperate climate conditions, permafrost climate conditions, glacial climate conditions and submerged climate conditions. Figures over the solubility limits for the other studied elements can be found in Appendix F of reference [2].

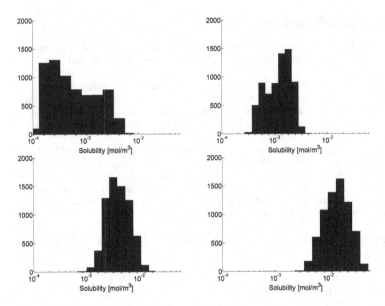

Figure 1. Solubility limits for Ra using groundwater data representing temperate conditions (upper left), permafrost conditions (upper right), glacial conditions (lower left) and submerged conditions (lower right). The y-axis shows the number of realisations in each bin.

Impact of uncertainty in thermodynamic data

The impact of uncertainty in thermodynamic data has an obvious effect on the solubility limits of the radioelement in this study (Np, Pb, Pu, Ra, Se, Th, U and Zr). Since radium has been discussed earlier in this paper, the solubility limits for radium with varying equilibrium constants compared to unvarying equilibrium constants are shown in Figure 2. Figure 2 is also a representative illustration of the behaviour of the other elements in this study, and complete figures of the solubility limits for all the studied elements can be found in Appendix F of reference [2]. Only lead seems to have a somewhat different pattern, as shown in Figure 3.

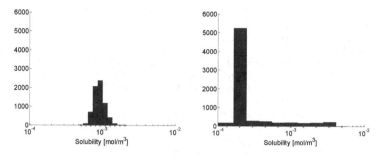

Figure 2. Solubility limits for Ra using groundwater data from temperate conditions. Fixed groundwater composition and varying thermodynamic data (left). Varying groundwater composition and fixed thermodynamic data (right). The y-axis shows the number of realisations in each bin.

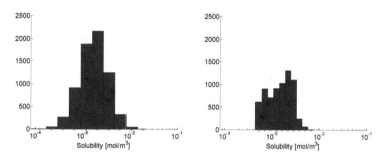

Figure 3. Solubility limits for Pb using groundwater data from temperate conditions. Fixed groundwater composition and varying thermodynamic data (left). Varying groundwater composition and fixed thermodynamic data (right). The y-axis shows the number of realizations in each bin.

CONCLUSIONS

The relative importance of variability in groundwater composition compared to uncertainty in thermodynamic data was evaluated by keeping one of them constant. The results showed that uncertainty in thermodynamic data has a bigger impact on the results for almost every radioelement. The sole exception to this is radium, which happens to be the most safety critical element, where variability in water composition has a somewhat larger impact.

REFERENCES

1. SKB. *Long-term safety for the final repository for spent nuclear fuel at Forsmark. Main report of the SR-Site project.* SKB TR-11-01, Svensk Kärnbränslehantering AB (2011).

2. SKB. *Radionuclide transport report for the safety assessment SR-Site.* SKB TR-10-50, Svensk Kärnbränslehantering AB (2010).
3. @RISK, *Risk analysis and simulation add-in for Microsoft Excel, Palisade Corporation,* USA. www.palisade.com
4. Grivé M, Domènech C, Montoya V, Garcia D and Duro L. *Simple Functions Spreadsheet tool presentation.* SKB TR-10-61, Svensk Kärnbränslehantering AB (2010).
5. Hummel W, Berner U, Curti E, Pearson F and Thoenen T. NAGRA/PSI Chemical Thermodynamic Data Base 01/01, Universal Publishers (2002).
6. Duro L, Grivé M, Cera E, Domenech C and Bruno J. *Update of a thermodynamic database for radionuclides to assist solubility limits calculation for performance assessment.* SKB TR-06-17, Svensk Kärnbränslehantering AB (2006).
7. Duro L, Grivé M, Cera E, Gaona X, Domènech C and Bruno J. *Determination and assessment of the concentration limits to be used in SR-Can.* SKB TR-06-32, Svensk Kärnbränslehantering AB, (2006).
8. Grivé M, Domènech C, Montoya V, Garcia D and Duro L. *Determination and assessment of the concentration limits to be used in SR-Can. Supplement to TR-06-32.* SKB R-10-50, Svensk Kärnbränslehantering AB (2010).

Mater. Res. Soc. Symp. Proc. Vol. 1518 © 2013 Materials Research Society
DOI: 10.1557/opl.2013.67

Glass-Iron-Clay interactions in a radioactive waste geological disposal: a multiscale approach

Diane Rébiscoul[1*], Emilien Burger[1], Florence Bruguier[1], Nicole Godon[1], Jean-Louis Chouchan[1], Jean-Pierre Mestre[1], Pierre Frugier[1], Jean-Eric Lartigue[2], Stephane Gin[1].

[1]CEA, DEN, (DTCD/SECM/LCLT) –Marcoule, F-30207 Bagnols-sur-Cèze Cedex, France
[2]CEA DEN, (DTN SMTM LMTE) – Cadarache, F- 13108 Saint-Paul Les Durance, France
*diane.rebiscoul@cea.fr

ABSTRACT

In France, nuclear glass canisters arising from spent fuel reprocessing are expected to be disposed in a deep geological repository using a multi-barrier concept (glass/canister/steel overpack and claystone). In this context, glass - iron or corrosion products interactions were investigated in a clayey environment to better understand the mechanisms and driving forces controlling the glass alteration. Integrated experiments involving glass - metallic iron or magnetite - clay stacks were run at laboratory scale in anoxic conditions for two years. The interfaces were characterized by a multiscale approach using SEM, TEM, EDX and STXM at the SLS Synchrotron. Characterization of glass alteration patterns on cross sections revealed various morphologies or microstructures and an increase of the glass alteration with the proximity between the glass and the source of iron (metallic iron or magnetite) due to the consumption of the silica coming from the glass alteration. In case of magnetite, the silica consumption is mainly driven by a sorption of silica onto the magnetite. For experiments containing metallic iron, the silica consumption seems to be strongly driven by silicates precipitation including Fe and Fe/Mg when the Fe is not enough available. Moreover, in addition to Fe-silicates observed at the surface of the gel layers, iron is incorporated within the gel probably as nanosized precipitates of Fe-silicates which could affect its physical and chemical properties. Those results highlighted the impact of the distance between glass and iron source and the nature of the iron source which drive the process consuming the silica coming from the glass alteration.

INTRODUCTION

In the French high level radioactive waste management strategy, it is expected to dispose around 40,000 nuclear glass canisters arising from spent fuel reprocessing in a deep geological repository using a multi-barrier concept: nuclear glass is poured into a stainless steel canister and the resulting system is placed in a low-alloy steel overpack, directly stored in a 100 m thick clayey host rock located 500 m below the surface [1]. The lifetime of this borosilicate glass must exceed several thousand years to limit the radiological impact of disposal. At this time scale, the prediction of glass durability implies a mechanistic approach, requiring an accurate description of the physicochemical mechanisms controlling glass dissolution [2]. Consequently, source term resulting from interactions between the nuclear glass, the groundwater and the near-field materials (iron, corrosion products, claystone) must be assessed.

In this study, glass – iron or corrosion products interactions were investigated in a clayey environment to better understand the mechanisms and driving forces controlling the glass alteration. Integrated experiments involving glass - metallic iron or magnetite - claystone stacks were run at laboratory scale in anoxic conditions for two years. The interfaces were characterized by a multiscale approach using scanning and transmission electron microscopy (SEM, TEM), energy-dispersive X-ray (EDX) and scanning transmission X-ray microscopy (STXM) at the SLS Synchrotron. We specifically focused on the influence of the glass – iron source distance on the morphology and chemistry of glass alteration layers, and the speciation of iron valence state of iron in the different zones of the glass / iron source interface.

EXPERIMENTAL DETAILS

Two experimental systems consisting in vertical stacks of three materials in contact were made (Figure1 (b)). From bottom to top: (i) a Callovo-Oxfordian (COx) claystone core (sampled at the Bure site (Meuse/Haute-Marne, France)), (ii) magnetite powder (Puratronic Alfa Aesar, purity of 99,997 %) or two drilled disks (5 mm thick, 39 mm diameter, Armco, 99.8 % purity) of iron filled with iron powder (1 μm to 6 μm size fraction, 98 % purity, Goodfellow) to minimize the dead volume and allow the water circulation, and (iii) SON68 powder, the inactive surrogate of French R7T7 glass (63 to 125 μm size fraction and a specific surface area of 0.06 ± 0.01 $m^2 \cdot g^{-1}$). During this experiment performed at 50°C, the synthetic solution at thermodynamic equilibrium with COx clay (-190 mV and pH= 6.73 at 25°C) was vertically flow through the materials stack using a pump at 15.10^5 Pa to maintain a constant flow rate (around 2 mL/month and a Peclet number below 1) and an another pump to maintain a confinement pressure of 26.10^5 Pa around the system to ensure the vertical flow of the solution (Figure1 (a)).

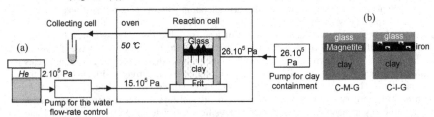

Figure1: Schematic views of (a) the experimental leaching setup (NFM : near-field materials) and (b) the materials stacks of the two experiments (C-M-G: clay - magnetite - glass; C-I-G: clay - iron - glass).

At the end of the experiments, i.e 782 days for the C-M-G and 756 days for the C-I-G experiments, the leaching cells were cooled to 25°C, frozen at −17°C, dismantled, and placed in a freeze-drier for 48 days (condenser at −80°C, P = 0.18 mbar) to eliminate the water. To limit oxidation, the materials stack was degassed for 48 hours in a high vacuum and then impregnated with a very fluid epoxy resin (EPOXY®) at 50 mbar. The sample was placed in an oven at 40°C for 24 hours to allow penetration of the resin and for 48 hours at 60°C to harden the resin. The block was cut and polished with SiC paper under ethanol. The final polishing step to grade 4000 was performed in nitrogen glove box to remove the first nanometers of oxidized iron and to avoid potential oxidation.

Polished cross-sections were imaged by SEM on a Zeiss Gemini Supra 55 FEG-SEM system operated at an acceleration voltage of 15 kV. EDX spectra were recorded with a SDD detector and quantitatively analyzed using commercial software (Bruker AXS Esprit™). TEM observations and EDX analyses of thin foils were carried out on a Technai G2 (FEI), equipped with a LaB$_6$ source operating at 200 kV and a scanning mode. The detectors were a Gatan CCD camera, a STEM BF-DF detector and an EDAX Genesis for the EDX analyses. The spot size for EDX was around 25 nm. Analyses by STXM coupled with Fe $L_{2,3}$-edge

($2p_{1/2,3/2} \rightarrow 3p$) X-ray absorption Near Edge Structure (XANES) were carried out on thin foils at the PolluX beamline of the Swiss Light Source (SLS, Villigen). The thin foils were entirely prepared into an anoxic glove box until their transfer into the microscope. XANES and TEM analyses were performed on the same thin foils. A great care was given to analyze the same points by comparing the morphological details on TEM microphotographs and X ray absorption images gathered above the Fe absorption edge by STXM. Samples were first mapped by taking images below the iron $L_{2,3}$-edge (~700 eV) in order to localize the areas of interest. Fe $L_{2,3}$-edge XANES spectra were then acquired to obtain qualitative data on Fe^{III} / Σ Fe [3]. A 1200 groove/mm grating was used. Entrance and exit slits were set at 0.5 and 0.2, providing a theoretical energy resolution of 140 meV, and a spatial resolution of about 30 nm. Because $L_{2,3}$ X-ray absorption spectra are sensitive not only to valence, but also to coordination and crystal field strength, a quantitative approach could not be performed here.

RESULTS

C-M-G system

Figure2 presents the SEM images and EDX maps of the altered glass located at the glass – magnetite interface and at a distance more than one millimeter from the magnetite. These results show any detectable alteration layer for the glass located far from the magnetite and an alteration layer around two to five micrometers when glass is in contact with the magnetite. The EDX maps highlight some alteration products containing Si, Al, Ca, O and Mg. The presence of Mg could be related to the formation of a magnesium silicates such as sepiolite $Mg_4Si_6O_{15}(OH)_2 6(H_2O)$ on top of gel as previously observed in COx solution at 90°C [4]. Moreover, silica is detected on magnetite, showing the affinity of the silica with iron. This result could be associated to a silica sorption process on magnetite [5] sustaining the glass alteration.

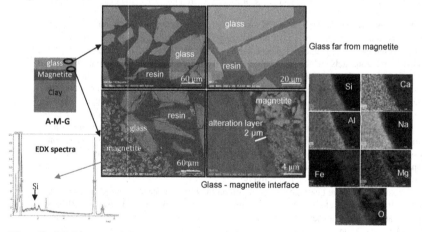

Figure2: SEM images of the altered glass at a distance higher than one millimeter from the magnetite and at the glass - magnetite interface with the associated EDX maps and EDX spectra highlighting the presence of silica on magnetite.

C-I-G system

The glass alteration patterns of the C-I-G experiment presented on Figure3, highlight also the impact of the proximity of the metallic iron on the microstructure and morphology of alteration products. First, while any phase is observed above the iron disk, there is high quantity of secondary phases such as calcite and Fe/Mg silicates between the glass grains when glass is located inside the iron disk notch close to the iron powder. Second, the presence of such phases is also correlated with thick alteration layer of around 2 to 5 μm containing Si, Al, Ca, Nd, O, and also Fe and Mg. For glass grains far from the iron source, alteration layer are thinner than 1 μm.

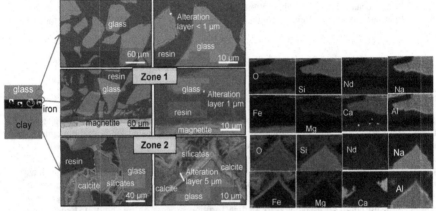

Figure3: SEM images with the associated EDX maps of the altered glass far from the iron, at the top of the iron disk (zone 1) and at the proximity of iron powder in the disk hole (zone 2).

At sub-microscopic scale, TEM-EDX analyses of a glass grain in the disk notch having one of its two sides in contact with iron powder show some stratified and polyphased alteration layers (Figure4 (a), (b) and (d)). These layers containing between 10 and 30 wt% of Fe, consist in porous gels, which are characterized by the presence of zirconium, an element provided by the glass and known for its low mobility during most of the glass alteration [6]. The thickness of these gels seems to depend on the iron local flow which is very geometry - and distance - dependent from the glass. As an example, a gel of 1.5 μm was observed on the grain side directly in contact with the iron powder whereas this gel is 50% thinner at the opposite side. Moreover, this phenomenon is associated with an important amount of Fe-silicates around the iron powder completely corroded (data no shown here), whereas iron is less corroded further. On both side of the grains, a 200 nm lamellar precipitate is seen at the top of the gel. This precipitate is different compare to the phase formed around iron powder with Fe:Si ratio of 1:1 vs 1:3-4, respectively.

In order to have qualitative information on the iron valence state in the different zones of the alteration products, STXM analyses (Figure4 (c) and (e)) were performed on FIB cross-section and compared to several references having various Fe(II)/Fe(III) ratio : maghemite (γ-$Fe^{III}_2O_3$), siderite ($Fe^{II}CO_3$), magnetite ($Fe^{II}Fe^{III}_2O_4$) and berthierine ($Fe^{II},Fe^{III},Al)_3(Si,Al)_2O_5(OH)_4$. The Fe L2,3-edge spectra taken on the pristine glass (Figure 5 (c) 1# and (e) 1# to 3#) show a single peak at 710 eV suggesting the predominance of Fe(III), which is consistent with the fabrication process [7]. Concerning the alterations products, for both side of the grain, two observations can be made. First, in the gel layer (Figure 5 (c) 2# to

4#and (e) 4# to 6#) the peak at 710 eV appears together with another peak at 708 eV, suggesting a mix of Fe(II) and Fe(III), indicating that most of the iron present in the gel comes from the solution rather than from the glass. Second, regardless of the total amount of iron in the gel, the relative intensity of each contribution is constant, indicating that the Fe(II)/Fe(III) ratios are quite close in the gel layer and in Fe-silicates (Figure 5 (c) 5# to 6#and (e) 7#). This observation could be consistent with the integration of iron within the gel as nanosized precipitates of Fe-silicates (unlike berthierine) having an iron local structure close to the one of Fe-silicates located at the top of the gel.

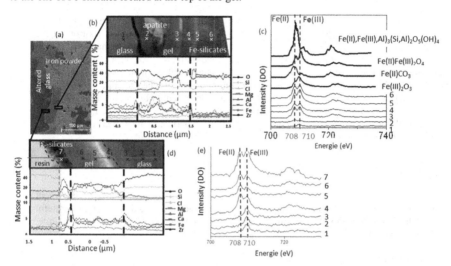

Figure4: Microstructure, morphology and composition of the alteration products of a glass grain of the C-I-G experiment having one of its two sides in contact with iron powder in the disk hole (a). (b)(d) microphotographs of the glass / resin - iron interface and corresponding EDX profiles; and (c)(a) $L_{2,3}$-edge XANES spectra extracted from different areas of the stack compared with the spectra of maghemite (γ-$Fe^{III}_2O_3$), siderite ($Fe^{II}CO_3$), magnetite ($Fe^{II}Fe^{III}_2O_4$) and berthierine ($Fe^{II},Fe^{III},Al)_3(Si,Al)_2O_5(OH)_4$.

DISCUSSION

Multi-scale characterizations of glass iron interactions allow us to make mechanistic interpretation in order to explain why and how iron increases glass alteration.

In case of magnetite, the silica consumption is mainly driven by a sorption of silica onto the magnetite. However, some simulations using GRAAL [9] show that silica sorption on magnetite is not the only mechanism driving the glass alteration, Fe-silicates precipitation or magnetite transformation into more oxidized phases like hematite or goethite that present higher silica sorption capacities could also occur and control the long-term rate.

For experiments with metallic iron, the silica consumption seems to be strongly driven by silicates precipitation including Fe and Fe/Mg. Moreover, in addition to Fe-silicates observed at the surface of the gel layers, iron is incorporated within the gel probably as nanosized precipitates of Fe-silicates. These observations suggest a diffusion of the dissolved iron through the gel porosity and a precipitation within the pores to form nanosized silicates. Similar nano-organized structures have been already observed in alteration products of Fe-rich basaltic glass, formed in aerated conditions [10] and in archeological glassy samples altered in clayey and anoxic medium [11]. Such local precipitation which could lead to a porosity clogging and thus limit the transport of species, could also play the same role as the external Fe-silicates, impacting the chemical affinity of glass dissolution. Such silica consumption, limited by the transport, does not allow the system to be saturated regarding the silica nor to form protective gel layer leading to higher glass alteration rate than without iron.

CONCLUSIONS

In this study, glass – iron or corrosion products interactions were investigated in a clayey environment to better understand the mechanisms and driving forces controlling the glass alteration. Integrated experiments involving glass - metallic iron or magnetite - clay stacks were run at laboratory scale in anoxic conditions for two years. Those results highlighted the impact of the distance glass – iron source and the nature of the iron source which drive the process consuming the silica coming from the glass alteration. Such silica consumption, limited by the transport, does not allow the system to be saturated regarding the silica nor to form protective gel layer leading to higher glass alteration rate than without iron. Future works will be focus on the modeling of these experiments including a first step of secondary phases determination.

ACKNOWLEDGMENTS

This work was partly funded by the French National Radioactive Waste Management Agency (ANDRA), whose financial support is gratefully acknowledged. We are also grateful for the local contact of Pollux Beamline at Swiss Light Source (Paul Scherrer Institute, Villigen, Swizerland).

REFERENCES

1. ANDRA, Synthesis – Evaluation of the feasibility of a geological repository in an argillaceous formation. Andra, Chatenay-Malabry, France (2005).
2. C. Poinssot, S. Gin, J. Nucl. Mater. 420, 182–192 (2012).
3. J.P. Crocombette, M. Pollak, F.Jollet, N. Thromat and M.Gautier-Soyer, Phys. Rev. B 52, 3143–3150 (1995).
4. P. Jollivet, P. Frugier, G. Parisot, J.P. Mestre, E. Brackx, S. Gin, S. Schumacher, J. Nucl. Mater. 420 (1-3), 508-518 (2012).
5. V. Philippini, A. Naveau, H. Catalette, S. Leclercq S., J. Nucl. Mater. 348, 60-69 (2006).
6. C. Cailleteau, F. Angeli, F. Devreux, S. Gin, J. Jestin, P. Jollivet, O. Spalla, Nat. Mater. 7, 978–983 (2008).
7. E. Pèlegrin, G.Calas, P. Ildefonse, P. Jollivet, L.Galoisy, J. Non-Cryst. Solids. 356, 2497–2508 (2010).
8. P. Frugier, S. Gin, Y. Minet, T. Chave, B. Bonin, N. Godon, J-E. Lartigue, P. Jollivet, A. Ayral, L. de Windt, G. Santarini, J. Nucl. Mat. 380 (1-3) 8-21 (2008)
9. N. Godon, S. Gin, D. Rébiscoul, P. Frugier, submitted to Procedia Earth and Planetary Science (2012).
10. Ph. Idefonse, R.J. Kirkpatrick, B. Montez, G. Calas, M. Flank, P. Lagarde, Clays and Clay Minerals 42 (3), 276–287 (1994).
11. A. Michelin, E. Burger, D. Rébiscoul, D. Neff, F. Bruguier, E. Drouet, P. Dillmann, S. Gin, submitted to Environmental Science and Technology (2012).

Mater. Res. Soc. Symp. Proc. Vol. 1518 © 2013 Materials Research Society
DOI: 10.1557/opl.2013.88

Suitability of Bioapatite as Backfill Material for Nuclear Waste Isolation

A.J. Finlay[1], A.E. Drewicz[1], D.O. Terry, Jr.[1], D.E. Grandstaff[1], and Richard D. Ash[2]

[1] Earth and Environmental Science, Temple University, Philadelphia, PA 19122 USA

[2] Department of Geology, University of Maryland, College Park, MD 20742 USA

ABSTRACT

Bioapatite, found in vertebrate bones and teeth, is highly reactive and may incorporate high concentrations of some radionuclides, including U, Pu, and Sr. Therefore, bioapatite may be useful in backfill or overpack materials in nuclear waste repositories. The dissolution rate for bioapatite is constant at pH > 4 and is about 5 times faster than fluorapatite. In terrestrial environments, bioapatite recrystallizes over periods of up to ca. 40 ka.

INTRODUCTION

Monazite (CePO₄), apatite, and other phosphate minerals can contain high concentrations of actinides, lanthanides, and other elements with anthropogenic short-lived isotopes (e.g., ^{90}Sr) found in nuclear waste. Because of their high capacity and durability, phosphate minerals or phosphate-silicate solid solutions have been proposed as waste-forms for nuclear waste disposal [1]. These characteristics have also led to proposals for phosphate minerals to be used as backfill or overpack materials in high-level nuclear waste repositories [2,3,4]. Bioapatite (dahllite), found in vertebrate bones and teeth, has also been investigated as a reactive barrier material to remove U, Pb, Cd, and other heavy metals from solution at contaminated sites [5,6,7]. We suggest that bioapatite may also be used in backfill or overpack materials in nuclear waste repositories.

Common apatite types include hydroxy- (HAP)[Ca₅(PO₄)₃(OH)], chlor- [Ca₅(PO₄)₃(Cl)], and fluorapatite (FAP)[Ca₅(PO₄)₃F], found in igneous and metamorphic rocks and carbonate fluorapatite (CFA), in sedimentary phosphorites. In CFA and bioapatite, CO_3^{2-} substitutes for PO_4^{3-}. The charge deficiency is usually compensated by omission of Ca, hydroxyl, or substitution of monovalent cations, such as Na, producing a defect structure [8]. Carbonate apatite crystals in bone are poorly crystalline, plate- or tablet-shaped and extremely small, with average dimensions of 50 x 25 x 2 to 4 nm and very large specific surface areas of ca. 240 m²/g [8,9].

The carbonate substitution, poor crystallinity, and defect structure make bioapatite more soluble and reactive than many other apatites [5,9]. In apatite-containing backfill, concentrations of nuclear waste species may be controlled either by solubility of their phosphate minerals or by sorption on apatite [7,10]. Therefore, the higher dissolved phosphate concentrations resulting from greater solubility of bioapatite would produce lower concentrations of radionuclides. In near-neutral pH solutions, measured sorption constants [3,4,6] between apatite and U, Pu, and rare earth elements (REE) range from ca. 5 x 10^5 to 1 x 10^7. Therefore, sorption could also significantly decrease dissolved waste concentrations. During diagenesis, adsorption and incorporation of fluoride and other trace elements and growth of larger crystals decreases apatite solubility and reactivity, and enable wastes to be sequestered for millions of years [6,11].

Bioapatite is more soluble and reactive than other apatite forms. These are desirable properties for repository performance as they potentially allow greater retention of radionuclides. However, the rate of bioapatite dissolution has not been measured under environmentally relevant conditions. After burial, bioapatite recrystallizes to form the more stable and less soluble CFA. The period of bioapatite recrystallization is important for repository performance as it potentially limits the time during which the greatest amount of sorption and radionuclide incorporation may occur. In this paper we investigate dissolution kinetics and duration of bioapatite sorption and uptake reactions to assess its suitability for use as backfill material in nuclear waste repositories.

METHODS

Bioapatite dissolution rate experiments were conducted at 38, 20, and 4°C with solution pH values between 2 and 8 in continuously stirred flow-through tank reactor (STR) systems. The specific surface area of the crushed bone was measured using a Monosorb single-point BET. Bioapatite dissolution rates were calculated based on Ca and P release rates. REE and trace element (TE) concentrations in fossil bones were determined using a Thermo-Finnigan Element 2 single collector high resolution magnetic sector ICP-MS, coupled with a New Wave UP 193 or 213 Laser Ablation System operating at ~ 2-3 J/cm^2 and calibrated with the NIST 610 glass standard. A 15 to 40 μm spot size was used for transects across bones. Concentrations were normalized to 55% CaO.

Millard and Hedges [4] suggested that U and other TE were incorporated into bone by diffusion from the bone surface. They developed a model of diffusion and adsorption, based on Fick's second law, to calculate periods of uranium uptake. We calculated bioapatite reaction/diffusion periods (t) from measured concentration profiles of uranium and selected lanthanides following the methods of Millard and Hedges [4] using:

$$t = \frac{t'\left(\dfrac{K_d}{\rho}+1\right)l^2}{\left(\dfrac{D_o VW}{\tau^2}\left(1-\dfrac{a}{A}\right)^2 F(a)\right)} \tag{1}$$

in which t' is a dimensionless parameter calculated from the concentration gradient, K_d the distribution coefficient between apatite and solution, ρ the specific bone porosity, l the bone radius, D_0 the diffusion coefficient in solution, V the volume porosity (0.35), W a soil saturation factor (0.50 ± 0.1), τ the tortuosity (0.35), a the hydrodynamic radius of the diffusing species, A the pore diameter (80Å), and F(a) the Faxen drag coefficient [4]. Values of sorption, diffusion, and other constants used in diffusion period calculations are given in Table I. Based on oxygen isotopes in bones from Toadstool Park, a temperature of 20°C was used in calculations [12].

Table I. Parameter values used in diffusion period calculations.

Element	log K_d	D_o x 10^6 cm^2 s^{-1}	D(bone) x 10^8 cm^2 s^{-1}	a Å
Uncertainty	$\pm$ 0.1	$\pm$ 0.06	$\pm$0.06	
La	5.85 (3)	6.17 (13)	9.27	6
Ce	5.8 (3)	6.16 (13)	9.26	6
Pr	5.9 (3)	6.15 (13)	8.72	7
Nd	6.0 (3)	6.14 (13)	8.71	7
Gd	5.7 (3)	5.9 (13)	7.88	8
Yb	5.28 (3)	5.82 (13)	7.78	8
U	5.7 (4)	6.8 (14)	10.1	8

SAMPLES

Modern bioapatite was obtained from the scapulae of a white-tailed deer (*Odocoileus virginianus*) found as dry and defleshed osseus remains in Eastern Pennsylvania. Bones were crushed to 75 to 106 and 106 to 246 μm and cleaned of residual organic matter with hydrogen peroxide. The bone composition was $Ca_{4.14}Mg_{0.07}Na_{0.18}(PO_4)_{2.76}(HPO_4)_{0.12}(CO_3)_{0.13}F_{0.01}$, with a = 9.368 $\pm$ 0.004Å and c = 6.884 $\pm$ 0.004Å, and an average BET surface area of 255 m^2 g^{-1}.

Fragments from five fossil Brontothere mammal bones were obtained from ca. 35 Ma smectitic overbank deposits from the terrestrial late Eocene Chadron Formation (White River Group) near Toadstool Geologic Park, northwest Nebraska.

RESULTS AND DISCUSSION

Fluids in all dissolution experiments were highly undersaturated based on bioapatite saturation constants [9]. During initial experiments, bioapatite dissolution was non-stoichiometric, with Ca released faster than P. However, as experiments continued, the release rates became essentially stoichiometric at molar Ca:P $\approx$ 1.44. Initial non-stoichiometric release may have resulted from the production of fine-grained particles and dislocations during crushing. Fine-grained "dust" particles are present in SEM photomicrographs of initial apatite surfaces; which are absent in micrographs of reacted bone. Our dissolution rates are based on Ca release. Bioapatite dissolution rates were pH-dependent at pH $\leq$ 4, but were pH-independent at higher values (figure 1). At 20°C the dissolution rate (R) may be described by:

$$R = k_1(H^+)^n + k_2 \qquad (3)$$

where n is the order of the reaction with respect to hydrogen ion activity and k_1 and k_2 are rate constants. In this equation: n = 1.03 $\pm$ 0.14, log k_1 = -5.68 + 0.10, and log k_2 = -9.38 $\pm$ 0.05 (R^2 = 0.79). Therefore, below ca. pH 4 the rate is directly dependent on hydrogen ion activity within experimental error. Some previous FAp dissolution experiments [15] have also found both pH-dependent and independent dissolution intervals, with a transition at ca. pH 6 and n $\approx$ 0.8. However, no such transition was found for CFA [15] at pH < 7 (figure 1).

Dissolution rates are also a function of temperature (figure 2). The calculated activation energy (E_a) is 18.1 $\pm$ 5.4 kJ m^{-1}. This is less than the activation energy found for FAp [14] (E_a =

34.7 ± 1 kJ). The lower value suggests a diffusion-controlled rather than a surface-controlled reaction. Since the bone consists of very small bioapatite grains within a porous structure, the reaction may be controlled by diffusion of ions into and out of the porous interior of the bone to the bone surface.

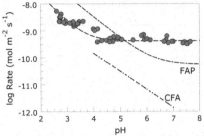

Figure 1. Bioapatite dissolution rates (○) in this study as a function of pH at 20°C and comparison with some previous results for CFA and FAP [13].

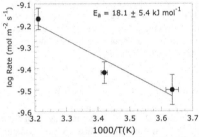

Figure 2. Arrhenius plot of bioapatite dissolution rate. The activation energy calculated is consistent with a diffusion controlled reaction.

    Concentrations of many trace elements are high at the bone surfaces and decrease inward toward the center (figure 3), consistent with models in which trace elements are introduced into the bone by diffusion from the surface. Diffusion periods for selected lanthanides (La, Ce, Pr, Nd, Gd, Yb) and U were calculated from concentration gradients in eight fossil bones using equation (1)[4]. An example of a 17.3 ± 1.8 ka diffusion profile fitted by non-linear regression to Nd concentrations from the bone edge through the outer cortical bone of sample F08-80 is shown in figure 4. Near the bone center, in the trabecular (spongy or cancellous) and inner-most cortical bone, Nd concentrations were somewhat higher than predicted by diffusion from the bone surface. The higher concentrations probably result from introduction of Nd, as well as other trace elements, by more rapid diffusion and advection through post-mortem cracks and the spongy trabecular bone. Therefore, diffusion may occur not only from the bone surface but also through cracks and pathways into the bone interior. Within individual bones, calculated diffusion period were generally the same within analytical error for the selected elements (figure 5). However, in some cases (e.g., F08-08, and F08-09) the diffusion period for uranium was greater than that for the lanthanides, possibly reflecting changes in redox state during diagenesis. Average diffusion

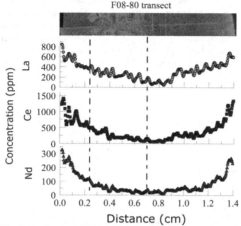

Figure 3. Concentrations of selected lanthanides in a transect (marked at top) across fossil mammal bone sample F08-80. The boundary between outer cortical and inner trabecular bone is indicated bv dashed lines.

periods in terrestrial environments (figure 5) range from ca. 1 ka [16] to as much as 40 ka, with most greater than 15 ka.

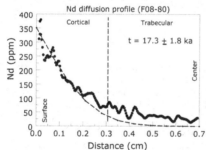

Figure 4. Example diffusion gradient for Nd in Eocene fossil mammal bone F08-80. Calculations indicate diffusion occurred over a period of 17.3 ± 1.8 ka.

IMPLICATIONS FOR REPOSITORY PERFORMANCE

Although the bioapatite dissolution rate is directly dependent on hydrogen ion activity below ca. pH 4, the rate is independent of pH under repository-relevant conditions. In contrast, FAp and CFA dissolution rates are slower [15] and CFA rates may be dependent on repository fluid pH [15]. For example, at 40°C in the pH-independent regions the bioapatite dissolution rate is 8×10^{-10} mol m^{-2} s^{-1} compared with 1×10^{-10} mol m^{-2} s^{-1} for FAP [15]. However, because the bioapatite specific surface area is much greater than FAP, the reactivity of bioapatite would be

much greater and smaller amounts would be required to retard radionuclide migration.
After burial, bioapatite recrystallizes to form more stable CFA. Concentration gradients indicate that this recrystallization occurs over periods of up to 40 ka. Recrystallization periods may be longer in drier repository environments. Therefore, bioapatite will persist for periods longer than those previously required for U.S. repository licensing. Even after recrystallization, the CFA formed will continue to retard and sequester radionuclides.

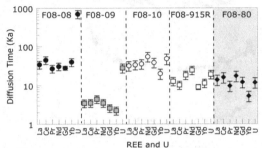

Figure 5. Calculated diffusion periods for selected REE and U in five Eocene fossil mammal bones. All of the samples, except F08-80, were collected from one bone bed.

ACKNOWLEDGMENTS

We thank William Lukens for field assistance in South Dakota. Funding was provided by Temple University and the U.S. Forest Service (05-PA-11020700).

REFERENCES

1. Ewing R.C. and Wang, L., 2002. Reviews in Mineralogy and Geochemistry, 48, 673-699.
2. Oelkers E.H., and Montel, J-M., 2008. Elements, 4, 113-116.
3. Koeppenkastrop D. and DeCarlo E.H., 1992. Chemical Geology 95, 251-263.
4. Millard A.R. and Hedges R.E.M., 1996. Geochim. Cosmochim. Acta 60, 2139-2152.
5. Bostick, W.D., Jarabek, R.J., Bostick,D.A., and Conca, J., 1999. Adv. Envtl. Res. 3, 488-498.
6. Wright, J., and Conca, J., 2002. Amer. Chem. Soc. 42, 117-122.
7. Raicevic, S., Wright, J.V., Veljkovic, V., Conca, J.L., 2006. Sci. Total Envt. 355, 13-24.
8. Wopenka, B. and Pasteris, J.D., 2005. Materials Science & Engineering C, 25, 131-143.
9. Berna, F., Matthews, A., and Weiner, S. 2004. Jour. Archaeological Sci., 31, 867-882.
10. Johannesson, K.H., Lyons, W.B., Stetzenbach, K.J., and Byrne, R.H., 1995. Aquatic Geochemistry 1, 157-173.
11. Grandstaff, D.E. and Terry, Jr., D.O. Applied Geochemistry, 24, 733-745.
12. Zanazzi, A., Kohn, M.J., MacFadden, B.J., and Terry, D.O., 2007. Nature, 445, 639-642.
13. Li, Y-H, and Gregory, S., 1974. Geochim. Cosmochim. Acta 38, 703-714.
14. Kerisit, S., and Liu C., 2010. Geochim. Cosmochim. Acta, 74, 4937-4952.
15. Guidry, M.W., and Mackenzie, F., 2003. Geochim. Cosmochim. Acta, 67, 2949-2963.
16. Suarez, C.A., Macpherson, G.L., González, L.A., and Grandstaff, D.E., 2010. Geochim. Cosmochim. Acta 74, 2970-2988.

Mater. Res. Soc. Symp. Proc. Vol. 1518 © 2012 Materials Research Society
DOI: 10.1557/opl.2012.1568

Collocation and Integration of Back-End Fuel Cycle Facilities with the Repository: Implications for Waste Forms

Charles Forsberg[1]
[1]Department of Nuclear Science and Engineering, Massachusetts Institute of Technology, 77 Massachusetts Avenue, Room 24-207B, Cambridge, MA 02139-4307, U.S.A.

ABSTRACT

The organization of the fuel cycle is a legacy of World War II and the cold war. Fuel cycle facilities were developed and deployed without consideration of the waste management implications. This led to the fuel cycle model of an isolated single-purpose geological repository for disposal of wastes shipped from distant processing facilities. There is an alternative: collocation and integration of reprocessing and other backend facilities with the repository. Such an option alters waste form functional requirements by reducing storage and transport requirements. This, in turn, broadens the choice of waste forms by relaxing the incentives to minimize waste volumes. Waste forms can be chosen primarily on meeting two goals: repository performance and minimizing costs. Less restrictive waste volume constraints enable termination of safeguards on all wastes, enable use of solubility-limited waste forms, and reduce radiation damage as a waste form limitation. The implications of such changes in waste form requirements are discussed.

INTRODUCTION

In most industries that generate significant wastes or where disposal of wastes has significant costs, waste disposal is associated with the production facilities. This is true of mining, coal-fired power plants, steel mills, and many other industries. This historical norm also implies (1) coupling of jobs, taxes and waste management and (2) a public understanding and acceptance of this coupling.

The unique history of nuclear energy, starting with development and building of nuclear weapons in World War II, created a different industrial model. Fuel cycle facilities would be located based on other considerations and the wastes would be shipped long distances to separate disposal facilities. This model exists even though shipping many radioactive wastes is expensive because the package often weighs more and takes up more volume than the wastes being shipped. This large shipping penalty is because radiation shielding is required to ship many wastes (spend nuclear fuel [SNF] and high-level waste [HLW]) and accident protection is required to ship other wastes (transuranic) with low gamma radiation levels.

The consequences of this industrial model are that waste forms have to meet multiple requirements that have little or nothing to do with safe disposal of waste. Volumes must be small to control shipping costs. Waste forms have to meet shipping requirements that have nothing to do with safe disposal.

Recent studies [1, 2] have recommended integration of the fuel cycle with the repository because of potential economic, safety, repository performance, nonproliferation, and institutional incentives. One option for integration is collocation and integration of all backend facilities at the repository site. The near-term incentive is to develop a successful repository by coupling benefits (jobs, taxes, etc.) with waste management. The longer term incentive [3] is that such integration

(Fig. 1) may result in significant economic savings because a repository consists of three major facilities that can multiple users.

- *SNF receiving facilities.* The repository will have the nation's largest SNF receiving and storage facilities. The U.S. navy inspects all SNF before disposal to better estimate long-term performance for nuclear fuel in its reactor fleet. Utilities inspect some fuel for the same reasons. The economic location for such inspection facilities is at the repository to avoid the need to build separate SNF receiving and waste packaging facilities. If the nation adopts a closed fuel cycle, locating the reprocessing and recycle facilities at the repository avoids the need to build separate SNF receiving and storage facilities.
- *Waste packaging facilities.* The repository will have the largest waste packaging facility. Collocation of any major waste generator at the site such as a reprocessing plant avoids duplication of facilities.
- *Underground disposal facilities.* For any major generator of radioactive wastes, collocation with the repository enables waste forms to be chosen to meet only disposal criteria, not disposal and transport criteria.

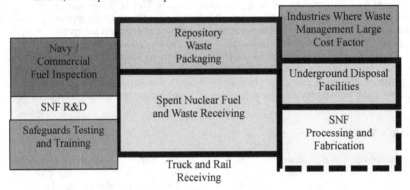

Fig. 1. Integrated Fuel Cycle and Repository

IMPLICATIONS FOR PROCESSING AND WASTE TREATMENT

There are large economic benefits by integrating reprocessing, fabrication, and waste disposal into a single backend fuel-cycle facility. Table I shows the cost breakdown for a commercial PUREX reprocessing plant [4, 5]. Some functions such as receiving are required by reprocessing facilities and repositories; thus, collocation can avoid facility duplication.

Equally important, less than 7% of the cost of a reprocessing facility is associated with separation of fissile and fertile material from the SNF—the purpose of the reprocessing plant. About half of the cost of reprocessing is associated with waste management—either processing wastes or storing wastes. Major improvements in reprocessing economics are only likely with major changes in waste management requirements because that is where most of the costs are. Many of the historic design choices have been driven by the need to minimize waste volumes. There are alternative processes [6] and the potential for more economic systems if the design constraints, particularly with waste management, are changed. An example can clarify this.

TABLE I. Cost Breakdown of Reprocessing LWR SNF

Area	% Cost	Subarea	% Cost
Receiving	7.8		
Front End	25.5	Mechanical Feed Preparation	13.00
		Tritium Confinement	3.65
		Dissolution	8.16
		Feed Preparation	0.69
Offgas	5.74	Dissolver Off Gas	4.17
		Vessel Off Gas	1.22
		Head-end Off gas	0.35
Separations	6.59	Solvent Extraction Uranium	1.39
		Solvent Extraction Plutonium	1.56
		Solvent Treatment	1.04
		Acid and Waste Recovery	1.91
		Low-Enrichment Uranium Purification	0.69
High-Level Waste	10.42	HLW concentration	1.04
		Intermediate Level Waste Concentration	1.39
		HLW Solution Storage	3.13
		HLW Solidification	4.86
Product Conversion	6.6	Low-Enriched Uranium Conversion	3.99
		Plutonium Conversion	2.26
		Plutonium Storage	0.35
SNF/HLW Storage	26.9		
Clad Storage	10.4		

SNF cladding is typically a third of the mass of a SNF assembly; thus, removal of the clad (part of mechanical feed preparation) and storage of clad are major cost components of reprocessing plants. Clad can be separated from fuel materials by mechanical or chemical methods. Chemical decladding of zircaloy-clad SNF has been done on an industrial scale for defense SNF at the Hanford site using the Zirflex process [6]. However, the higher waste volumes have made chemical decladding non-viable for commercial facilities that ship wastes offsite to repositories. If waste volumes are not a constraint, chemical decladding of some fuels becomes a viable option with reductions in reprocessing plant capital costs and simpler operations. Immediate on-site disposal would eliminate most or all clad storage costs independent of the choice of process or choice of clad material.

In the cold war, the Purex reprocessing plant at the Hanford site in the United States processed 7000 tons of SNF per year [7]. In comparison the commercial French LaHague reprocessing plant (a much larger facility and the largest commercial reprocessing plant in the world) has a capacity of 1700 tons per year. The biggest difference is that the Hanford defense complex used on-site disposal that simplified and reduced the cost of reprocessing with processes such as chemical decladding of SNF. Because of decisions to minimize short-term waste management costs, a massive cleanup effort is underway at the Hanford site. However, this and other lines of evidence [3] suggest large savings for reprocessing if waste volume constraints are relaxed. This is only viable if the reprocessing plant is collocated with the repository to allow appropriate disposal without long-distance transport of wastes. It does require new repository waste forms to match the process.

WASTE FORM IMPLICATIONS

Collocation and integration reduces the waste volume constraints. This greatly expands the potential waste forms because there are many excellent waste forms that have been rejected in the past because of their low waste loadings. However, it also creates waste forms that can meet much higher performance standards that are not possible to meet with a waste form with a high waste loading. Three such options have been identified.

Solubility-limited waste forms. The release rates of many radionuclides from a repository are limited by the solubility of the specific radionuclide in groundwater. If the specific radionuclide is diluted by a factor of a thousand with the non-radioactive isotopes of that element, its concentration in groundwater is reduced by a factor of a thousand which should lead to a commensurate reduction in radionuclide releases to the environment. Isotopic dilution is the most direct way to improve repository performance for solubility-limited radionuclides.

For example, radioactive carbon-14 can be removed from reprocessing off gas streams by scrubbing with calcium hydroxide to create calcium carbonate [8]. The calcium carbonate can be incorporated into cement with non-radioactive calcium carbonate. Because the actual mass of most radioactive isotopes in SNF is small, high isotopic dilution factors are possible ($>10^3$). The same can be done for other isotopes such as iodine by isotopic dilution with nonradioactive iodine.

Limiting radiation damage. Waste forms with high concentrations of radionuclides are degraded by (1) long term radiation damage to the waste form and (2) change in the chemical composition of the wastes caused by the decay of radionuclides into different elements. Both effects are reduced by using waste forms with low waste loadings. Increasing volumes by a factor of ten reduces the cumulative radiation dose to the waste form per unit volume by a factor of ten.

Terminating safeguards on wastes. If wastes contain significant quantities of plutonium and other fissile materials, there is a requirement for multi-generational long-term repository safeguards. However, dilution of such wastes can make the fissile materials "not practically recoverable" [9, 10] and safeguards can be terminated before disposal. Required levels of dilution in various waste matrixes have been defined [11, 12] by the International Atomic Energy Agency for safeguards termination. The economic requirement is cositing facilities so one can afford waste forms with lower waste loadings.

REPOSITORY VOLUME CONSTRAINTS

Repositories dispose of two waste classes [13]: high-decay-heat wastes and low-decay-heat wastes. High-decay-heat wastes include SNF and HLW. To avoid excessive repository temperatures that can degrade the waste form, package, and geology, these wastes are spread over parallel tunnels to enable decay heat to be conducted from the waste form through the waste package, and through the repository environment to the earth's surface. Integrating reprocessing and repositories may not change the treatment or disposal of HLW but does change the treatment and disposal strategy for all other back-end radioactive wastes—the low-heat wastes.

Low-decay-heat wastes can be disposed of in large engineered caverns that have low incremental disposal costs [13] as has been demonstrated in several operational facilities. The low disposal costs for low-heat wastes are a major factor in the economics of collocation and integration of reprocessing and the repository. While there is no operating geological repository for SNF or HLW, there are operating geological repositories for the disposal of low-heat transuranic and chemical wastes (Table II).

TABLE 2
Operational Geological Repositories

Repository Type	Chemical	Radioactive
Facility	Herfa Neurode (Germany)	Waste Isolation Pilot Plant (U.S.)
Operational	1975	1999
Capacity	200,000 tons/y	175,570 m³(Lifetime) (~ 350,000 tons)
Hazard Lifetime	Forever	>10,000 years

There are other types of operational high-volume underground facilities for the disposal of radioactive wastes. In 1988 Sweden opened the *Final Repository for Short-lived Radioactive Wastes* (SFR) [14] for low and intermediate level radioactive wastes. It is located under the Baltic seabed in granite about a kilometer off the Forsmark Nuclear Power Plant site with access by tunnel. Low activity wastes are disposed of in large mined caverns while higher activity wastes are disposed of in concrete silos with bentonite clay barriers between the silos and rock. Silo diameters are ~26 meters with heights of ~50 meters. Waste packages up to 100 tons in weight are placed in silos. A cement grout is used to create monolithic structures with low water permeability. The low surface to volume ratio minimizes the groundwater that can contact the silo and the wastes in the silo. The silos are high-performance low-cost packages for disposal of high-volume low-heat wastes.

Wastes with high decay heat have higher disposal costs because the wastes must be spread out over large areas underground to limit peak temperatures that can degrade waste forms and geologies. There have not been studies on the tradeoffs between processing costs and repository costs of adopting high-heat waste forms with lower fission product loadings with collocated facilities.

CONCLUSIONS

For half a century the fuel cycle has been defined as separate facilities at separate sites that treat and package wastes that are then transported to a repository. That system architecture results in waste form requirements that emphasized high waste loadings to minimize transportation costs and associated storage costs.

There is alternative system architecture: collocation and integration of backend facilities at the repository. Changing the system architecture changes waste form requirements. High waste loadings to meet transportation and storage requirements no longer driver waste form choices. Total waste system performance and minimizing repository costs drive waste form requirements.

It is uncertain today which system architecture will be adopted although there is increasing evidence that the second architecture may be required for successful repository siting. The second option requires rethinking the backend of the fuel cycle from institutional structures to waste forms. Because waste processing and waste forms are a major part of this system, this will require rethinking by the materials community of the waste form options. It also implies a need to begin to map out what the waste form options in this alternative future.

REFERENCES

1. Blue Ribbon Commission on America's Nuclear Future, *Report to the Secretary of Energy*, http://brc.gov/ (2012).
2. M. Kazimi, E. Moniz, C. W. Forsberg, et. al., *The Future of the Nuclear Fuel Cycle*, Massachusetts Institute of Technology (2011).
3. C. W. Forsberg, "Coupling the Backend of Fuel Cycles with Repositories, *Nuclear Technology*, **180** (2) November (2012).
4. M. J. Haire, "Nuclear Fuel Reprocessing Costs," *Advances in Nuclear Fuel Management III*, Hilton Head, South Carolina, American Nuclear Society Topical Meeting, October 5-8, 2003.
5. M. J. Haire, G. L. Ritter, and R.E. Tomlinson, "The Economics of Reprocessing Alternative Nuclear Fuels," *National American Institute of Chemical Engineers Meeting*, San Francisco, California, November 25-29, 1979.
6. J. T. Long, *Engineering for Nuclear Fuel Reprocessing*, Gordon and Breach, Inc. (1967).
7. M. S. Gerber, *A Brief History of the Purex and UO₃ Facilities*, WHC-MR-0437 (1993)
8. C. W. Forsberg, *Theoretical Analysis and Preliminary Experiments on the Feasibility of Removing CO_2 Containing C-14 Selectively with a $Ca(OH)_2$ Slurry from a ^{85}Kr-Contaminated HTGR Reprocessing Plant Off-Gas Stream*, ORNL/TM-5825, Oak Ridge National Laboratory, Oak Ridge, Tennessee (1977).
9. G. Linsley and A. Fattah, "The Interface Between Nuclear Safeguards and Radioactive Waste Disposal: Emerging Issues," *IAEA Bulletin*, International Atomic Energy Agency, Vienna, February (1994).
10. D. C. Christensen and M. A. Robinson, *Development and Implementation of Attractiveness Level E Criteria and the Plutonium Disposition Methodology*, LA-13425-MS Los Alamos National Laboratory, Los Alamos, NM (1998).
11. International Atomic Energy Agency, *Guidance for the Application of an Assessment Methodology for Innovative Nuclear Energy Systems: INPRO Manual—Proliferation Resistance*, IAEA-TECDOC-1575/Vol. 5 (2007).
12. International Atomic Energy Agency, *Consultants' Report on Meeting for Development of Technical Criteria for Termination of Safeguards for Material Categorized as Measured Discards*, STR-251 (Rev. 2) (1990).
13. C. W. Forsberg, "Rethinking High-Level Waste Disposal: Separate Disposal of High-Heat Radionuclides (^{90}Sr and ^{137}Cs)," *Nuclear Technology*, **131** (2): pp. 252-268, August 2000.
14. SFR: Swedish Repository for Short-Lived Radioactive Wastes: SKB website: http://www.skb.se/Templates/Standard____25485.aspx.

Mater. Res. Soc. Symp. Proc. Vol. 1518 © 2012 Materials Research Society
DOI: 10.1557/opl.2012.1569

Processing and Disposal of Radioactive Waste: Selection of Technical Solutions

Michael I. Ojovan and Zoran Drace
Department of Nuclear Energy, International Atomic Energy Agency, Vienna International
Centre, PO Box 100, Vienna,
1400, Austria

ABSTRACT

An overview of selection criteria for waste processing and disposal technologies is given.
A systematic approach for selection of an optimal technology is proposed. Optimal selection of a
technical processing and disposal option is case specific to the waste management needs. Waste
streams considered are from nuclear applications, research, power generation, nuclear fuel cycle
activities and decommissioning of nuclear facilities as well as for NORM-containing waste.

INTRODUCTION

Waste management is recognized as an important link for public acceptance of nuclear
energy and its applications. Technical options and technologies are crucial for safe management
of radioactive waste. A wealth of information is currently available about a multitude of waste
management technologies and their technically novel and alternative designs, as well as about
emerging technologies, which require further development and/or validation. Selection among
available options and technologies can be done on a national level either by waste generators or
by waste management organizations. The selection principals may vary from organizational
preference, collected or known experience or following an optimization procedure. In any case,
because of the costs involved, the potential complexity of technical and environmental
considerations, as well as the necessity to assure adequate performance, the selection mechanism
will always require rather clear criteria, to address waste management needs. Some criteria will
be fairly general and applicable to almost any waste management system. Others may apply to
specific waste categories or to particular waste management steps. The aim of this paper is to
discuss a systematic approach for selection of optimal solutions.

SELECTION BASIS

The IAEA publications [1-3] form the basis for establishing appropriate strategies and
infrastructure for the management of radioactive waste. The infrastructure requires selection of
an optimized technology/option because of the variety of processes and techniques available for
different waste streams at specific waste management steps. The technologies selected for
different waste management steps should be then combined in an integrated strategy to optimize
the overall waste management system [4]. The selection of waste technologies for each specific
waste stream/category shall be based on an evaluation process with the following elements: (i)
Identification and nature of specific radioactive waste inventories and associated properties; (ii)
Consideration and review of various options for the management of that waste; (iii) Evaluation
of the advantages and disadvantages of each option using multi-attribute utility analysis [5] or

any other suitable methodology that compares safety, technological status, cost-effectiveness and social and environmental factors; (iv) Selection of the best available technology(ies) not entailing excessive cost and satisfying all regulatory requirements [6]. Approval (via licensing, authorization) of the selected technology(ies).

WASTE ROUTING, CLASSIFICATION AND CATEGORIZATION

Nuclear materials are typically routed depending on their further usefulness. The main routes envisaged are as follows (Fig. 1): (a) Clearance from regulatory control, which assumes unrestricted disposal of waste and unrestricted reuse of useful materials; (b) Authorized release, which assumes authorised discharge of waste to the environment and authorised reuse of useful materials; (c) Regulated disposal of waste and regulated transfer of useful materials to other practices.

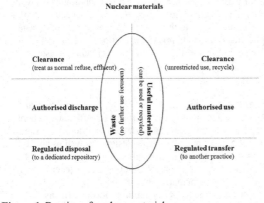

Figure 1. Routing of nuclear materials.

There is great diversity in the types and amounts of radioactive waste in different countries. Technologies for management of the waste are also diverse, although the main technological approaches are likely to be similar everywhere. Adequate processes and technologies can be identified based on detailed information of the current or forecasted waste e.g. waste classification, categorization, properties and inventory.

The IAEA provided an internationally accepted waste classification system [7] which defines the following classes according to the activity and half-lives of radionuclides in waste: (i) Exempt waste (EW); (ii) Very short lived waste (VSLW); (iii) Very low level waste (VLLW); (iv) Low level waste (LLW); (v) Intermediate level waste (ILW); (vi) High level waste (HLW). The IAEA classification is based primarily on long term safety and therefore is oriented on selection of most appropriate disposal routes (end-points) for solid or solidified waste.

Management of waste in different steps prior to disposal (e.g. pre-treatment, processing, storage) requires complementary information on the waste properties relevant to particular activities. Categorization of waste is used to provide a consistent approach to waste processing and storage. Categorization of waste has to include information such as origin, physical state, types, properties and process options [8].

WASTE MANAGEMENT STEPS

Waste management is typically divided into predisposal and disposal steps. Predisposal comprises all the steps in the management of radioactive waste from its generation up to disposal, including processing (e.g. pretreatment, treatment and conditioning), temporary (interim) storage and transport. Disposal envisages permanent emplacement of waste in an appropriate facility without the intention of retrieval. Radioactive waste is prepared for disposal by processing technologies primarily intended to produce a waste form that is compatible with the selected or anticipated disposal option. For evaluation of a particular process or technology it is necessary to review the availability of selected options to meet waste processing, storage and disposal requirements. Technical options for waste management have been described in many publications [9-19]. The life cycle of radioactive waste consists of a number of steps: (1) Pretreatment includes any operations prior to waste treatment, to allow selection of technologies that will be further used in processing of waste (treatment and conditioning), such as: collection; segregation; decontamination; chemical adjustment and fragmentation. (2) Treatment of radioactive waste includes those operations intended to improve safety or economy by changing the characteristics of the radioactive waste. The basic objectives of treatment are: (a) volume reduction; (b) radionuclide removal from waste; and (c) change of physical and chemical composition. Some treatment may result in an appropriate waste form. However in most cases the treated waste requires further conditioning either by solidification, immobilization or encapsulation. (3) Conditioning covers those operations that produce a waste package suitable for handling, transportation, storage and/or disposal. It may include: immobilization of the waste; enclosure of the waste in containers; and, if necessary; provision of an overpack. Immobilization refers to the conversion of waste into a waste form by solidification, embedding or encapsulation. Common immobilization matrices include cement, bitumen and glass. (4) Storage of radioactive waste involves maintaining the radioactive waste such that retrievability is ensured and confinement, isolation, environmental protection and monitoring are provided during storage period. (5) Transportation refers to the deliberate physical movement of radioactive waste in specially designed packages from one place to another. For example, raw waste maybe transported from its collection point to centralized storage or processing facility. Conditioned waste packages may be transported from processing or storage facilities to disposal facilities. (6) Disposal envisages emplacement of waste in an appropriate facility without the intention of retrieval. Note that in some countries controlled discharge of effluents to the environment is often considered as a regulated disposal option.

Characterization of radioactive waste is important at every stage of waste management. It involves determination of the physical, chemical and radiological properties of the waste to establish the need for further adjustment, treatment, conditioning, or its suitability for further handling, processing, storage and disposal. Up-front characterization as part of the pretreatment stage is essential for technical decision-making involving the selection of the most efficient treatment process. Methods of radioactive waste characterization and the methodology of characterization including sampling procedures have been described in detail [8].

METHODOLOGIES FOR TECHNOLOGY SELECTION

The selection of pretreatment, processing, storage and disposal technologies is necessarily bound up with the overall strategy for the management of the waste under consideration, and this is turn may be part of a larger scheme embracing many waste types. To achieve a satisfactory waste management strategy, waste management components must be complementary and compatible with each other [1]. Although many aspects have to be addressed, the challenge is to achieve the optimal solution in a logical, structured and justified way. Furthermore is important to ensure that all three basic waste management routes (e.g. clearance, discharge or regulated disposal) are taken into account and evaluated in defining waste management strategies. Strategies for all waste streams generated at facilities or sites need to be considered rather than options for individual waste streams. In addition, most national regulators now demand an impact assessment of proposed technologies and a justification for the selected technology. The process of selection of a waste management technology typically starts by collecting and assessing available data, by considering all potentially influencing factors such as applicable regulations, waste properties, waste routes and associated good practice indicators (Table I).

Table I. Technology-related criteria and attributes.

	Criteria	Good practice attributes
1	**National policy and strategies**	Compliance with the intent of national polices and strategies. In the case of insufficient national policies and strategies, compliance with international 'good practice'
2	**Regulatory framework**	Compliance with the requirements of the regulatory framework. In the case of insufficient regulatory framework, compliance with international 'good practice'. Clearance levels are set up. Mechanism for authorized discharged is established
3	**Funding and cost**	Both direct and indirect costs (e.g. stakeholder involvement and public acceptance) addressed. Total cost of the viable technology evaluated or compared and technology selected/eliminated in terms of main cost factors. Adequate financial resources or financial security and funding mechanisms available for the funding of viable technology.
4	**Health, Safety and Environmental (HSE) impact**	HSE impacts of viable technologies known and considered in the selection of technologies; HSE impact optimized by reducing exposure of the workforce and the members of the public. The need for transportation of radioactive material is minimized.
5	**Waste characterizatio n**	Identification of all sources of waste generation. Waste characterization developed and can be implemented at all stages of the waste management process.
6	**Waste management system**	Waste management system exists and can support the newly introduced technology. Storage/disposal facilities available. Operational waste generation control programme in place.
7	**Human resources**	Availability of suitably qualified and experienced personnel. Consideration of lessons learned from implementation of other technologies.

8	Social impacts and stakeholder involvement	Technologies discussed with stakeholders and considered in a transparent way. All stakeholders involved in the selection of a technology and reasonable consensus reached.
9	Technical factors	All technical factors affecting the selection of a technology (e.g. maturity, robustness, complexity and maintainability etc.) are taken into account.
10	Physical infrastructure	Physical structure is available and can support the newly introduced technology.

A set of possible technological options is then devised together with a preliminary waste management plan for implementing each option. These plans can be relatively brief at this stage but still sufficiently well defined, so that the associated major hazards and risks can be visualized. The next step is to perform technology selection studies. During this process, formal decision aiding techniques and 'workshop' discussion sessions can be employed. Selection of a preferred or optimized waste processing technology is best achieved through the evaluation of the general criteria and constraints in terms of their attributes for a specific waste stream or facility. This evaluation can benefit from the use of formal decision-aiding techniques that address the influencing factors and associated good practice indicators. When evaluating the various influencing factors for a specific technology option, a simple 'decision-tree' approach could be adopted, in which the various factors are evaluated. The limitations of a linear approach are that influencing factors may only be considered one at a time, and in descending order of priority. Project selection decisions require multiple, generally non-linear, objectives to be simultaneously optimized. In addition, factors that are mutually influential cannot be considered in combination.

MULTI-ATTRIBUTE UTILITY ANALYSIS

Multi-Attribute Utility (MAU) analysis [5] is a powerful tool accounting for many criteria and constraints involved in the technology selection process. It is an effective and efficient way of showing the impact of each technology option in terms of good practice attributes, and of reaching conclusions that address all of the influencing factors. Such analysis involves assigning numerical ratings and weightings to the factors considered, followed by comparison of the obtained total scores for the options. If necessary (i.e., when two options have very close scores), a sensitivity analysis can be performed to check whether or not the preferred option is a right choice. A simple scoring of the criteria for a given option allows any option to be discarded or considered for further evaluation. Regardless of the approach it is necessary to produce a justifiable and auditable solution for selected options.

MAU method makes clear to all involved the basis on which the alternatives are being evaluated. It offers quantifiable principals for choosing options. This is particularly important in-group decision-making situations in which many different points of view and alternative decisions have to be reviewed and taken into account. The attributes needed for evaluation of options must be identified. They are assigned a weight that reflects their importance to the decision. A value of 3, 2, or 1 might be assigned to each attribute, depending on its importance. Alternatively 100 points can be assigned and distributed over the attributes according to their importance. A score to each of the alternatives for each attribute can be given. A scale of 1-10 may be given. Each alternative's score for each attribute is then multiplied by the weight of that attribute, and the total is calculated. That total represents the value of that option, which can then be compared to the same calculation for the others. If it is a group process, each member of the group scores the attributes for each option and the group's ratings can be totalled or averaged. The final result of this example analysis would be a relative, numerical ranking of the options (the score for each option).

Furthermore, various criteria such as non-safety related matters could also be considered in the process of selecting an option. Where relevant, safeguards related issues should also be considered in optimizing both safety and resources in the decision making process. The costs of maintenance, surveillance and physical protection for the waste management facility should also be taken into account. It shall be ensured that the selected option meets all the applicable safety requirements. A MAU model can be used to further explore the consequences of changing the attributes, their weights, or the scores they received. Since the criteria are transparent, it is possible to make several changes and review the results. For example, if it appears that some attributes are too important in determining the results, the weights could be adjusted to produce more realistic results. Workshop sessions (sometimes called brainstorming sessions or decision conferences) can provide a practical and motivating way forward. In such sessions a panel of relevant experts agrees on the list of influencing factors and then assesses the impact of these factors on each of the technological options, assisted by the use of decision aiding techniques. It is important to produce a report of the workshop sessions, describing the technique adopted, the considerations addressed and the results obtained. This report can be a valuable aid in support of the waste management plan and the associated safety justification.

The processes of selecting a preferred technology and the subsequent detailed strategy are best approached by ensuring that the team clearly understands the underlying safety logic. This logic must be applied to each of the candidate options (at an appropriate level of detail), as part of the process of selecting a preferred option. The key point is to ensure that there is a demonstrable connection between the characteristics and amounts of radioactive waste at generation, the proposed technologies and associated risks in implementing these technologies, the safety management arrangements, and the costs. E.g. analysis of the risks involved logically determines the requirements for key aspects such as additional or modified equipment, staff training, procedures, work instructions, maintenance and security arrangements.

CONCLUSIONS

The selection of a technology is based on the evaluation of all relevant criteria and constraints. To address complex waste management needs it is essential to analyse the waste generation and understand properties, types and volumes of waste before selection of a particular technology. Furthermore it is necessary to comply with regulatory regime, and to consider disposal options available assuming that legal and regulatory infrastructure exists or is going to be established. A forthcoming IAEA publication "Selection of Technical Solutions for the Management of Radioactive Waste" will give a detailed consideration of selection process.

REFERENCES

1. Policies and Strategies for Radioactive Waste Management, IAEA Nuclear Energy Series No. NW-G-1.1, IAEA, Vienna (2009).

2. Methodology for Establishing an Inventory of Radioactive Waste and for Assessing the Subsequent Management Needs. IAEA, Vienna (2012, to be published).
3. Economics of Radioactive Waste Management. IAEA, Vienna (2012, to be published).
4. Predisposal Management of Radioactive Waste, IAEA Safety Standards Series No. GSR Part 5, IAEA, Vienna (2009).
5. A. Rahman. Multi-attribute Utility Analysis — a Major Decision Aid Technique, Nuclear Energy, 42, No 2, April, 87–3 (2003).
6. Fundamental Safety Principles, IAEA Safety Standards Series No. SF-1, IAEA, Vienna (2006).
7. Classification of Radioactive Waste, General Safety Guide No. GSG-1, IAEA, Vienna (2009)
8. Strategy and Methodology for Radioactive Waste Characterization, IAEA-TECDOC-1537, IAEA, Vienna (2007).
9. M. Ojovan, W.E. Lee. An Introduction to Nuclear Waste Immobilisation, Elsevier, Amsterdam, 315pp. (2005).
10. M. Ojovan. Handbook of advanced radioactive waste conditioning technologies. Woodhead, Oxford, 512 p. (2011).
11. Concepts for the Conditioning of Spent Nuclear Fuel for Final Waste Disposal, Technical Reports Series No. 345, IAEA, Vienna (1992).
12. Spent Fuel Reprocessing Options, IAEA-TECDOC-1587, IAEA, Vienna (2008).
13. Naturally Occurring Radioactive Material (NORM V), Proceedings of an international symposium, Seville, Spain, 19–22 March 2007.
14. Waste Forms Technology and Performance: Final Report. National Research Council. 340 p., The National Academies Press, Washington, D.C. (2011).
15. W.E. Lee, M.I. Ojovan, M.C. Stennett, N.C. Hyatt. Immobilisation of radioactive waste in glasses, glass composite materials and ceramics. *Advances in Applied Ceramics*, **105** (1), 3-12 (2006).
16. Interim Storage of Radioactive Waste Packages, Technical Reports Series No. 390, IAEA, Vienna (1998).
17. Scientific and Technical Basis for the Near Surface Disposal of Low and Intermediate Level Waste, Technical Reports Series No. 412, IAEA, Vienna (2002).
18. Scientific and Technical Basis for the Geological Disposal of Radioactive Wastes, Technical Reports Series No. 413, IAEA, Vienna (2003).
19. J. Ahn, M.J. Apted. Geological repository systems for safe disposal of spent nuclear fuels and radioactive waste. Woodhead, Cambridge, 792 p. (2010).

Mater. Res. Soc. Symp. Proc. Vol. 1518 © 2013 Materials Research Society
DOI: 10.1557/opl.2013.70

Assessment of the evolution of the redox conditions in the SKB ILW-LLW SFR-1 repository (Sweden)

L. Duro[1], C.Domènech[1], M. Grivé[1], G.Roman-Ross[1] and J.Bruno[1] and K. Källström[2]
[1] Amphos 21 Consulting, S.L., P. Garcia Faria 49-51, 1-1, Barcelona, E-08019, Spain.
[2] Svensk Kärnbränslehantering AB, Avd. Låg- och medelaktivt avfall, Box 250, 101 24 Stockholm, Sweden.

ABSTRACT

The evaluation of the redox conditions in the Swedish ILW-LLW repository, SFR-1, is of high relevance in the performance assessment. The SFR-1 repository contains heterogeneous types of wastes, of different activity levels and with different materials in the waste and in the matrices and packaging. Steel and concrete-based materials are ubiquitous in the repository. The assessment presented in this work is based on the evaluation of the redox conditions and of the reducing capacity in 15 individual and representative waste package types in SFR-1. A combination of the individual models is used to determine the redox evolution of the different vaults in the repository. The results of the model indicate that in the initial time after repository closure, O_2 is consumed through degradation of organic matter and metal corrosion during the initial time after repository closure. Afterwards, the system is kept under reducing conditions for long time periods, and $H_2(g)$ is generated due to the anoxic corrosion of steel forming magnetite as main corrosion product. The time at which steel is depleted varies with the amount and characteristics of steel and ranges from 5,000 to over 60,000 years. After complete steel corrosion, the reducing capacity of the system is mainly given by magnetite. The calculated redox potential under the chemical conditions imposed by the massive amounts of cements in the repository is in the order of -0.75 V (at pH 12.5). In case of assuming that the Eh of the system is controlled by the interaction between Fe(III)/Magnetite as a result of groundwater/magnetite interactions, redox potentials in the range -0.7 to -0.01V are calculated, considering the uncertainty in the pH prevalent in the system If the absence of oxic disturbances the Eh of the repository system would be kept reducing. In the event of oxidising and diluted glacial meltwater intrusion, magnetite would gradually convert into Fe(III) oxides, buffering the redox potential of the system and preventing it from oxidation for long time periods.

INTRODUCTION AND OBJECTIVE

The SFR1 repository at Forsmark (Sweden) is used for the final disposal of ILW-LLW produced by the Swedish nuclear power programme, industry, medicine and research. The repository is composed of different vaults: Silo, BMA, 1&2BTF and BLA.

The repository contains large amounts of cementitious materials, metals (containers or as scrap), organic materials (bitumen and cellulose among others), sand and bentonite. Wastes are conditioned in cement or bitumen and in some cases they are not conditioned.

The objective of the work summarised here was to develop a methodology to calculate the evolution of the redox state of SFR1 with deposition time. According to the Swedish regulations, the time frame for the assessment must span over 100,000 y after the repository closure.

MODEL DEVELOPMENT

Table I briefly describes the 15 waste package types whose redox evolution has been calculated in the model. The reader is referred to [1] for a more complete description of the packages.

Table I. Waste type representative for Silo and BMA, BLA, 1&2BTF vaults.

	Type	Package	Waste	Matrix		Type	Package	Waste	Matrix
	B.06	Steel	I.E.resins	Bit.		B.05	Steel	I.E.resins	Bit.
	F.18	Steel	I.E.resins	Bit.		F.05	Steel	I.E.resins	Bit.
SILO	O.02	Conc.	I.E.resins	Bit.		F.17	Steel	I.E.resins	Bit.
	R.16	Steel	I.E.resins	Bit.	BMA	F.23	Steel	Scrap metal	Conc.
	O.24*	Conc.	Scrap metal	Conc.		R.01	Conc.	I.E.resins	Conc.
BTF	B.07/O.07	Conc.	I.E.resins	Conc.		R.15	Steel	I.E.resins	Conc.
	S.13	Steel	ashes	Conc.		O.23	Conc.	Scrap metal	Conc.
BLA	O.12	ISO-cont.	Scrap metal	--					

Conc.: concrete; Bit.:bitumen; I.E.resins: ion-exchange resins. *Considered equal to O.23.

Calculations are done with the code PHREEQC [2] for a total simulation time of 100,000 y. Temperature is considered constant and equal to 25°C and the system is considered to be water saturated. No mechanical changes are assumed to affect the integrity of the system. Neither the transport or diffusion of gases through the container wall nor the mechanical effects on the system that the presence of these gases can produce are considered. Gases generated from the different chemical reactions are left to accumulate in solution and leave the system once their pressure exceeds 6.87 atm (hydrostatic pressure at 70 m depth).

The relevant redox chemical processes considered in the model are:

1) Steel corrosion.

Steel is considered homogeneous C-Steel. Its corrosion is kinetically controlled. Corrosion rates are listed in Table II. Initial oxygen trapped in the repository due to the construction and operation periods generates goethite as steel corrosion product (r.1). Once the initial oxygen is consumed and anoxic conditions are achieved ($[O_2] < 10^{-6}$ M), steel corrosion continues at expenses of water reduction (r.2) following an anoxic rate and being magnetite and H_2 the main corrosion products. Between both periods, there is a transient period in which the amount of goethite oxically formed is transformed into magnetite (r.3).

$$4 Fe^0 + 3 O_2 + 2 H_2O \rightarrow 4 FeOOH \qquad \text{r.1}$$
$$3 Fe^0 + 4 H_2O \rightarrow Fe_3O_4 + 4H_2 \qquad \text{r.2}$$
$$2 FeOOH + Fe^0 + 2 H_2O \rightarrow Fe_3O_4 + H_2 \qquad \text{r.3}$$

2) Microbial activity.

O_2- and SO_4-reducing bacteria contribute to the degradation of bitumen, cellulose, acetate and organic matter in general. Their biological activity has been kinetically modeled. For the sake of simplicity, a threshold of $[SO_4^{2-}]_{total}=10^{-6}$ M under which SO_4-reducing bacteria is considered negligible has been fixed [3, 4, 5].

3) Degradation of organic matter, bitumen, acetate and cellulose.

Organic matter, bitumen and acetate are kinetically degraded under hyperalkaline and neutral conditions. The degradation rate of these organic compounds is controlled by the growth rate of aerobes and SO_4-reducing bacteria. Under abiotic conditions, bitumen and organic matter are assumed to degrade following a constant rate while degradation of acetate is not considered.

Under oxic and hyperalkaline conditions cellulose degrades kinetically to acetate and carbonate. In the presence of aerobes and sulphate reducing bacteria, cellulose degradation rate is controlled by the growth of these communities. Once abiotic conditions are developed in the system, cellulose produces ISA according to the kinetic law shown in [6].

All processes in the system are assumed to occur under hyperalkaline conditions due to the presence of concrete materials in the waste matrices and/or the container materials, except for waste package type O.12, where circumneutral conditions are considered. In the numerical model, portlandite has been taken as a proxy for concrete. Kinetic rates implemented in the model are listed in Table II. Thermodynamic equilibrium is assumed for the rest of reactions.

The contribution that Al and Zn corrosion can have to the development of anoxia in the system has been conservatively neglected; sludge and other inorganic materials are not included in the model because they appear in very low amounts when comparing to those of steel and organic matter.

The groundwater considered to interact with the wastes is in equilibrium with portlandite and initially equilibrated with atmospheric O_2.

Table II. Kinetic rates used in the calculations. Rates in mol dm^{-3} s^{-1}, except when indicated.

Material	Rate	Comments
Steel	Oxic Constant rate	0.1 µm/y (hyperalkaline) [7], 60 µm/y (neutral) [8]
	Anoxic constant rate	0.05 µm/y (hyperalkaline) [9], 2.8 µm/y (neutral) [10]
Bitumen and organic matter	Biotic rate	$rate_{B/OM}^{biotic} = \sum_{i=O_2,SO_4} Biomass_i \; K_{B/OM}^i \frac{i}{k_{1/2}^{B/OM,i} + i}$
	Abiotic constant rate	1.0×10^{-12} mol dm^{-3} s^{-1} (two orders of magnitude lower tan biotic rate)
Acetate	Biotic rate	$rate_{acetate}^{biotic} = \sum_{i=O_2,SO_4} Biomass_i \; K_{acetate}^i \frac{i}{k_{1/2}^{acetate,i} + i}$
Cellulose	Biotic rate	$rate_{cellulose}^{biotic} = \sum_{i=O_2,SO_4} Biomass_i \; K_{cellulose}^i \frac{i}{k_{1/2}^{cellulose,i} + i}$
	Abiotic rate	$celdeg_{cellulose} = \left(1 + e^{-k_h t} \frac{k_1}{k_t}\right) G_{t\,0} \left(1 - e^{-k_h t}\right) - 1$ [6]
O_2-consumer bacteria		$rate_{O_2-consumer} = - \sum_{j=acet,BOM \atop cellulose} Y_j^{O_2} rate_j^{biotic,O_2} - b_{O_2}[Biomass_O]$
SO_4-consumer bacteria		$rate_{SO_4-consumer} = - \sum_{j=acet,BOM \atop cellulose} Y_j^{SO_4} rate_j^{biotic,SO_4} - b_{SO_4}[Biomass_{SO4}]$

$K_j^{O_2}$=1.0×10^{-3} mol$_i$ mol^{-1}$_{Biomass}$ s^{-1} [11]; $K_j^{S(VI)}$=6.9×10^{-3} mol$_i$ mol^{-1}$_{Biomass}$ s^{-1} [12]; $k_{1/2}^{j,O_2}$=1.0×10^{-6} mol dm^{-3} [11]; $k_{1/2}^{j,S(VI)}$=1.7×10^{-4} mol dm^{-3} [11]; b_{O_2}=2.3×10^{-6} s^{-1} [4]; b_{SO4}=1.3×10^{-5}s^{-1} [11]; $Y_j^{O_2}$=0.05 mol$_{Biomass}$/mol$_j$ [4]; $Y_j^{SO_4}$=0.085 mol$_{Biomass}$/mol$_j$ [4]; (G$_t$)$_o$=5.6×10^{-4} [6]; k$_1$ =(2.0±0.3)×10^{-3}h^{-1} [6]; k$_t$ =(3.3±0.5)×10^{-4} h^{-1} [6]; k$_h$ =(2.9±0.3)×10^{-8} h^{-1} [6].

Several relevant uncertainties of the model developed must be considered, such as:
- the uncertainty in the parameter values used in biotic degradation rates are not expected to largely affect the results because of the short-term impact that the degradation of organic materials will have in the overall redox behaviour of the repository;
- given the discrepancies in the literature concerning bitumen degradation, a sensitivity case has been run by neglecting the role of bitumen in oxidant consumption (not shown here);
- One unique type of cellulose has been considered in the system. Different type of cellulose materials may present higher degradation rates, thus higher ISA concentrations;

-the acetate concentration calculated to accumulate in the system; it will not importantly affect the behaviour of radionuclides;

-the possible reduction of organic ligands under the very low Eh values developed in the system has not been considered;

- probabilistic analyses on the corrosion rate of steel indicate that this uncertainty will not have a large effect on the capacity of the system to buffer an oxidant intrusion because magnetite can buffer it.

-The oxidation of metals other than steel during the oxic period would be very fast and its contribution to the generation of $H_2(g)$ is not expected to be very relevant;

- the degradation of concrete and, therefore, the decrease of the pH of the system due to the formation of different CSH may buffer the pH at lower values and therefore may result in higher redox potentials of the system than the ones calculated at the pH of portlandite;

-considering a continuous inflow of water equilibrated with atmospheric oxygen content to simulate a glacial event after 60 ky of the repository closure; clearly an over-pessimistic assumption as the reducing capacity of the fracture materials and the soils is not considered. It is also very unlikely that the melting water occurs continuously for 40 ky (from 60 ky after the closure until the 100 ky of the time frame of the assessment).

SUMMARY OF RESULTS

Figure 1 summarises the evolution of the redox potential in SFR1. Biotic degradation of bitumen, organic matter, acetate and cellulose is only significant during the early times after repository closure. After 5 y a reducing environment has been developed in all packages (Figure 2a).

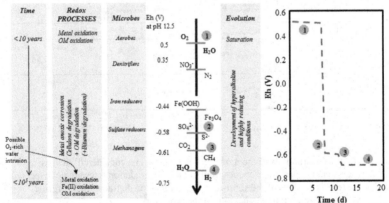

Figure 1. Sketch of the evolution of the redox potential in SFR1 in non disturbed conditions, depending on the redox couple governing the Eh at a pH 12.5 (portlandite equilibrium).

During the short oxic period, steel corrodes, goethite is the corrosion product, and $Fe(II)$ the major aqueous Fe species (Figure 2b). As Eh decreases, goethite transforms to magnetite which becomes the stable corrosion product of steel under anoxic conditions (Figure 2b).

Magnetite precipitation continues until steel is totally corroded (Figure 2c). The time at which steel is exhausted ranges from 5,000 to over 60,000 y, depending on the amount and properties of the steel in each waste package type. After complete steel corrosion, the reducing capacity of the system is mainly provided by magnetite. The redox potential achieved is imposed by the steel corrosion and by the production of $H_2(g)$ (which is considered reactive) in the order of -0.75 V at pH 12.5 (Figures 1, 2a).

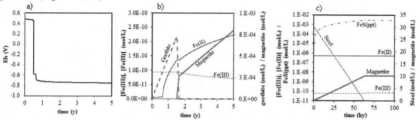

Figure 2. a) Short term evolution of Eh calculated for R.15 waste package type. b) Short term and c) long term evolution of Fe(III) and Fe(II) aqueous concentration, and of steel, goethite, magnetite and FeS(ppt) calculated for O.O2 waste package type.

Neither concrete degradation nor the formation of other corrosion products other than goethite and magnetite have been considered in the calculations. Both processes can, however, have an important control on the pH or the Eh finally achieved in the repository. Figure 3 shows the Eh/pH field framed by the different redox couples more likely to exert the redox control in the repository at the different pH values that porewater can reach due to concrete degradation. In all cases reducing redox potentials are developed. Only in the case of considering Ferrihydrite (HFO) as the oxidation product of magnetite (extreme case considering the modeling time), mildly reducing potentials could be attained.

	pH	13.5	12.5	10.5
	H_2O/H_2	-0.807	-0.749	-0.633
Eh(V)	Fe_3O_4/Fe_2O_3	-0.721	-0.663	-0.547
	$Fe_3O_4/FeOOH$	-0.700	-0.642	-0.526
	Fe_3O_4/HFO	-0.190	-0.132	-0.016

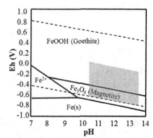

Figure 3. Range of redox potential calculated depending on the active couple considered to control the system at the pH stages established by the degradation of cementitious materials.

Figure 4 shows the effect of a fast intrusion of glacial groundwater rich in O_2 into the SFR-1 repository. O_2 added to the system, due to a melting glacial water intrusion occurring 60ky after repository closure is buffered by the oxidation of the magnetite previously formed through steel corrosion. The amount of magnetite in the system provides enough reducing capacity as to counteract an O_2 intrusion for more than 50 ky. After this period Eh (V) increases.

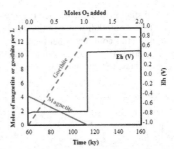

Figure 4. Evolution of the moles of magnetite and goethite in S.13 waste package type and the value of the redox value. Time = 60ky indicate the time at which the glacial groundwater is simulated to intrude in the system.

CONCLUSIONS

Results of the model have shown that the reducing capacity of the repository wil be initially provided by organic matter and by steel and its corrosion products. After 100 ky of simulation time, the amount of magnetite present in the vaults and the remaining steel provide a large reducing capacity to buffer eventual oxidant intrusions. Under undisturbed conditions the reducing capacity of the system is kept stable once all steel has been corroded. In the case of an oxidant intrusion the reducing capacity provided by magnetite is calculated to be sufficiently high as to buffer the redox potential to reducing values for long time periods.

REFERENCES

1. L. Almkvist and A. Gordon A. SKB report R-07-17 (2007).
2. D. L. Parkhurst and C. A. J. Appelo USGS, Water resources investigations report 99-4259 (1999).
3. C. A. J. Appelo and D. Postma *Geochemistry, groundwater and pollution*. A.A. Blakema Publishers, the Netherlands, 649 pp (2005)
4. J. S. Small, M. Nykyri, M. Helin, U. Hovi, T. Sarlin and M. Itävaara *Applied Geochemistry*, 23 (6), 1383-1418 (2008).
5. P. B. McMahon and F. H. Chapelle *Groundwater* 46 (2), 259-271 (2008).
6. M. A. Glaus and L. R. Van Loon Environ. Sci. Technol., 42, 2906–2911 (2008).
7. D. J. Blackwood, L. J. Gould, C. C. Naish, F. M. Porter, A. P. Rance, S. M. Sharland, N. R. Smart, M. I. Thomas and T. Yates AEA Technology Report AEAT/ERRA-0318 (2002).
8. D. Kuron, H. Gräfen, H. P. Batroff, K. Fäßler and R. Münster 1985. *Werkstoffe und Korrosion*, 36, 68-79 (1985).
9. N. R. Smart, D. J. Blackwood, G. P. Marsh, C. C. Naish, T. M. O'Brien, A. P. Rance and M. I. Thomas M I, 2004. *AEAT/ERRA-0313*, United Kingdom Nirex Limited (2004).
10. R. Schenk 1988. *Nagra Technical Report 86-24*, Wettingen, Switzerland (1988).
11. C. Sena PhD. Thesis. Universidade de Aveiro, Portugal (2009)
12.0Y. H. Lin and K. K. Lee. *Journal of environmental engineering*, February 119-126 (2001).

Mater. Res. Soc. Symp. Proc. Vol. 1518 © 2013 Materials Research Society
DOI: 10.1557/opl.2013.73

Characterization of Radionuclide Retaining Properties of Backfill Materials for Near Surface Repositories for Low and Intermediate Level Radioactive Wastes

Elizaveta E. Ostashkina[1], Galina A. Varlakova[2], Zoya I. Golubeva[1]
[1]Scientific and Industrial Association "Radon", 119121 Moscow, Russia, the 7-th Rostovsky Lane, 2/14
[2]Joint Stock Company "A.A.Bochvar High-technological Research Institute of Inorganic Materials", 123098, Moscow, Russia, the Rogov Str, 5a.

ABSTRACT

The sand of glaciolacustrine origin is offered as a major component of the backfill in repositories for solid or solidified low and intermediate level waste (LILW). Clinoptilolite, hematite, and magnesium oxide are offered as additives for increasing of sorption. In this work was carried investigation of sorption properties of sand, clinoptilolite, hematite and magnesium oxide and mixtures of sand with these mineral additives.

INTRODUCTION

Backfill is very important element of engineered barriers of repositories for solid or solidified low and intermediate level waste (LILW) [1, 2]. It must retain radionuclide migration from repository, restrict the access of water to waste packages and do not complicate necessary actions for closure of repository after operating period.

In various concepts of repositories bentonite clay (bentonite), concrete or sand are proposed to use as a backfill [3]. Bentonite swells after contact with water and forms a water impermeable barrier [4]. The concrete engineered barriers with time form voids and cracks. Groundwater after the contact with concrete has alkaline reaction. In case, when concrete and bentonite are used in the same multiple barrier, where is an exchange of ions of (H^+, Na^+) for Ca^{2+} in the structure of montmorillonite, and so bentonite loses plasticity and ability to swell [5].

In this paper, some compositions of sand of glaciolacustrine origin with mineral additives are offered for use as a backfill [6-8]. The sand consists of such basic components as quartz, feldspar, glauconite and illite [6, 12] and has good physical and mechanical properties. The sand has also good sorption properties [9, 10] due to mineral layer on the surface of quartz grains [11-13]. Mineral additives can increase sorption properties of the sand.

EXPERIMENTAL DETAILS

The work studies a degree of sorption of ^{137}Cs, ^{90}Sr, ^{60}Co, U and Pu from aqueous solutions on sand, clinoptilolite, hematite and magnesium oxide and on mixtures of sand with these mineral additives.

The experimental procedure was the as follow as. Solid phase contacted with radionuclide aqueous solutions with or without complexing compounds and competing ions until the of solid-liquid equilibrium reached. During the experiment, time changes of radionuclide concentration in the liquid phase were determined. The time when radionuclide concentration reached a constant value was considered as equilibrium time.

Series of sorption experiments were performed in presence of complexing agents and competing ions. The degree of sorption (C, %, further in the text as sorption) was calculated by formula 1.

$$C = \left[\left(A_0 - A_{ocm}\right) / A_0\right] \cdot 100 \qquad (1)$$

Where A_o is an initial activity of solution, Bq/mL; A_{ocr} is final activity of solution, Bq/mL.

Table I. Conditions of experiments.

Solid phase	- Sand of glaciolacustrine origin from homeland rock; - clinoptilolite $((Na, K_2)O \bullet Al_2O_3 \bullet 10SiO_2 \bullet 8H_2O)$ - natural zeolite of supercrust rock origin from one of the Russian's deposits; - hematite (Fe_2O_3) – chemical agent; - magnesium oxide (MgO) - chemical agent.
Size of particles	<2 mm (<5 mm for clinoptilolite)
Liquid phase	Distilled water with complexing compounds and competing ions
Solid/liquid ratio	$1 \div 2$; $1 \div 4$
Radionuclide concentration	^{137}Cs - 105 Bq/mL; ^{90}Sr - 532 Bq/mL; ^{60}Co - 23,5 Bq/mL; U - 2,64 Bq/mL, Pu - 87 Bq/L, 1 Bq/mL
Temperature	22 °C

The experiments were conducted in polypropylene vessels and centrifugal beakers (capacity 60 ml and 25 ml respectively). Solid phase was separated in a centrifuge (20 min, 5000 rpm).

DISCUSSION

Sorption of ^{137}Cs, ^{90}Sr and ^{60}Co in backfill materials

To evaluate influence of mineral salts on sorption of ^{137}Cs, experiments were conducted in presence of competing $NaNO_3$ or stable Cs NO_3 (0.01 M) in liquid phase. To study the sorption of ^{90}Sr in clinoptilolite, sand and hematite experiments were conducted in presence of competing $Ca(NO_3)_2$ or stable $Sr(NO_3)_2$ (0.01 M).

Figure 1 shows that sorption of ^{137}Cs in sand and clinoptilolite in presence of competing ions Na^+ was 99.91 % (the final concentration of ^{137}Cs in liquid phase was 0.032-0.053 Bq/mL).

Sorption of ^{137}Cs in clinoptilolite in presence of competing ions of stable Cs^+ was 99.97 % (the final radionuclide concentration in liquid phase was 0.032 Bq/mL). In that experiment concentration of stable Cs (0.13 M) corresponded to $4.3 \cdot 10^9$ Bq/L, which at least by $1 \cdot 10^5$ times more than concentration of radioactive ^{137}Cs ($1.05 \cdot 10^5$ Bq/L). Supposing that all Cs was radioactive and in respect that no difference in sorption of stable and radioactive Cs, it can be seen that sorption capacity of clinoptilolite is very high.

In presence of stable Cs sorption of ^{137}Cs in sand decreased to 28.6 % (the final concentration of ^{137}Cs in liquid phase was 74.7 Bq/mL) (Fig. 1).

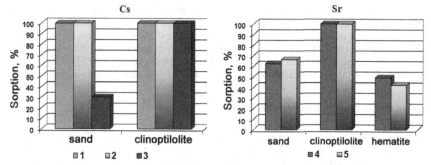

Figure1. Sorption of ^{137}Cs and ^{90}Sr in clinoptilolite, sand and hematite from aqueous solutions. (Liquid phase: 1 – distilled water, 2 – NaNO$_3$ (0.01 M), 3 – CsNO$_3$ (0.01 M), 4 – Ca(NO$_3$)$_2$ (0.01 M), 5 – Sr(NO$_3$)$_2$ (0.01 M).

Sorption of Sr^{90} in clinoptilolite was 99.94 % (the final concentration of Sr^{90} in liquid phase was 0.33 Bq/mL) and did not depend on the presence of competing ions in solution.

Sorption of Sr^{90} increased in line hematite<sand<clinoptilolite. Sorption of Sr^{90} in hematite amounted to 42-49 %. Sorption of Sr^{90} in sand varied from 62 to 66 % (the final radionuclide concentration in liquid phase was 271-320 and 160-217 Bq/mL respectively).

Mixtures of sand and clinoptilolite (20, 30, 50 wt %) characterized by a high sorption of Sr^{90} and ^{137}Cs (98.7 – 100 %) in the experimental conditions. Sorption did not depend on amount of clinoptilolite in a mixture and on presence of competing ions.

To study the sorption of ^{60}Co in sand and mineral additives the experiments were conducted in presence of stable Co(NO$_3$)$_2$ (0.01 M) and complexing agents in form of ethylenediaminetetraacetic acid (EDTA) (0.005 M). These complexing agents are conventional in liquid wastes arriving to SIA "Radon".

All analyzed materials, except hematite, adsorbed well ^{60}Co in form of cation. In that experiment liquid phase did not contain competing ions or complexing agents, what can be seen from figure 2.

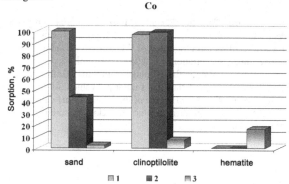

Figure 2. Sorption of ^{60}Co in clinoptilolite, sand and hematite from aqueous solutions. (Liquid phase: 1 – distilled water, 2 – Co(NO$_3$)$_2$ (0.01 M), 3 – EDTA (0.005 M)

219

Sorption of ^{60}Co decreased in presence of stable Co and complexing agents. Sorption of ^{60}Co in presence of stable Co amounted to 43.5 % in sand and 98.8 % in clinoptilolite (the final concentration of ^{60}Co in liquid phase was 13.06-13.56 and 0.268-0.306 Bq/mL respectively). Sorption of anionic Co-EDTA complexes in analyzed materials amounted to 2.82 – 16.8 % (for hematite). In this case the minimal concentration of ^{60}Co in liquid phase was 19.57-19.83 Bq/mL. When adding 20-50 wt % of clinoptilolite to sand, sorption of ^{60}Co slightly increased up to 10 %.

Sorption of [235]U and [239]Pu in backfill materials

In nature, uranium can be found in different forms, depending on environmental conditions [14]. Sorption of uranium was studied in its form of cation UO_2^{2+} (liquid phase - distilled water and 0.01 M $NaNO_3$) and in form of anionic complexes $[UO_2(CO_3)_3]^{4-}$ (liquid phase – 0.005 M Na_2CO_3). The results of these studies are shown in figure 3.

U

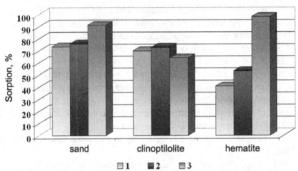

Figure 3. Sorption of [235]U in clinoptilolite, sand and hematite from aqueous solutions. (Liquid phase: 1 – distilled water, 2 – $NaNO_3$ (0.01 M), 3 – Na_2CO_3 (0.005 M))

In experimental conditions, sand adsorbed [235]U better than all the backfill materials (sorption varies from 74 to 92 %; final concentration in liquid phase was 0.15-0.73 Bq/mL respectively). Clinoptilolite adsorbed [235]U well regardless of its form. Sorption of [235]U in form of anionic complex in hematite amounted to 98 % (final concentration was 0.04-0.05 Bq/mL).

Sorption of [239]Pu in backfill materials was studied in the pH range from 8 to12. Infiltrated water near the concrete waste packages and barrier materials had similar pH values. All the backfill materials showed maximal sorption (≈ 92 %) at pH≈10. For the boundary pH values, sorption decreased by 1-6.5 %. Thus, it can be seen from fig. 4 what all backfill materials showed a very good sorption abilities to [239]Pu.

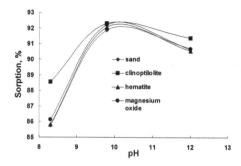

Figure 4. Sorption of [239]Pu in backfill materials in the pH range from 8 to 12

Waste which coming to SIA "Radon" for processing commonly contain small quantities such materials as paper, cardboard, plastic, rubber, wood, etc. Significant amounts of CO_2 released during degradation of these materials, results in decreasing pH and increasing solubility of actinides. In the light of that, it is proposed to use magnesium oxide as another mineral additive in backfill. It is known [15] that MgO sorbs good undesirable CO_2. Due to the buffer properties of magnesium oxide, it maintains alkaline pH, and can increase sorption and retention of Pu and other actinides.

Batch experiments were performed to evaluate sorption properties of sand-clinoptilolite-magnesium oxide mixtures with different ratios of components (Table II).

Table II. Ratio of components in mixtures

Number of mixtures	Ratio of components, wt %		
	sand	clinoptilolite	magnesium oxide
1	50	5	45
2	50	10	40
3	50	15	35
4	50	20	30
5	50	25	25
6	50	30	20
7	50	35	15
8	50	40	10
9	50	45	5

The mixtures had constant content of quartz sand and variable contents of clinoptilolite and magnesium oxide. Sorption of ^{137}Cs in these mixtures was 99.2-99.9 %, [90]Sr 87.6-99.8 % and [235]U 95.6-99.4 %. Sorption of [239]Pu in these mixtures was 59-64 %. The liquid phase with [239]Pu before the contact with mixtures had pH 2. After the contact with mixtures, pH increased to 8 – 10.5. This confirms the ability of mixtures with minimal content of magnesium oxide (5-10 wt %) to maintain alkaline pH and to increase sorption and retention of [239]Pu (Figure 5).

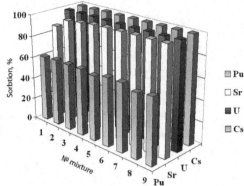

Figure 5. Sorption of radionuclides in sand-clinoptilolite-magnesium oxide mixtures

Thus, the introduction of magnesium oxide into mixtures did not decrease sorption of ^{137}Cs and ^{90}Sr, increased sorption of [235]U and stabilized sorption of [239]Pu at enough high level.

CONCLUSIONS

Sand of glaciolacustrine origin is offered as a main component of backfill for near surface repositories for solid or solidified low and intermediate level waste (LILW). The sand has optimal physical and mechanical properties. Using of quartz sand of local origin reduces cost of repository. Additives of clinoptilolite, hematite and magnesium oxide improve sorption properties of backfill mixtures.

ACKNOWLEDGMENTS

The authors would like to thank A.S.Barinov for scientific maintenance and assistance in incubation of this work.

REFERENCES

1. GOST P 52037-2003. *Near surface repositories for disposal of radioactive waste. General technical requirements. Edition of standards,* (Moscow, 2003).
2. NP-069-06. *Near surface disposal of radioactive waste. Safety requirements. Federal rules and regulations in the field of atomic energy,* (Moscow, 2006).
3. IAEA-TECDOC-1255, International Atomic Energy Agency, *Performance of engineered barrier materials in near surface disposal facilities for radioactive waste. Results of a co-ordinated research project,* (IAEA. Vienna, 2001), p.56.
4. O. Karnland, T. Sanden in *Long Term Test of Buffer Material at Aspo Hard Rock Laboratory, Sweden,* (Mater. Res. Symp. Proc. **608**, 2000) pp. 173-178.
5. F.P. Glasser in *Characterisation of the barrier performance of cements,* (Mater. Res. Symp. Proc. **713**, 2002) pp.721-732.

6. G.A. Varlakova, A.S. Barinov, E.E. Ostashkina, Z.I. Golubeva in *Selection of Backfill materials for near surface repository for low and intermediate Level Waste,* (Migration'11, Beiging, China, 2011) pp. 233-234.

7. G.A. Varlakova, Z.I. Golubeva, S.V. Roschagina et.al. in *Development of backfill mixtures for near surface repository for radioactive waste,* ("Radiochemistry-2009" Moscow, 2009) pp. 253-254.

8. G.A. Varlakova, E.E. Ostashkina, Z.I. Golubeva. et.al., RU Patent No. 2101200 (17 December 2010).

9. B.E. Serebryakov, E.A. Ivanov, A.P. Shchukin, Atomic Energy, **100**, 3, 220-225 (2006).

10. K.J. Cantrell, R.J. Serne, G.V. Last in *Hanford Contaminant Distribution Coefficient Database and Users Guide,* (PNNL-13895 Rev. 1, June 2003).

11. V.A. Kuznetsov, V.A. Generalova, Radiochemistry, **42**, 2, 154-157 (2000).

12. M.N. Sabodina, E.V. Zakharova, S.N. Kalmykov et.al., Radiochemistry, **50**, 1, 81-86 (2008).

13. S.A. Dmitriev, A.S. Barinov, G.A. Varlakova et al. in *Distribution and migration of radionuclides in host rock surrounding a shallow ground of vitrified radioactive waste,* (Migration'2005, Avignon, France, 2005) p. 201.

14. M.P. Gorbacheva, S.A. Kulyuhin, Radiochemistry, **51**, 3, 251-254 (2009).

15. J.L.Krumhansl, H.W. Papenguth, P.C. Zhang in *Scientific Basis for Nuclear Waste Management XXIV,* (Mater. Res. Symp. Proc. **608**, 2000). pp. 155–160.

Mater. Res. Soc. Symp. Proc. Vol. 1518 © 2012 Materials Research Society
DOI: 10.1557/opl.2012.1715

Some Experiments on Sorption Behavior of Iodide ions into CSH Gel under the Condition Saturated with Saline Groundwater

Yuichi Niibori, Taihei Funabashi and Hitoshi Mimura
Dept of Quantum Science and Energy Engineering, Tohoku University,
6-6-01-2, Aza-Aoba, Aramaki, Aoba-ku, Sendai, 980-8589 Japan.

ABSTRACT

The main hydrate of cement is calcium silicate hydrate (CSH). Such a cement-based material is essential for constructing the geological disposal system of TRU radioactive wastes including I-129 in Japan. So far, the sorption behavior of iodine on CSH gel has been examined by using the CSH samples dried once. However, the Japan's repository would be constructed under water table. Therefore, we must focus on also the interaction of altered cementitious material and iodine under the condition saturated with saline groundwater.

In this study, the sorption behavior of iodide ions into CSH gel, formed without dried processes, was examined in imitated saline groundwater. Ca/Si ratio was set to 0.4, 0.8, 1.2 and 1.6, and NaCl concentration of each sample also was set to 0.6 M, 0.06 M or 0.006 M. These samples were synthesized with CaO, SiO_2 (fumed silica), and distilled water in a given combination of 20 ml/g in liquid/solid ratio. A NaI solution was added after curing the CSH gel (hereinafter referred to as the "Surface sorption sample") for 7 days, setting the initial concentration of NaI to 0.5 mM in sample tube. The values of Eh and pH of each sample showed iodide ions as the chemical species of iodine in the sample tube. Furthermore, this study prepared the "Co-precipitation sample" of CSH gel with iodide ions. Here, the NaI solution was added before curing the CSH gel. For all samples, the contact time-period of the CSH gel with iodide ions was set to 7 days. After each contact time-period, each sample for analyses was separated into the solid and the liquid phases by 0.20 μm membrane filter. In the liquid phase, the concentrations of Ca, I, Si and Na ions in the liquid phase were measured by ICP-AES. Besides, the Raman spectra were obtained from the solid phases of the surface sorption sample and the co-precipitation sample without dried process.

The results showed that the sorption of iodide ions into CSH gel strongly depends on the amount of water included in the CSH gel. Such a sorption behavior was confirmed in both co-precipitation samples and the surface sorption samples, even if the Ca/Si ratio is low. This means that iodide ions can be easily immobilized through the water-molecular of CSH gel. Besides, Na concentration did not so much affect the sorption behavior of iodide ions into CSH gel. In addition, the Raman spectra showed that the degree of polymerization of SiO_4 tetrahedrons in CSH gel was unaffected with increasing Na ions concentration. These results suggest that the CSH gel saturated with groundwater would retard the migration of iodide ions, even if the groundwater includes salinity.

INTRODUCTION

It is well-known that cement material used in the construction of the geological disposal system would alter the groundwater up to 13 in pH in the surroundings of the repository (e.g., [1-

4]). Such a highly alkaline groundwater may produce calcium-silicate-hydrate (CSH) with relatively low Ca/Si mol-ratio as a secondary mineral. Narita et al. [4] have pointed out that the dried process limits the flexibility of CSH structure, strongly affecting the sorption behavior of the surrounding ions. Since as in many countries the repository system in Japan is saturated with groundwater, we must examine the sorption behavior of CSH gels also without dried processes. Besides, [129]I is a key nuclide in the performance assessment of the repository system of transuranium (TRU) wastes eliminated mainly from the reprocessing facility. Iodine undergoes iodide ions under such a reducing condition saturated with high pH groundwater. This study examined the sorption behavior of iodide ions into CSH gel formed without dried processes, using imitated saline groundwater (considered potential sites for disposal along the coastal areas).

EXPERIMENT

Samples

Table I shows a given combination to synthesize each sample with CaO, SiO_2, and distilled water. The liquid/solid weight ratio was uniformly set to 20 ml/g. For observing sorption behavior of iodine onto CSH, the two types of CSH samples were prepared by the following procedures: (1) a NaI solution is added to adjust the sample to 0.5 mM in initial [I⁻] after curing the CSH gel (hereinafter referred to as the "Surface Sorption sample"), (2) a NaI solution is added to adjust the sample to 0.5 mM in initial [I⁻] before curing the CSH gel (hereinafter referred to as the "Co-Precipitation sample"). Furthermore, this study prepared CSH samples with the Ca/Si molar ratio set to 0.4, 0.8, 1.2, and 1.6. NaCl concentration of each sample also was set to 0.6 M, 0.06 M or 0.006 M. SiO_2 (fumed silica, AEROSIL 300) was obtained from Japan AEROSIL Ltd. The specific surface area of BET (N_2 gas) was 300±30 m²/g. The other chemicals were obtained from Wako Pure Chemical Industries Ltd., and were used without further purification.

Table I. Materials used to synthesize CSH gel.

Ca/Si molar ratio	CaO / g	SiO_2 / g	distilled water / ml
0.4	0.408	1.092	30.0
0.8	0.641	0.859	30.0
1.2	0.792	0.708	30.0
1.6	0.898	0.602	30.0

Procedures

Figure 1 shows the details of the experimental procedures in which the sample does not undergo dried processes. These main processes were based on the procedures already reported by the authors [5, 6]. The synthesis of each CSH gel sample was conducted in a glove box saturated with nitrogen gas, in order to avoid contact with air. The temperature was kept constant at 298 K. "The surface sorption sample" was cured for 7 days in the sample tube (sealed and gently shaken with 120 strokes/min) before doping the iodine solution. The curing time-period was determined so that both concentrations of Ca and Si reached equilibrium in the solution. After 7 days for curing "the surface sorption sample", a NaI solution was added into the tube containing "the surface sorption sample" of CSH gel. The contact time-period was set to 7 days. On the other hand, for the "co-precipitation sample" already containing NaI in the synthesis process of CSH gel, its curing time-period (the contacting time) was consistently set to 7 days. After each contact

226

time-period, the samples for analyses were centrifuged with 7,500 rpm during 10 min. The liquid phase was filtrated by 0.20 μm membrane filter. Then, the concentrations of iodine, Ca, Na and Si were also measured by inductively-coupled plasma atomic emission spectrometry (ICP-AES). Using the pH-Eh diagram considering Na ions concentration, it was confirmed that iodine in each sample exists mainly as iodide ions. The Eh values measured in each sample were in the range of +159 mV to +368 mV and the values of pH were in the range of 9.7 to 12.6. After each contact time-period, the solid phase was observed by Raman spectroscopy (JASCO, NRS-3000, the irradiation with light of 532 nm wavelength) without dried process. Additionally, X-ray diffraction pattern was examined by using each vacuum-dried solid phase.

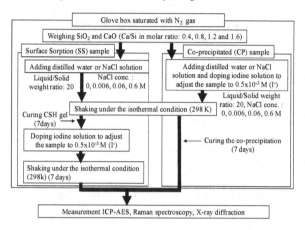

Figure 1. Procedures of the experiment.

RESULTS AND DISCUSSION

Figure 2 shows both the amount of iodine in the solution and the volume of liquid phase filtrated by 0.20 μm membrane filter in each co-precipitated sample after the contacting time. In Fig. 2(a), the difference from the initial value means the sorption amount of iodine. In the same way, Fig. 2(b) shows the water volume used to form CSH gel or adsorbed to CSH gel. While the concentration of I⁻ in the liquid phase did not decrease so much compared to the initial concentration, these results suggest that the hydrated water to form CSH gel plays a role to stabilize I⁻ ions. However, this behavior of CSH gel may be limited to co-precipitated samples.

Figure 3 shows the results of "surface sorption samples". As shown in this figure, even if a NaI solution is added after curing CSH gel, the CSH gel contributes the stabilization of I⁻. Noshita et al. [7] reported by using dried samples that iodine is sorbed on CSH by electrostatic adsorption reaction. Also in the CSH gel without dried processes, an electrostatic reaction such as ≡SiOCa⁺ + I⁻ → ≡SiOCaI might occur in the CSH gel with iodide ions through the water-molecular interlayer formed in the hydrate [8]. Besides, as shown in Figs. 2 and 3, the Ca/Si molar ratio did not strongly affect the sorption behavior of iodide ions, while Na ions slightly increased the sorption amount. This suggests that the silicate chain of CSH (≡SiO⁻) directly influences the sorption behavior.

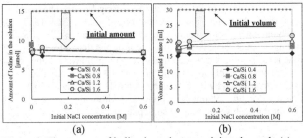

(a)   (b)

Figure 2. The amount of iodine in each co-precipitated sample (a) and its volume of liquid phase (b).

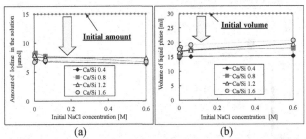

(a)   (b)

Figure 3. The amount of iodine in each surface-sorption-sample (a) and its volume of liquid phase (b).

Figure 4 shows the Ca/Si molar ratio of CSH gel measured after 7 days contacting with I⁻. Here, each ratio was evaluated by using the Ca and Si concentrations in the liquid phase measured after 7 days. While the Ca/Si ratio slightly decreased with increment of initial Na ions concentration, the initially set Ca/Si ratio was maintained. Besides, it was confirmed that Na ions apparently decreased, particularly when the Ca/Si ratio was relatively small. Na ions contribute to stabilize the form of CSH with relatively small Ca/Si <0.8 [8]. Furthermore, Si concentration remarkably increased when Ca/Si ratio was 0.4. In general, such a CSH gel tends to raise the Ca/Si ratio by releasing Si to the liquid phase, while Na ions as in 0.6 M decrease the solubility of Si because water activity decreased due to the hydration of Na ions [9].

Figure 5 shows the Raman spectra of co-precipitated samples. Here, the peaks of 663 cm⁻¹, 875 cm⁻¹ 1019 cm⁻¹ and 445 cm⁻¹ correspond to the Q^2 symmetrical bending band (SB $Q^{2)}$), the Q^1 symmetrical stretching bands (SS Q^1), the Q^2 symmetrical stretching bands (SS Q^2) and the symmetric stretching v2[SiO₂], respectively. As shown in Fig. 5 (a), the peak of 875 cm⁻¹ perfectly disappeared in the CSH gel samples of relatively small Ca/Si ratio. This means that such a CSH gel would undergo the polymerization of silicate chain structures. On the other hand, relatively large Ca/Si ratio samples clearly showed the peak of 875 cm⁻¹, indicating that the degree of the polymerization was relatively small.

To compare these spectra, this study focused on the intensity ratio of SS Q^1 and SS Q^2 (shown in Fig. 5(b)). Figure 6 is the intensity ratio of SS Q^1 to SS Q^2. The result shows that CSH gel structure does not change so much due to sodium or iodide ions, while Borrmann et al. [10]

assumed nano-structured calcium silicate hydrate functionalized with iodine. Furthermore, it was confirmed that the intensity ratios both of the co-precipitated sample and the surface sorption sample are almost similar in the dependencies on the Ca/Si molar ratio and Na ions concentration. Furthermore, the X-ray diffraction patterns of the CSH gel samples (dried) clearly showed the structure of calcium-silicate hydrate as in Tobermorite and Jennnite [11], when the samples of Ca/Si exceed 0.4.

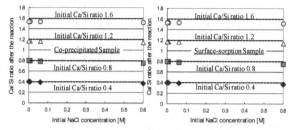

Figure 4. Ca/Si molar ratio after 7 days contacting with iodide ions ((a): in co-precipitated (CP) samples, (b): in surface sorption (SS) samples).

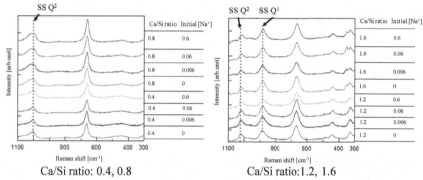

Ca/Si ratio: 0.4, 0.8   Ca/Si ratio:1.2, 1.6

Figure 5. Raman spectra of co-precipitated samples.

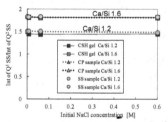

Figure 6. The intensity ratio of Q^1/Q^2 to initial NaCl concentration. (CP sample: co-precipitated, SS sample: surface sorption sample)

CONCLUSIONS

In this study, the sorption behavior of iodide ions into CSH gel, formed without dried processes, was examined in imitated saline groundwater. Also considering CSH deposited as a secondary mineral around the repository, this study adjusted Ca/Si molar ratio to 0.4, 0.8, 1.2 and 1.6. As the results, even if the Ca/Si ratio is relatively small (<1.0), the iodide ions stabilized both in the co-precipitated sample and the surface-sorption sample. That is, the hydrated water might play a role to incorporate anion such as iodide ions through water-molecular interlayer of the CSH structure under the condition saturated with groundwater (not dried condition). In the Raman spectra, the characteristic peaks of CSH gel, such as the Q_1 symmetrical stretching bands, the Q_2 symmetrical bending band, were confirmed in the samples of 1.2 and 1.6 in Ca/Si molar ratio even if in the condition after adding the iodine solution into the CSH samples. Furthermore, when the samples of Ca/Si exceeded 0.4, the X-ray diffraction patterns of any CSH gel samples (dried) clearly showed the CSH structure as in Tobermorite and Jennnite. These suggest that CSH gel incorporates the iodide ions into the CSH gel, of which structure is not changed so much by the sorption of Na and/or I ions. Since the use of cementitious materials to construct the repository alters the surrounding groundwater up to >10 in pH for a long time-period such as 10^4 years [1, 2], it is expected that a secondary mineral such as CSH with a low Ca/Si molar ratio also retard the migration of radionuclides. To understand more details of the mechanism under the underground condition limiting the amount of water compared to the volume of solid phase, we need to examine also the influences of the liquid/solid ratio on the sorption behavior of iodide ions to CSH gel by considering both the hydrated water and the adsorption water in CSH gel.

ACKNOWLEDGMENTS

This study was supported by Japan Society for the Promotion of Science, Grant-in-Aid for Scientific Research (B) No. 21360460.

REFERENCES

1. A. Atkinson, AERE-R 11777, UKAEA (1985).
2. FEPC (Federation of Electric Power Companies of Japan) and JNC (Japan Nuclear Cycle development institute), JNC TY1400 2005-013, FEPC TRU-TR2-2005-02 (2005).
3. T. Chida, Y. Niibori, K. Tanaka and O. Tochiyama, Applied Geochemistry 22, 2810 (2007).
4. M. Narita, Y. Niibori, H. Mimura, A. Kirishima, J. Ahn, Proc. of WM2010 Conference, Paper No. 10096 (2010).
5. K. Shirai, Y. Niibori, A. Kirishima, H. Mimura, Proc. of ASME 13th ICEM, Paper No. 40089 (2010).
6. Funabashi et al., Proc. of WM2012 Conference, Paper No. 12145 (2012).
7. K.Noshita et al. in Scientific Basis for Nuclear Waste Management XXIV, edited by K. P. Hart, and G. R. Lumpin, (Mater. Res. Soc. Symp. Proc., 663, Pittsburgh, PA, 2001), pp. 115-123.
8. I.G. Richardson, Cement and Concrete Research, 38, 137 (2008).
9. R. O. Fournier and W. L. Marshall, Geochimica et Cosmochimica Acta, 47, 587 (1983).
10. T. Borrmann et al., Journal of Colloid and Interface Science, 339, 175 (2009).
11. R. J. Kirkpatrick et al., Advanced Cement Based Materials, 5(3), 93 (1997).

Mater. Res. Soc. Symp. Proc. Vol. 1518 © 2013 Materials Research Society
DOI: 10.1557/opl.2013.74

Sorption Behavior of Nickel and Palladium in the Presence of $NH_3(aq)/NH_4^+$

Taishi Kobayashi[1], Takayuki Sasaki[1], Ken-you Ueda[1] and Akira Kitamura[2]
[1]Graduate School of Engineering, Nuclear Engineering, Kyoto University, Yoshida-honmahi, Sakyo-ku, Kyoto, Japan
[2]Radionuclide Migration Research Group, Geological Isolation Research and Development Directorate, Japan Atomic Energy Agency, Muramatsu, Tokai, Ibaraki, Japan

ABSTRACT

It is necessary to assess the impact of nitrate salts and their reduction products (e.g. $NH_3(aq)/NH_4^+$) contained in low-level radioactive waste generated from nuclear reprocessing process for the safety assessment of geological disposal of the waste. In the present study, sorption behavior of Ni and Pd on pumice tuff was investigated in the presence of $NH_3(aq)/NH_4^+$. Under various $NH_3(aq)/NH_4^+$ concentration, pH and ionic strength conditions, distribution coefficient (K_d) of Ni and Pd on pumice tuff was determined by a batch experiment. For Ni system, the K_d values showed no significant dependence on initial NH_4^+ concentration ($[NH_4^+]_{ini}$ < 1 M) in neutral pH region, which agreed with the prediction from thermodynamic data. For Pd system, the K_d values decreased with an increase of $[NH_4^+]_{ini}$, suggesting the formation of stable ammine complexes ($Pd(NH_3)_m^{2+}$ (m: 1 – 4)). The obtained K_d values for Ni and Pd were analyzed using a surface complexation model. By taking complexes predicted by thermodynamic data into account, sorption behavior of Ni and Pd in the presence of $NH_3(aq)/NH_4^+$ were well explained.

INTRODUCTION

Some of low-level waste generated from nuclear reprocessing process contain considerable amount of nitrate salts which may affect the migration behavior of several radionuclides. Nitrate ion (NO_3^-) and its reduced components, ammonia ($NH_3(aq)$) and ammonium ion (NH_4^+) may form stable complex with radionuclides. Since nickel (Ni-63) as radioactivated product and palladium (Pd-107) as fission product potentially form stable ammine complexes [1], it is important to evaluate the impact of $NH_3(aq)/NH_4^+$ on the migration behavior of Ni and Pd under relevant conditions. The sorption behavior of Ni and Pd, thus, may be influenced by the formation of ammine complexes. In the present study, we focused on the sorption behavior of Ni and Pd on pumice tuff in the presence of $NH_3(aq)/NH_4^+$. Distribution coefficient (K_d) value for Ni and Pd on pumice tuff was determined by a batch experiment at various pH, ionic strength, and initial NH_4^+ concentrations and the obtained K_d values were analyzed using a surface complex model and values based on the thermodynamic data in order to understand and predict the sorption behavior in the presence of $NH_3(aq)/NH_4^+$.

EXPERIMENT

A batch sorption experiment was carried out to determine the K_d ($m^3 \cdot kg^{-1}$) value of Ni and Pd on pumice tuff. The pumice tuff was purchased from Nichika Inc. and ground into a powder using a stainless steel mortar and pestle. The powder with a particle size of $63 - 125$ μm was then obtained by sieving. The chemical and mineralogical characterization of the pumice tuff was performed in the previous study [2]. Stock solutions of non-radioactive Ni and Pd were prepared from the 1,000 ppm standard solutions (Wako, Japan). Aliquots of Ni, or Pd stock solution, sodium and ammonium perchlorate solutions and 300 mg of pumice tuff powder were introduced to a sample tube, and pH was adjusted to 6, 8, 11 and 12 with dilute NaOH solution. The amount of each sample solution was 30 mL and the initial Ni and Pd concentration ([M]$_{ini}$, M = Ni, Pd) was set to 10^{-6} mol/dm^3 (M). The NH_4ClO_4 concentration was set from 0 to 1.0 M and ionic strength was adjusted from 0.01 to 1.0 by $NaClO_4$. All the chemicals used were of reagent grade. The sample tube was kept in an Ar-filled glove box at $25 \pm 1°C$ for given periods, and mildly shaken by hand at intervals. After contacting a given period, the hydrogen ion concentration exponent (pH$_c \equiv -\log$ [H$^+$]) of each sample solution was measured using combination glass electrodes (9611-10D, Horiba). The internal solution of the electrode constituted 3.6 M NaCl + 0.4 M $NaClO_4$ instead of saturated KCl solution, to prevent precipitation of $KClO_4$ at the junction between the electrode and the sample solution and the electrode was calibrated against pH buffers (pH 4, 7, Horiba). Part of the sample solution (1 dm^3) was filtrated through a syringe filter (0.45 μm pore size, Advantec) and ultrafilters (3 and 10 kDa; ca. 2 and 3 nm pore size, respectively, Millipore). The filtrate was diluted into 0.1 M HNO_3 and the Ni and Pd concentrations ([M]$_{liquid}$) were determined by ICP-MS (ELAN DRC2, PerkinElmer). The detection limit was around 10^{-8} M for both Ni and Pd. The K_d value was obtained by the equation; $K_d = ([M]_{ini} - [M]_{liquid}) / [M]_{liquid}$.

DISCUSSION

Sorption behavior of Ni and Pd

Figure 1 shows the pH$_c$ dependence of log K_d value for Ni on pumice tuff at different $NaClO_4$ and NH_4ClO_4 concentrations, obtained after 10 kDa filtration. Since there was no significant difference of the data with the aging time of 4 weeks from those with more than 4 weeks, it was considered that the equilibrium state was achieved after 4 weeks. The log K_d values determined after 3 kDa and 0.45 μm filtrations showed similar to those after 10 kDa filtration, indicating no significant effect of colloids. The log K_d values were almost constant under neutral pH condition (pH = 6 and 8). In alkaline solution, it was considered that Ni precipitates as $Ni(OH)_2(s)$ at $[Ni]_{ini} = 10^{-6}$ M from the thermodynamic data [1], and the K_d value was not reliable and therefore removed in this study. At $[NH_4^+]_{ini} < 1$ M in neutral pH region, no significant dependence of the log K_d value on $[NH_4^+]_{ini}$ was found. This agrees with the prediction from the thermodynamic data [1], where Ni^{2+} is considered to be as a dominant species. At $[NH_4^+]_{ini} > 1$ M, $Ni(NH_3)_m^{2+}$ was considered to be a dominant species [1], and the decrease of log K_d value suggests less sorption ability of the $Ni(NH_3)_m^{2+}$.

Figure 2 shows the pH$_c$ dependence of log K_d value for Pd on pumice tuff at different $NaClO_4$ and NH_4ClO_4 concentrations after 10 kDa filtration. Similar to the Ni system, equilibrium state was achieved after 4 weeks. No significant effect of colloids was found through various filtrations. The log K_d value slightly decreases with increasing pH$_c$ and $[NH_4^+]_{ini}$. From the thermodynamic data [1], the dominant species were considered to be $Pd(OH)_2(aq)$ and

Pd(OH)$_3^-$ at low [NH$_4^+$]$_{ini}$, and Pd(NH$_3$)$_m^{2+}$ at [NH$_4^+$]$_{ini}$ > 10^{-4} M. In the absence of NH$_4^+$, the log K_d values for Pd are about one order of magnitude higher than those for Ni, suggesting that Pd(OH)$_2$(aq) has larger sorption capacity than Ni^{2+}.

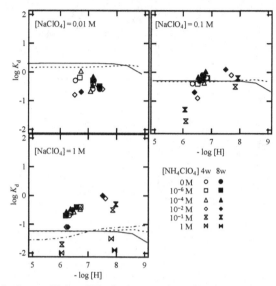

Figure 1. Distribution coefficient (K_d) of Ni on pumice tuff in the presence of NH$_3$(aq)/NH$_4^+$ obtained after 4 and 8 weeks. The solid, dotted and broken curves represents the calculated values for [NH$_4^+$]$_{ini}$ = 10^{-6}, 10^{-2}, and 1 M, respectively.

<u>**Surface complex model for Ni and Pd**</u>

A surface complexation model has been adopted to describe the apparent sorption behavior at the anionic sorption reaction sites ($\equiv$S-O$^-$), depending on pH, ionic strength, and the concentration of sorption species [3, 4]. In our previous study, the model has been applied to understand the sorption behavior of Cs onto pumice tuff, and the behavior was well explained by assuming one sorption site [2]. In the present study, we apply the model to describe the sorption behavior of Ni and Pd onto pumice tuff. Assuming one major sorption site, the reactions relating the sorption behavior of Ni can be expressed by the following equations:

$$S\text{-}OH + H^+ \Leftrightarrow S\text{-}OH_2^+ \qquad K_1 = \frac{[S-OH_2^+]}{[S-OH][H^+]} \qquad (1)$$

$$S\text{-}OH \Leftrightarrow S\text{-}O^- + H^+ \qquad K_2 = \frac{[S-O^-][H^+]}{[S-OH]} \qquad (2)$$

$$NH_4^+ \Leftrightarrow NH_3(aq) + H^+ \qquad K = \frac{[NH_3(aq)][H^+]}{[NH_4^+]} \qquad (3)$$

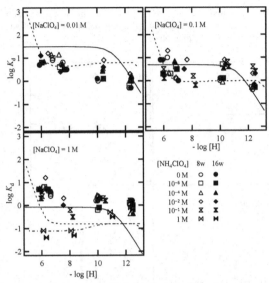

Figure 2. Distribution coefficient (K_d) of Pd on pumice tuff in the presence of $NH_3(aq)/NH_4^+$ obtained after 8 and 16 weeks. The solid, dotted and broken curves represents the calculated values for $[NH_4^+]_{ini} = 10^{-6}$, 10^{-2}, and 1 M.

$$Ni^{2+} + n\,OH^- \Leftrightarrow Ni(OH)_n^{(2-n)+} \qquad \beta_{1,n,0} = \frac{[Ni(OH)_n^{(2-n)+}]}{[Ni^{2+}][OH^-]^n} \quad (n = 1-4) \tag{4}$$

$$Ni^{2+} + m\,NH_3(aq) \Leftrightarrow Ni(NH_3)_m^{2+} \qquad \beta_{1,0,m} = \frac{[Ni(NH_3)_m^{2+}]}{[Ni^{2+}][NH_3(aq)]^m} \quad (m = 1-4) \tag{5}$$

$$S\text{-}O^- + Ni^{2+} \Leftrightarrow S\text{-}ONi^+ \qquad K_3 = \frac{[S\text{-}ONi^{2+}]}{[S\text{-}O^-][Ni^{2+}]} \tag{6}$$

$$S\text{-}O^- + Ni(OH)_n^{(2-n)+} \Leftrightarrow S\text{-}ONi(OH)_n^{(1-n)+}$$

$$K_{(n+3)} = \frac{[S\text{-}ONi(OH)_n^{(1-n)+}]}{[S\text{-}O^-][Ni(OH)_n^{(2-n)+}]} \quad (n = 1, 2) \tag{7}$$

$$S\text{-}O^- + Ni(NH_3)_m^{2+} \Leftrightarrow S\text{-}ONi(NH_3)_m^+$$

$$K_{(m+5)} = \frac{[S\text{-}ONi(NH_3)_m^+]}{[S\text{-}O^-][Ni(NH_3)_m^{2+}]} \quad (m = 1-4) \tag{8}$$

$$S\text{-}O^- + Na^+ \Leftrightarrow S\text{-}ONa^0 \qquad K_{10} = \frac{[S\text{-}ONa^0]}{[S\text{-}O^-][Na^+]} \tag{9}$$

$$S\text{-}O^- + NH_4^+ \Leftrightarrow S\text{-}ONH_4^0 \qquad K_{11} = \frac{[S\text{-}ONH_4^0]}{[S\text{-}O^-][NH_4^+]} \ . \tag{10}$$

The number of surface reaction sites (n_s) is given by,

$$n_s = \{[S-OH^0]+[S-OH_2^+]+[S-O^-]+[S-ONi^+]+[S-ONiOH^0]+[S-ONi(OH)_2^-]$$
$$+[S-ONiNH_3^+]+[S-ONi(NH_3)_2^+]+[S-ONi(NH_3)_3^+]+[S-ONi(NH_3)_4^+] \qquad (11)$$
$$+[S-ONa^0]+[S-ONH_4^0]\} \times \frac{V}{1000\,W}$$

where V is the volume (dm^3) of the solution and W is the weight (g) of the solid sample. Total nickel concentration ($[Ni]_0$), representing $[Ni]_{ini}$ in this study, and the K_d are defined as

$$[Ni^{2+}]_0 = [S-ONi^+]+[S-ONiOH^0]+[S-ONi(OH)_2^-]+[S-ONiNH_3^+]$$
$$+[S-ONi(NH_3)_2^+]+[S-ONi(NH_3)_3^+]+[S-ONi(NH_3)_4^+]$$
$$+[Ni^{2+}]+[NiOH^+]+[Ni(OH)_2(aq)]+[Ni(OH)_3^-]+[Ni(OH)_4^{2-}] \qquad (12)$$
$$+[NiNH_3^{2+}]+[Ni(NH_3)_2^{2+}]+[Ni(NH_3)_3^{2+}]+[Ni(NH_3)_4^{2+}]$$

$$K_d = \frac{\left\{\begin{array}{l}[S-ONi^+]+[S-ONiOH^0]+[S-ONi(OH)_2^-]+[S-ONiNH_3^+] \\ +[S-ONi(NH_3)_2^+]+[S-ONi(NH_3)_3^+]+[S-ONi(NH_3)_4^+]\end{array}\right\}}{\left\{\begin{array}{l}[Ni^{2+}]+[NiOH^+]+[Ni(OH)_2(aq)]+[Ni(OH)_3^-]+[Ni(OH)_4^{2-}] \\ +[NiNH_3^{2+}]+[Ni(NH_3)_2^{2+}]+[Ni(NH_3)_3^{2+}]+[Ni(NH_3)_4^{2+}]\end{array}\right\}} \cdot \qquad (13)$$

From the Eqs. (1)-(13), the K_d value for Ni is described as a function of pH_c, $[NH_4^+]_{ini}$, and $[Na^+]$. The K_d value for Pd was expressed by the same type equations. The values in Figs. 1 and 2 were analyzed in a least-square fitting by treating $\log K_l$ (l = 1-11) as model parameters. **Table 1** shows the obtained fitting parameters, together with fixed parameters. The n_s, $\log K_1$, $\log K_2$, $\log K_{10}$ and $\log K_{11}$ were taken from the previous study [2] and treated as fixed parameters.

The calculated curves of $\log K_d$ for $[NH_4^+]_{ini} = 10^{-6}$, 10^{-2}, and 1 M were drawn in Figs. 1 and 2 as solid, dotted and broken lines, respectively, using the obatained parameters in Table 1. For Ni system, the experimental values are reproduced at $I = 0.1$, however, not well fitted at $I = 0.01$ and 1.0. This might be caused by the fixed parameters of $\log K_{10}$ and $\log K_{11}$ [2], which represent the dependence of $\log K_d$ on ionic strength. For Pd system, the dominant solution species was $Pd(NH_3)_m^{2+}$ in neutral pH region at $[NH_4^+]_{ini} > 10^{-4}$ M, large K_d values under such conditions indicate strong sorption of $Pd(NH_3)_m^{2+}$ onto the surface of pumice tuff.

CONCLUSIONS

We determined the $\log K_d$ values of Ni and Pd on pumice tuff under different $NH_3(aq)/NH_4^+$ concentration, pH_c and ionic strength conditions by a batch sorption experiment. For Ni system, the K_d values showed no significant dependence on the initial NH_4^+ concentration in neutral pH region at $[NH_4^+]_{ini} < 1$ M. For Pd system, the K_d values decreased with an increase of $[NH_4^+]_{ini}$, suggesting the formation of stable ammine complex ($Pd(NH_3)_m^{2+}$) at $[NH_4^+]_{ini} > 10^{-4}$ M. The obtained K_d values for Ni and Pd were explained by a surface complexation model by taking these complexes into account.

235

Table 1 Parameters for Ni, Pd and pumice tuff fitted by surface complexation model. Values with bold type were determined in the analysis.

Reaction	Parameter	Ni	Pd	Ref.
	n_s	1×10^{-6}	1×10^{-6}	[2]
$S\text{-}OH + H^+ \Leftrightarrow S\text{-}OH_2^+$	$\log K_1$	2	2	[2]
$S\text{-}OH \Leftrightarrow S\text{-}O^- + H^+$	$\log K_2$	-3.65	-3.65	[2]
$S\text{-}O^- + M^{2+} \Leftrightarrow S\text{-}OM^+$	$\log K_3$	**2.50 ± 0.13**	–	
$S\text{-}O^- + MOH^+ \Leftrightarrow S\text{-}OMOH^+$	$\log K_4$	–	–	
$S\text{-}O^- + M(OH)_2(aq) \Leftrightarrow S\text{-}OM(OH)_2^0$	$\log K_5$	–	**3.69 ± 0.28**	
$S\text{-}O^- + MNH_3^{2+} \Leftrightarrow S\text{-}OMNH_3^+$	$\log K_6$	**2.91 ± 0.46**	**9.61 ± 2.76**	
$S\text{-}O^- + M(NH_3)_2^{2+} \Leftrightarrow S\text{-}OM(NH_3)_2^+$	$\log K_7$	2.91	**7.60 ± 0.53**	
$S\text{-}O^- + M(NH_3)_3^{2+} \Leftrightarrow S\text{-}OM(NH_3)_3^+$	$\log K_8$	2.91		
$S\text{-}O^- + M(NH_3)_4^{2+} \Leftrightarrow S\text{-}OM(NH_3)_4^+$	$\log K_9$	2.91	**3.02 ± 0.15**	
$S\text{-}O^- + Na^+ \Leftrightarrow S\text{-}ONa^0$	$\log K_{10}$	1.79	1.79	[2]
$S\text{-}O^- + NH_4^+ \Leftrightarrow S\text{-}ONH_4^0$	$\log K_{11}$	1.79	1.79	[2]
$M^{2+} + OH^- \Leftrightarrow MOH^+$	$\log \beta_{1,1,0}$	4.5	13.35	[1]
$M^{2+} + 2\,OH^- \Leftrightarrow M(OH)_2(aq)$	$\log \beta_{1,2,0}$	10.0	24.89	[1]
$M^{2+} + 3\,OH^- \Leftrightarrow M(OH)_3^-$	$\log \beta_{1,3,0}$	12.3	27.8	[1]
$M^{2+} + 4\,OH^- \Leftrightarrow M(OH)_4^{2-}$	$\log \beta_{1,4,0}$	11.0	–	[1]
$M^{2+} + NH_3(aq) \Leftrightarrow MNH_3^{2+}$	$\log \beta_{1,0,1}$	2.23	9.61	[1]
$M^{2+} + 2\,NH_3(aq) \Leftrightarrow M(NH_3)_2^{2+}$	$\log \beta_{1,0,2}$	3.95	18.51	[1]
$M^{2+} + 3\,NH_3(aq) \Leftrightarrow M(NH_3)_3^{2+}$	$\log \beta_{1,0,3}$	5.46	26.02	[1]
$M^{2+} + 4\,NH_3(aq) \Leftrightarrow M(NH_3)_4^{2+}$	$\log \beta_{1,0,4}$	6.47	32.83	[1]

ACKNOWLEDGMENTS

This work was performed in the Project on "Combined development of nitrate salt removal technology and an assessment system for the impact of nitrate on the co-locational disposal of TRU waste and HLW (FY 2011)" funded by Agency for Natural Resources and Energy, Ministry of Economy, Trade and Industry of Japan.

REFERENCES

[1] A. Kitamura, K. Fujiwara, R. Doi, Y. Yoshida, M. Mihara, M. Terashima and M. Yui, JAEA-Data/Code 2009-024 (2010).
[2] M. Rajib, T. Sasaki, T. Kobayashi, Y. Miyauchi, I. Takagi, H. Moriyama, J. Sci. Nucl. Technol. **48**, 950 (2011).
[3] R. J. Silva, L. V. Benson, A. W. Yee, G. A. Parks, "Theoretical and experimental evaluation of waste transport in selected rocks," in Waste Isolation Safety Assessment Program Task 4: Collection and Generation of Transport Data, PNL-SA8571, Lawrence Berkeley Laboratory (1979).
[4] D. A. Dzombak, F. M. M. Morel, Surface Complexation Modeling; Hydrous Ferric Oxide, John Wiley, New York (1990).

Mater. Res. Soc. Symp. Proc. Vol. 1518 © 2013 Materials Research Society
DOI: 10.1557/opl.2013.89

Sensitivity Analysis for the Scenarios on Deterioration or Loss of Safety Functions Expected in Disposal System Due to Human Error on Application of Engineering Technology

Seiji Takeda[1], Yoshihisa Inoue[1] and Hideo Kimura[1]
[1] Nuclear Safety Research Center, Japan Atomic Energy Agency (JAEA), Tokai, Ibaraki, 319-1195 Japan

ABSTRACT

The sensitive analysis of radionuclide migration for the scenarios on deterioration or loss of safety functions expected in HLW disposal system due to the human error (initial defective scenarios) is performed in this study. Release rates for Cs-135 and Se-79 are estimated from Monte Carlo-based analysis. Maximum release rates of Se-79 and Cs-135 from natural barrier in initial defective scenarios for vitrified waste and overpack are approximately equivalent to that in normal scenario on all safety function working. Maximum release rate of Se-79 in initial defective scenario of buffer under the condition of colloidal migration is about 30 times as high as that in normal scenario. Maximum release rate of Cs-135 in initial defective scenario of plugs is about two orders of magnitude higher than that in normal scenario. These results especially indicate the need to understand the feasibility on two types of initial defective scenario, leading to the loss of restraint for colloidal migration in buffer and the loss of restraint with plugs from short-circuit migration.

INTRODUCTION

Nuclear Safety Research Center of Japan Atomic Energy Agency has carried out regulatory research to provide technical support for safety review of the license application for candidate sites of final disposal of high-level radioactive waste (HLW), in which Nuclear Waste Management Organization of Japan (NUMO) as an implementer will propose through the stepwise setting process. NUMO will concretely provide a repository design and engineering technology of repository construction tailored to a specific site. From a viewpoint of safety regulation, it is important to review suitably the progress of engineering technical development applied for stepwise implementation of safety geological disposal towards a future safety review and to grasp the safety function of the engineered barrier materials, namely vitrified waste, overpack, sand-bentonite buffer material and sealing plugs, affected by various factors. The human error on application of engineering technology is one of important factors.

The sensitive analysis of radionuclide migration for the scenarios on deterioration or loss of expected safety functions of the engineered barrier materials due to the human error on the application of engineering technology (hereinafter referred to as "initial defective scenarios") is performed in this study. The analysis specifies the initial defective scenario, which should be especially focused toward future safety review.

ANALYTICAL METHOD

Scenario description

In this analysis, disposal site is not specific, and geological disposal system is basically the same design as in the H12 Project report [1] of Japan Nuclear Cycle Development Institute. The repository is constructed in stable granitic bedrock at a depth of 1,000 m. The rock is described in terms of two major hydraulic units; fractured zone and rock mass. It is reasonable to assume that the repository is constructed in a stable rock mass having enough distances from fractured zones, in order to avoid the occurrence of a short path of groundwater from the repository to the biosphere. The engineered barrier system is composed of borosilicate vitrified waste form, carbon steel overpack and bentonitic buffer material (a mixture of 70 wt% of bentonite and 30 wt% of sand). Cementitious materials are used for the grouting and support of tunnel. Sealing plugs of a sand-bentonite is applied to high permeability area of tunnels with the excavation disturbed zone.

Two types of initial defective scenarios for vitrified waste are selected. The first scenario is that the crack of vitrified waste increases due to the external force to the waste caused by falling in handling process, and resulting enhanced glass dissolution rate. The second scenario assumes that the formation of molybdenum oxides or molybdates in vitrified waste, known as yellow phase [2], results from the human error of the fluid adjustment work in a vitrified-waste manufacturing. Additionally, high glass dissolution rate in yellow phase and the formation of oxidizing condition in engineered barrier due to molybdenum oxides are considered in this initial defective scenario.

In initial defective scenario of overpack, the human errors such as defect of quality management in manufacturing, transformation of overpack caused by falling in handling process, tiny crack of its surface and remaining stress occurred during welding operation have an effect on the corrosion progress of overpack. It is conservatively assumed that the function of no infiltration of groundwater inside the overpack loses immediately after post-closure of disposal site. Early high temperature (about 90 degrees) in the engineered barrier leads to the increase of glass dissolution rate, the increase of effective diffusion coefficient and the decrease of distribution coefficient in the buffer. Additionally, the radiolysis of groundwater, which is caused by gamma and alpha radiation with early migration of radionuclides into engineered barrier, brings about increased redox potential by radiolytic oxidants, and resulting enhanced solubility limit for radionuclide. Such a series of deterioration and loss of the functions is considered in the initial detective scenario of overpack.

Two types of the initial defective scenario of the sand-bentonite buffer are considered in this analysis. The heterogeneous decrease of density and the change of porosity structure in the buffer material result from crack or exfoliation of the buffer block, poor quality, corrosion and outflow of the buffer material, and heterogeneous swelling. The reasons are considered to be some kinds of human errors in quality management of materials, manufacturing of buffer blocks, construction work of the buffer, processing work of groundwater flowed in a tunnel and so on. The deterioration of buffer properties due to human errors leads to the increase of effective diffusion coefficient and the loss of restraint of advection in the buffer. In the case that the advective transport of radionuclides is dominant, the advection of groundwater decreases the silica concentration in the porewater of buffer, and resulting increased glass dissolution rate. The above description is one of the initial defective scenarios for the buffer. In another scenario of the buffer, it is assumed that the loss of restraint of colloidal migration in the buffer is due to the heterogeneous decrease of density and the change of porosity structure in the buffer, and resulting colloidal migration of radionuclide in natural barrier. Additionally, the enhanced solubility limit for radionuclide occurs owing to the formation of colloid-radionuclide complex. Basically, same types of the human errors for the buffer are considered for the sealing plugs of a sand-bentonite. In the initial defective scenario of the plugs, the loss of restraint with the plugs from migration through dominant pathway in the tunnels and their vicinity results from the human errors. It is assumed that the pathway of radionuclide migration is short-circuit through the excavation disturbed zone under high permeability and hyperalkaline groundwater derived from the degradation of cementitious materials affects the sorption of radionuclide in the pathway. Additionally, in the normal scenario, it is assumed that no inadvertent events are anticipated and radionuclides are transported in the intact geological disposal system.

Model and code

The probabilistic safety assessment code, GSRW-PSA [3], has been developed based on the Monte Carlo calculation. This code estimates 1-D migration of radionuclides in the engineered barrier and geological media such as the release from the vitrified waste form, migration in the buffer materials and fractured rock mass by diffusion and/or advection, migration of colloidal-radionuclide complex and so on. Same models of radionuclide migration used in the H12 Project report [1] basically apply to this study. However, the boundary condition to calculate the flux from the engineered barrier is zero concentration at rock mass, which is different from mixing sell model in the H12 Project report. The effect of parameter uncertainties to the fluxes of radionuclides from the engineered and natural barriers is evaluated on a basis of probabilistic distribution functions for parameters in GSRW-PSA.

Parameter settings

The parameters on the design of the engineered barrier system are referred from the H12 Project report [1]. The uncertainties on main parameters used in this analysis for the normal scenario and six types of initial defective scenarios are shown in **Table I**. The target radionuclides for this analysis are Cs-135 and Se-79, which are classified as key radionuclides on radiological effect in previous analysis [4].

RESULTS AND DISCUSSION

The sensitivity analysis is conducted for 1,000 parameter sets with Monte Carlo technique of GSRW-PSA. **Figure 1** shows a comparison among complementary cumulative probabilities of Cs-135 and Se-79 maximum release rates from the sand-bentonite buffer per a vitrified waste canister. The 97.5th percentile of maximum release rate from the buffer corresponding to the upper endpoint of the 95th percentile confidence interval is applied to an index for comparing the calculation results.

In the result of Cs-135 for the buffer as shown in **Figure 1** (a), the 97.5th percentiles of three initial defective scenarios on vitrified waste (increasing glass surface area and yellow phase) and overpack are two or three times as high as that of normal scenario on all safety function working. The 97.5th percentiles of two initial defective scenarios of buffer (advective transport and colloidal transport) are same and the highest flux.

Table I Uncertainties on main parameters for normal scenario and initial defective scenarios

Parameters		Min[*1]	Max[*1]	Unit	Comments
Normal scenario					
Glass dissolution rate		1E-04	1E-01	g/m²/d	Based on statistical analysis for published data of long-term glass dissolution rate [5], the value is revised at 60 degrees.
Solubility	Se	1E-07	6E-07	mol/L	Solubility variation of Se(cr) calculated by PASOL [6], considering uncertainties of thermodynamic data of Se represented in the NEA TDB [7].
	Cs	–			Since Cs is a highly soluble element, the solubility limit for Cs is not considered.
Distribution coefficient in the buffer	Se	1E-06	3E-02	m³/kg	The variation is estimated on a basis of the average of measurements for compacted bentonite obtained from the in-diffusion method and the deviation of measurements from batch sorption tests.
	Cs	1E-03	1E+00		
Effective diffusion coefficient in the buffer	Se	1E-12	1E-10	m²/s	The valiation is estimated from the statistical analysis for diffusion coefficient data picked up in the range of the density specification of montmorillonite gel. The values are revised at 60 degrees.
	Cs	3E-11	8E-10		
Groundwater velocity in the fracture		5E-02	5E+01	m/y	H12 Project report [1].
Distribution coefficient in natural barrier	Se	3E-04	4E-02	m³/kg	Statistical analysis for published distribution coefficient data for a granite.
	Cs	2E-03	2E+00		
Initial defective scenario of vitrified waste (increasing glass surface area)[*2]					
Glass surface area		17	170	m²	Variation of 10 to 100 times the external surface area of vitrified waste is assumed.
Initial defective scenario of vitrified waste (formation of yellow phase)[*2]					
Ratio of radionuclide inventory including in yellow phase		0.01	1	–	Variation of 1/10 to 10 times the ratio of 0.1 is assumed.
Solubility	Se	–		mol/L	No solubility limit under oxidizing condition due to molybdenum oxides.
	Cs	–			Cs is a highly soluble element.
Initial defective scenario of overpack[*2]					
Glass dissolution rate		3E-04	3E-01	g/m²/d	Glass dissolution rates in normal scenario are revised at 90 degrees.
Solubility	Se	1E-07	6E-07	mol/L	No solubility limit under the enhanced redox potential by radiolytic oxidants until 1,000y after the disposal. After 1,000y, solubity of Se(cr) is set up in normal scenario.
	Cs	–			Cs is a highly soluble element.
Distribution coefficient in the buffer	Se	1E-06	3E-03	m³/kg	It is supposed that increased temperature (about 90 degrees) has a little effect on the sorption of radionuclides in the buffer. However, the variations of distribution coefficients are 0.1 times as low as those for normal scenario and restricted to more than 1E-6 m³/kg.
	Cs	1E-04	1E-01		
Effective diffusion coefficient in the buffer	Se	2E-12	2E-10	m²/s	Effective diffusion coefficients in normal scenario are revised at 90 degrees.
	Cs	4E-11	1E-09		
Initial defective scenario of sand-bentonite buffer (advective transport of radionuclide)[*2]					
Groundwater velocity in the buffer		1E-04	1E+00	m/y	From the results of 2-D groundwater flow and trajectory analysis for advective transport case [4], the range of pore velocity in the buffer is determined.
Glass dissolution rate		4E-03	1E+00	g/m²/d	Under the assumption of increased dissolution rate of the glass matrix correlated with the advective transport, it is supposed that the maximum of the dissolution rate is conservatively equivalent to the level of the first-order dissolution rate.
Distribution coefficient in the buffer	Se	1E-06	3E-03	m³/kg	Under the assumption of decreased density of montmorillonite gel, the variations of distribution coefficients are 0.1 times as low as those for normal scenario and restricted to more than 1.0E-6 m³/kg.
	Cs	1E-04	1E-01		
Effective diffusion coefficient in the buffer	Se	1E-11	4E-09	m²/s	Under the assumption of the change of porosity structure in the buffer, maximum values are conservatively given by the effective diffusion coefficients in free water.
	Cs	2E-10	4E-09		
Initial defective scenario of sand-bentonite buffer (loss of restraint of colloidal migration)[*2]					
Solubility	Se	–		mol/L	It is assumed that the enhanced solubility limit for radionuclide occurs owing to the formation of colloid-radionuclide complex. No solubility limit is set up for Se.
	Cs	–			Cs is a highly soluble element.
Ratio of velocity of colloidal migration to groundwater velocity		1.3		–	Grindrod(1993) [8]. Variation of velocity of colloidal maigration is estimated from this ratio and the variation of groundwater velocity.
Concentraton of colloid in natural barrier		1E-04	1E-02	kg/m³	Concentration of bentonite colloid is measured to be about 1ppm(=0.001kg/m³) [9]-[11]. Variation of 0.1 to 10 times 1ppm is assumed.
Distribution coefficient for colloid	Se	1E+02	1E+04	m³/kg	Distribution coefficient of all radioactive elements for colloide is 1,000m³/kg in H12 Project report [1]. Variation of 0.1 to 10 times this value is assumed.
	Cs	1E+02	1E+04		
Initial defective scenario of sand-bentonite plugs[*2]					
Groundwater velocity in the pathway of short-circuit		0.1	10	m/y	Groundwater velocity in the pathway of short-circuit is estimated to be 1m/y from the result of groundwater flow analysis for a rock mass and excavation damaged zone along disposal tunnel under no sealing plugs condition [12]. Variation of 0.1 to 10 times this value is assumed.
Distribution coefficient in the pathway of short-circuit	Se	3E-05	4E-03	m³/kg	Radionuclide transport is forcused in excavation damaged zone and/or fractured cementitious support of tunnel. Distribution coefficient data are compared between concrete and rock under high pH condition, and lower values are selected.
	Cs	3E-05	1E+00		

(*1) The minium and maximum values are treated as the values of 0.1 percentile and 99.9 percentile in a log-normal distribution. However, the variation of solubilities of Se(cr) is the 95% confidence intervals calculated by PASOL.

(*2) The values of parameters shown in each initial defective scenario are different from those in normal scenario. The parameter uncertainties for advection of radionuclide in buffer are considered in the calculation for scenario on loss of restraint of colloidal migration. The other parameter valuses in initial defective scenario are same as normal scenario.

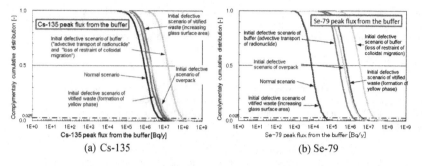

(a) Cs-135　　　　　　　　　　　　　(b) Se-79

Figure 1 Results of Cs-135 and Se-79 maximum release rates from the buffer

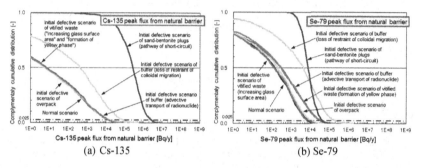

(a) Cs-135　　　　　　　　　　　　　(b) Se-79

Figure 2 Results of Cs-135 and Se-79 maximum release rates from natural barrier

In the result of Se-79 for the buffer as shown in **Figure 1** (b), there is little difference of 97.5[th] percentile for Se-79 between normal scenario and initial defective scenario of vitrified waste (increasing glass surface area). However, the 97.5[th] percentile for formation of yellow phase is calculated to be about two orders of magnitude higher than for normal scenario because of no solubility limit of Se caused by molybdenum oxides. The 97.5[th] percentiles of Se-79 for overpack and buffer (advection transport case) are 30 or 80 times higher than for normal scenario.

The results of maximum release rates from natural barrier are shown in **Figure 2**. In **Figure 2** (a), the 97.5[th] percentile of Cs-135 maximum release rate from natural barrier for initial defective scenario of buffer (colloidal transport case) is about 6 times higher than for normal scenario. The initial defective scenario of sealing plugs (pathway of short-circuit) indicates the highest flux of Cs-135, which is about two orders of magnitude higher than for normal scenario. For the other initial defective scenarios, their results from natural barrier are approximately equivalent to that for normal scenario. In **Figure 2** (b), the deterioration or loss of safety functions, which are caused by initial defective scenarios of vitrified waste (increasing glass surface area) and overpack, have little effect on enhanced release rate of Se-79 from natural barrier. Maximum release rates of Se-79 for two types of initial defective scenarios, buffer (colloidal transport case) and sealing plugs (pathway of short-circuit), tends to be higher than for the other scenarios. The 97.5[th] percentile of Se-79 for initial defective scenario of buffer (colloidal migration) is the highest and about 30 times as high as that for normal scenario.

CONCLUSIONS

Release rates per a vitrified waste for important radionuclides, Cs-135 and Se-79, are estimated from Monte Carlo-based analysis for 6 types of initial defective scenarios. Maximum release rates of radionuclides for initial defective scenarios, buffer (colloidal transport case) and sealing plugs (pathway of short-circuit), are about one or two order magnitude higher than that normal scenario. From a viewpoint of future safety review, these results especially indicate the need to understand the feasibility on two types of initial defective scenario for the buffer and sealing plugs, leading to the loss of restraint of colloidal migration in the buffer and the loss of restraint with the plugs from migration through dominant pathway in tunnels and their vicinity. It is suggested that adequate plan and implement are important for decreasing the human errors in quality management of materials, manufacturing of sand-bentonite blocks, construction work of the buffer and sealing plugs and so on, related to two initial defective scenarios.

ACKNOWLEDGMENTS

This research is funded by the Secretariat of Nuclear Regulation Authority, Nuclear Regulation Authority, Japan.

REFERENCES

1. Japan Nuclear Cycle Development Institute (JNC), "H12: Project to Establish the Scientific and Technical Basis for HLW Disposal in Japan - Supporting Report 3: Safety Assessment of the Geological Disposal System –", JNC TN410 2000-004 (2000).
2. IAEA, Chemical Durability and Related Properties of Solidified High-Level Waste Forms, Technical Reports Series No.257 (1985).
3. S. Takeda and H. Kimura, JAERI-Research 2002-014 (2002).
4. S. Takeda, T. Yamaguchi, H. Nagasawa, M. Watanabe, Y. Sekioka et. al., JAEA Research-2009-034 (2009).
5. Y. Sekioka, S. Takeda and H. Kimura, JAEA-Research 2009-062 (2010).
6. S. Takeda and H. Kimura, JAEA-Research 2006-069 (2006).
7. A. Olin, B. Noläng, E. G. Osadchii,L. O. Öhman and E. Rosén, "Chemical Thermodynamics of Selenium (Chemical Thermodynamics, Vol.7)", Elsevier Science B.V., Amsterdam, (2005).
8. P. Grindrod, Journal of Contaminant Hydrogy, 13, pp.167-181 (1993).
9. K. Matsumoto, K. Iijima and K. Tanai, JAEA-Research 2008-097 (2009)
10. Y. Kuno, K. Morooka, H. Sasamoto and M. Yui, Journal of Nuclear Fuel Cycle and Environment, 15, No.2, pp.117-129 (2009).
11. E. Wieland, J. Tits and M. Bradbury, Applied Geochemistry, 19, pp.119-135 (2004).
12. Y. Sugita, Y. Takahashi, M. Uragami, K. Kitayama, T. Fujita, et. al., NUMO-TR-05-02 (2005).

Fukushima Daichi

Mater. Res. Soc. Symp. Proc. Vol. 1518 © 2012 Materials Research Society
DOI: 10.1557/opl.2012.1713

Decontamination Pilot Projects: Building a Knowledge Base for Fukushima Environmental Remediation

Kaname Miyahara, Takayuki Tokizawa and Shinichi Nakayama
Fukushima Environmental Safety Center, Headquarters of Fukushima Partnership Operations,
Japan Atomic Energy Agency (JAEA),
2-2-2,Uchisaiwai-cho, Chiyoda-ku, Tokyo, 100-8577, Japan

ABSTRACT

After the Fukushima Dai-ichi nuclear accident, Japan Atomic Energy Agency (JAEA) was chosen by the Government to conduct decontamination pilot projects at selected sites in the contaminated area of Fukushima. Despite tight boundary conditions in terms of timescale and resources, the projects provide a good basis for developing recommendations on how to assure clean-up efficiency and worker safety and reduce time, cost, subsequent waste management and environmental impact. The results of the project can be summarised in terms of site characterisation and data interpretation, clean-up and waste minimisation and storage.

INTRODUCTION

The damage to the Fukushima Dai-ichi nuclear power plant (NPP) by the Great Tohoku earthquake and tsunami resulted in considerable radioactive contamination, both on- and off-site of NPP. After decay of shorter-lived nuclides, the contamination is now dominated by radiocaesium ($^{134, 137}$Cs), which is the focus for clean-up actions. Caesium tends to bind strongly to soil, especially clays. Dose rates are generally low with a few exceptional locations, resulting predominantly from external gamma irradiation and from continual reduction by washoff and soil mixing.

After the accident, staged clean-up of contaminated areas was initiated – first in populated areas (especially sensitive areas such as schools and playgrounds) and then, following stabilisation of the reactors and decay of shorter-lived nuclides, extending into evacuated zones.

The overarching "Special Measures" laws to manage radioactive contamination from this incident and establish an overall policy for decontamination were promulgated on 30th August and 11th November 2011. These specified responsibility for conducting a range of decontamination pilot projects to examine applicability of clean-up technologies to the higher levels of contamination within the evacuated zone. Based on such projects, the Government will develop the technical basis for efficient and effective clean-up technologies, assuring worker safety, establishing a regional remediation plan and advancing to stepwise implementation of decontamination, with a special focus on reducing dose rates and allowing evacuees to return to re-establish their normal lifestyles as quickly as possible.

Japan Atomic Energy Agency (JAEA) was chosen by the Government to conduct decontamination pilot projects at model sites. The first project included 2 residential sites with lower contamination levels, which ran from August 2011 until March 2012. The second project was implemented from September 2011 until June 2012 at 16 sites in 11 municipalities, including highly contaminated sites in evacuated zones. Despite tight boundary conditions in terms of timescale and resources, the decontamination pilot projects provided a good basis for developing recommendations on how to assure clean-up efficiency and worker safety and reduce time, cost, subsequent waste management and environmental impact.

The regional decontamination presently being initiated must be well planned, rigorously implemented and clearly presented to all stakeholders. However, decontamination on such a regional scale in a highly populated region has never been attempted before. Main challenges to implement full-scale decontamination are lack of both real-world examples and also experience for planning and implementing decontamination technology appropriate to Japanese boundary conditions. Therefore, the decontamination pilot projects played a key role to support drafting of guidelines and manuals that can be used as a source of reference by the national government, local municipalities and the contractors performing regional decontamination. This paper discusses this application of the decontamination pilot projects, focusing on those carried out in the evacuated zones.

DEVELOPMENT OF THE PILOT PROJECTS

The evacuated zone is quite typical of Northeast Japan, *i.e.*, comprising a narrow coastal plain and valleys leading into a spine of densely wooded mountains. Along the coast and in the plain, the population density is relatively high with agriculture being an important industry. In the more mountainous areas, population is mainly confined to narrow valleys although, even here, agriculture is important, as is tourism.

The locations of the sites selected for the decontamination pilot projects, specific constituents and features requiring clean-up, the level of contamination and project site grouping are summarised in Figure 1. As can be seen, these are representative of the challenges that will be faced in the regional remediation and allow for different approaches to remediation to be compared. Further, the locations include both urban and rural areas in different terrain (mountainous, hilly, plain). JAEA selected 3 contractor Joint Ventures to carry out the work, while JAEA managed, supervised and evaluated the overall program. Documentation of the decontamination pilot projects has been published in Japanese [1, 2].

Procedure

A tailored remediation plan was developed for each of the demonstration sites selected. The plan involves initial characterisation of the distribution of contamination, setting priorities and deciding details of remediation techniques to be implemented (*e.g.* extent of surface soil removal based on depth profiles of radiocaesium concentration). Such analysis also allowed first estimates of the volume and radioactivity of wastes to be expected and hence the requirements for temporary storage facilities to be determined. The remediation plan also explicitly considered operator safety, which involves consideration of both radiological and conventional labour hazards. Further requirements for the remediation plan were to consider minimisation of environmental impact and a process for communication to establishing dialogue with stakeholders.

Following preliminary remediation planning, implementation for each of the demonstration sites proceeded in the following steps:
- Radiation survey before remediation (establish maps of radionuclide distributions; particularly useful to guide remediation planning and determine depth profiles to allow assessment of benefits of different soil remediation approaches)
- Establish remediation implementation plan based on evaluation of radiation survey data
- Apply remedial measures

> Evaluate effectiveness of remedial measures
> Review effectiveness and assess input for remediation guidelines.

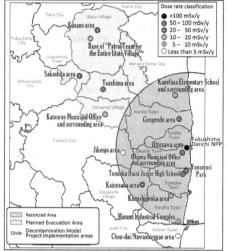

Group /Municipalities	Object of decontaminating (Total about 209 ha)	
	Main constituents and features	Extent
Group A Minami Soma City	Farmland, Building(Elementary School), Road, Forest, Residential area	About 13 ha
Kawamata Town	Forest, Farmland, Road, Residential area	About 11 ha
Namie Town	Building(Junior High School, etc.), Forest, Residential area, Road	About 5 ha
Namie Town	Building(Station / Track, Library, etc.), Private house, Road, Farmland	About 13 ha
Iitate Village	Building(Iitate-home, etc.), Farmland, Private house, Residential area, Forest, Road	About 17 ha
Group B Tamura City	Farmland, Forest, Residential area, Road	About 15 ha
Katsurao Village	Forest, Building (Elementary School, Municipal office), Residential area, Road	About 6 ha
Tomioka Town	Building (Junior High School, Playing field, etc.), Residential area, Forest, Road (Road bordered with cherry-tree)	About 9 ha / About 3 ha
Futaba Town	—	—
Group C Hirono Town	Building(Municipal office, Elementary School, Junior High School, Playing field), Residential area, Forest, Road	About 33 ha
Okuma Town	Building(Municipal office, Public hall, Park), Residential area, Road	About 6 ha
	Farmland, Forest, Residential area, Road	About 17 ha
Naraha Town	Farmland, Residential area, Forest, Road	About 4 ha
	Building(Factory, etc.), Road	About 37 ha
Kawauchi Village	Farmland, Forest, Private house, Road	About 23 ha

Figure 1. The sites and targets for the decontamination pilot projects

Site characterisation and data interpretation

Measurement approaches for site characterisation involved both modification of existing technology and development of new methods - measuring total dose rate, surface contamination or radiocaesium concentration. When linked to appropriate data loggers, these provided rapid and convenient electronic maps of radiocaesium distributions.

Maps were particularly useful to guide remediation planning. In-situ measurements could be subdivided into 2 broad classes, local dose rate measurements and determination of radiocaesium contamination levels. The former integrates dose rate due to gamma radiation from all sources in the vicinity at a defined height above ground surface (usually 1m or 1cm). To provide more information than a simple integrated dose rate, near-surface measurements (1 cm) were made using GM detectors that are particularly sensitive to betas. All demonstration projects started from pre-existing digital maps and aerial images, which were integrated with any pre-existing radiological survey data (*e.g.* aerial gamma scans, point measurements) to derive a first conceptual model of initial site contamination. Novel scanning tools developed (incorporated into a remote-controlled helicopter, a buggy or a backpack to facilitate access to the complex terrain involved) could be considered only semi-quantitative, but have proven useful for establishing relative radioactivity distributions and finding hotspots.

Depth profiles of radiocaesium concentration allowed assessment of benefits of different soil remediation approaches. In general, 80% or more of radiocaesium in soil was present within about 5 cm of the topsoil. In dense asphalt pavements, most of radiocaesium was present within about 2 to 3 mm from the surface.

Options of clean-up methods could be assessed using a model to predict effective dose reduction. Because of the long range of gamma rays in air (the half-distance for [137]Cs 0.66-MeV gamma ray in air is about 70 m), assessing the net impact of decontamination on local dose rates is not straightforward. As a guide to planning, therefore, a calculation tool (Calculation system for Decontamination Effect; CDE) has been developed by JAEA (http://nsed.jaea.go.jp/josen/: in Japanese only). Basically, this applies a 5-m mesh to the area of interest and for each lattice cell specifies a surface radiocaesium concentration (derived from measured air doses, assuming a continuous flat surface) and the land use. For a given set of specific decontamination factors applied to the different objects in the site, the resultant change in the map of dose rate distribution can be calculated. Although the obtained quantitative output is associated with uncertainties, the model is a first step to assess the consequences of different decontamination strategies on the net dose distribution and hence tailor a general remediation plan to a specific site.

Radioactivity monitoring was continued during remediation actions, to provide feedback on effectiveness and quantify the characteristics of wastes generated. This then led to a more complete survey after remediation to form the basis for assessment of effectiveness of different methods.

In principle, the effectiveness of the entire clean-up operation could be assessed by comparison of dose rates at a number of specified points before and after clean-up. This provided a measure of the dose reduction achieved, but the ratio of before/after (Dose Rate Reduction Factor; DRRF) tended to underestimate effectiveness, as it included a background from the surroundings, which could actually contribute a large percentage of the final measured dose rate.

Measurements with a GM detector were also made before and after remediation of some surfaces. If suitable shielding / collimation was used, the measurement is proportional to the extent of superficial contamination (due to the short range of the betas counted) and, for the case where radiocaesium does not penetrate to any significant depth, the ratio of count rate before/after provides a direct measurement of the decontamination factor.

Reliable measurements must be a basis to assure clean-up efficiency and worker safety. Special care was taken to check the working temperature of detectors, as these are generally assumed to be used within the range of about 0 to 40°C. For measurements below 0°C, which was often encountered at a few sites, recalibration was needed. Measurement biases were observed due to difference of NaI scintillation survey meters with or without energy compensation circuit. Therefore, measured data obtained from the latter survey meters were appropriately corrected.

Clean-up

In each of the selected demonstration sites, decontamination targets were identified and different technologies were applied to specific targets, such as buildings, forest, farmland, etc. Although the majority of the effort involved manual washing and contaminated material removal using conventional technology, methods that might improve clean-up while decreasing volumes of waste were tested (some examples illustrated in Figure 2). To avoid generating secondary contamination, decontamination proceeded from topographically higher locations to lower ones, with clean-up of roads the final step.

Radiation exposure of clean-up workers was continuously monitored by ensuring that all workers wear a cumulative dosimeter and a pocket dosimeter and remained low during the course of the projects and well within the specified dose limit. For example, in the case of highly contaminated agricultural and residential areas, average exposure dose of clean-up workers is 2.4mSv over 108 days. The atmospheric radioactivity concentrations observed in decontamination work areas of this pilot project were not particularly high. Because the workers wore protective equipment, the internal exposure doses of all workers were below the limit of measurement (1 mSv).

1) Trees and forest

The main decontamination methodology used for forests was simple removal of contaminated material, including undergrowth, fallen leaves, humus/litter layer, topsoil and tree pruning. Removal of leaves was carried out both manually (where vegetation was swept up using rakes) and mechanically, using vacuum suction (small "car-based" and large "lorry-based" vacuums were tested). Reduction of volume using a chipper was important for woody materials, such as bamboo, small trees and pruned branches.

Figure 2. Images of some of the remediation methods used.

2) Farmland

Decontamination methodologies for agricultural land included vegetation removal (manually with a strimmer or mechanically using a grass cutting machine), soil inversion (manually with a spade, mechanically with a small mechanical rotavator or ploughing). Soil removal using a mechanical digger was, in some case, preceded by "soil solidification" using a resin spray.

For a typical example of soil inversion, field vegetation was first harvested using strimmers, followed by deep ploughing to 25-cm depth; this technique inverts the soil profile and greatly decreases dose rate above the ground surface due to shielding by uncontaminated topsoil. An alternative technique for deeper emplacement involved removal, excavation and backfilling: the top 5 cm of soil (containing most, if not all, of the contamination) being removed followed by a further 45 cm of subsoil, the contaminated surface soil is then layered in the hole and covered by uncontaminated material.

3) Buildings

For houses, techniques designed to loosen surface contamination on roofs and walls included manual wiping, cleaning with brushes and high pressure water cleaning (although here care had to be taken to avoid water penetrating roofs). All water used was carefully collected and decontaminated by filtration or ion-exchange before discharge to drainage. The decontaminated water was reused for cleaning in some cases. Although roof tiles are made from diverse materials, with the exception of weathered cementitious roof tiles, these techniques were generally effective for decontamination. A focus for roofs was cleaning gutters, which often represented contamination hotspots.

Large buildings, such as schools and factories, were treated in a similar manner, but novel techniques were tested to clean larger concrete surfaces where simple washing was ineffective.

These included pneumatic shot blasting with either small steel balls or dry ice. Eroded thin layers of the contaminated concrete surface were collected by vacuum for later disposal (with magnetic separation of steel shot for reuse). Both these methods have the advantage of avoiding use of water (and resultant decontamination), but require careful dust management to avoid workers internal exposure as well as careful control of external exposure due to gamma irradiation from resulting waste.

After decontamination of buildings, the surrounding environment was treated. Garden vegetation was cut back, usually by strimming, mowing or clipping, followed by soil turnover or complete removal of soil surface layers if required. Particular attention was paid to removal of hotspots, often found underneath the eaves of roofs or in drains collecting roof runoff. Unpaved surfaces and gravel were generally treated by high-pressure water cleaning and wastewater collected for subsequent treatment.

4) Roads and paved surfaces

Roads were decontaminated using a number of methods, which were generally "lorry based". Initially roads were sprayed with water with mechanical cleaning by large coarse rotating brushes. Alternative techniques tested for removing contamination from roads and pavements included high pressure water jets, very high pressure water (such as spin jet washing which could erode thin layers of the contaminated surfaces), shot blasting and complete removal of asphalt from roads. In all cases contaminated run-off was collected and pumped into tankers for later treatment or reuse, and any other wastes (especially dust) captured for treatment and/or packaging for storage.

Waste minimisation and waste storage

A specific goal of the demonstration projects was to optimise clean-up procedures in order to reduce waste volumes to the maximum extent possible. For reduction of the volume of soil requiring disposal, the main approach used involved using measured profiles to determine depth of penetration of radiocaesium and then using a technique which removed only the most contaminated material. Indeed, when low-level radiocaesium contamination was limited to near the surface, soil profile inversion variants are effective management options, which produce no waste and reduce radiation dose rate by the natural shielding of uncontaminated soil. At some sites waste was segregated according to its radioactivity level with scan-sorting equipment. In the demonstrations, this equipment was used for excavated soil and farm wastes, which were sorted to allow material with a specific radioactivity less than a reference value to be returned to the excavation site and only that exceeding this level to be packaged for storage.

The large quantity of organic waste being produced was a particular focus for volume reduction. Methods employed for this material included chipping, physical compaction and incineration – both high (> 800 °C) and low temperature (250-400°C) variants. For the latter options, cost and throughput have to be balanced against the volume and radioactivity concentrations in resultant ash.

Wastewater from decontamination activities was either filtered for reuse or pumped into holding tanks for treatment. Treatment methods included ion-exchange and scavenging by co-precipitation, with filtration of resulting fluids and drying of sludges and precipitates. All resulting contaminated solids were placed in flexible plastic containers for storage and purified water released to drains.

In most cases, solid waste was simply placed in flexible plastic containers, labelled and then transported to a temporary store. These flexible plastic containers have a volume of about 1m³ and are strong enough to be lifted even when full of wet soil. They are impermeable, but cannot be considered gas- or water-tight. They were labelled with either a robust conventional tag or an electronic readable chip, which contains a sample location code, date of packaging, description of contents, estimated radiocaesium content and surface dose rate.

A regulatory constraint on waste management was that all significantly contaminated material had to be placed in temporary storage at the site being remediated. The locations of these storage facilities were selected taking into consideration topography, land use, available areas of land, local government requests and required the explicit agreement of local communities and landowners. Several different design options for such temporary storage, either on the surface or in shallow pits (see figure 3), were developed with the key aim of assuring safety over the required period (defined as lasting not more than 3 years). The details of the design were tailored to the storage site topography, which resulted in design variants for flat areas (on surface or sub-surface) and for sloping sites (inclined and stepped). In all cases, the temporary storage facility included an impermeable base, surface cover and uncontaminated soil backfill to provide shielding. It was not expected to make such structures completely watertight, so drainage due to gravity flow was incorporated in all cases. Drainage was monitored and captured in a water collection tank for any required treatment. Because of the organic waste content, allowance is made also to allow gas venting to ensure pressurisation does not disrupt the engineered barriers.

The temporary storage options implemented during the demonstration sites were designed to meet the guidelines issued by Ministry of the Environment. The facilities at the demonstration sites are being carefully monitored to check performance is maintained and, in case of any problems, appropriate actions will be taken.

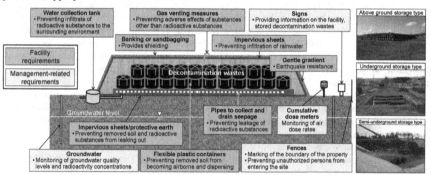

Figure 3. General concept for temporary storage and some illustrative examples

DISCUSSION

Wide area decontamination

The experience gained in the remediation projects has provided valuable input in terms of developing fast and efficient approaches to assessing the relative distribution of radioactivity (or dose rate) in any remediation area. Integration of electronic maps, geographical information (land use, etc.) and measured radiometric data has worked well. Figure 4 shows an example of results of dose rate reduction by wide area decontamination. Dose reduction here, based on average dose rates in each land use compartment, is between 40% and 80%.

In terms of estimated annual dose reduction to less than the initial evacuation level of 20 mSv/y has been demonstrated for areas with 20 - 30 mSv/y before decontamination. However, this goal could not be assured for areas exceeding 40 mSv/y before decontamination. In case of highly contaminated agricultural and residential areas, the dose rate was decreased by 70 %, but could not be reduced to below 50 mSv/y. In general, the fractional dose rate reduction was smaller in areas of relatively initial low contamination, compared with higher contaminated areas.

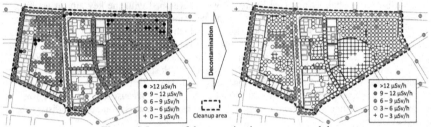

Figure 4. Impact of decontamination on measured doses

Recommended clean-up technologies

To facilitate use in tailoring to specific site conditions, the remediation toolkit requires pros and cons of different methods to be clearly identified. Most of the parameters involved were reasonably easy to assess during demonstration (*e.g.* cost, time and manpower requirements, volume of wastes generated). In principle, the effectiveness of the entire clean-up operation could be assessed by comparison of dose rates at a number of specified points before and after remediation. A pragmatic approach was to carry out remediation actions sequentially and compare dose rates measured at the target object before and after cleaning.

Table 1 shows comparison of forest decontamination methods. Recommended clean-up methods for each land use target were derived by comparing options in terms of dose reduction, speed, cost, waste management and environmental impact. In general, clean-up methods with higher dose reduction resulted in higher cost. For some clean-up methods, however, dose reduction was comparable but cost and/or efficiency was different.

The derived recommended clean-up methods for each land use target are listed in Table 2.

Lessons learned

The demonstration projects have served their primary purpose of development of a knowledge base to support more effective planning and implementation of stepwise regional remediation of the evacuated zone. A range of established, modified and newly developed techniques have been tested under realistic field conditions and their performance characteristics determined. This toolkit covers site characterisation, clean-up and waste storage.

Table 1. Characteristics of different remediation methods for wooded areas.
◎: highly effective; ○: effective; △: moderately effective; ▲: limited effect

Decontamination method		Removal of fallen leaves and humus (on flat ground)	Removal of fallen leaves and humus (on slopes)	Removal of fallen leaves, humus and topsoil (on flat ground)	Trees	
					Trunk washing	Branch trimming in the lower part
Proportions of radioactivity on evergreen trees (as of August - September 2011)			44-84 %		Trunks: 1-3%	Branches and leaves: 14-53%
Percentage dose reduction*		5-90 %	5-90%	20-80%	30-85 %	5-40 %
Volume of decontamination waste generated		0.2-0.9m³/m²	0.2-0.9m³/m²	1-2m³/m²	Small amount	2.7m³/m² (non-reducing waste volume)
Secondary contamination		Does not occur.	Does not occur.	Does not occur.	Occurs.	Occurs.
Effects on surrounding environments		On slopes, it is necessary to be careful not to cause erosion.			(Soil infiltration of droplet)	(Drop branches to forest floor)
Cost (JPY)		530/m²	760/m²	890/m²	3,390/tree	580/m²
Decontamination speed		510 m²/day (11 persons)	340 m²/day (11 persons)	220 m²/day (5 persons)	32 trees/day (4 persons)	150 m²/day (4 persons)
Applicability	Deciduous forests	◎	◎	○	▲	—
	Evergreen forests	◎	◎	○	▲	○

*Percentage dose reduction is calculated using the values of surface contamination density measured before and after decontamination. In the case of the branch trimming, the values of air dose rate are used.

Table 2. Comparative assessment of remediation options for different targets
◎ : highly effective, ○ : effective, △ : moderately effective, ▲ : limited effect

Land use classification			Comprehensive evaluation
Forest			◎Removal of fallen leaves and humus (on flat ground and slopes), ○Removal of fallen leaves, humus and topsoil (on flat ground), ▲Trunk washing, ○Branch trimming in the lower part (evergreen tree)
Farmland			◎Machine that strips off surface of soils, ○Backhoe (stripping off depth of 5 cm of the soil), ◎Reversal tillage (by tractor and plough), ○Ploughing to replace surface soil with subsoil (by backhoe)
Residential area	Roof		▲High pressure water, ○Brushing, ○Wiping, ▲Apply a remover
	Gutter		△High pressure water, ○Wiping
	Wall		○Brushing
	Topsoil		○Removal of topsoil
	Rubble		○Washing of the rubble, ○Removal of the rubble
	Turf		○Removal of the turf
	Garden tree		▲Clipping a garden tree
	Interlocking block		△High pressure water
Large structure	Concrete and Mortar surface		△Sanding machine with the dust-collection (Plane which scrapes concrete), ○Ultrahigh pressure water (Over 150Mpa), ○High pressure water (10-20Mpa), ○Iron shot blasting
	Roof floor	Concrete surface	○High pressure water (including brushing)
		Waterproof coating surface	○High pressure water (including brushing)
		Downpipe	○High pressure water(Maximum 50Mpa)
	Playing field		○Strips off surface of soils (Large mower+Sweeper), ○Strips off surface of soils (Road planers), ○Strips off surface of soils(Motor grader), ○Ploughing to replace surface soil with subsoil
	Swimming pool		○High pressure water
	Turf		○turf stripper
	Paved road		▲Road cleaners + Riding style road sweepers, △High pressure water (About 15MPa)+Brushing, △Car of a functional recovery drainage pavement, ○Ultrahigh pressure water (120~240MPa), ○Iron shot blasting, ○TS road planers

1) Site characterisation

Although the pilot projects were set up under extreme time pressure, which did not allow optimisation of procedures or standardisation of measurement protocols, short manuals were developed to guide sampling and field measurement and independent quality assurance checks were introduced. Improvements have been identified that will allow establishment of user-friendly sampling and measurement protocols which allow measurement uncertainties to be rigorously measured and minimise measurement biases.

Unlike conventional civil engineering work, it was difficult to visually assess the progress of large-scale decontamination work. For example, when stripping topsoil manually, it was difficult to visually judge if the entire contaminated surface layer had been removed. Therefore, during clean-up, the surface dose rate was often monitored to check the effectiveness of

253

procedures, particularly during removal of hotspots. Because of the complexity of sites and local movement/storage of wastes, such monitoring requires low background, well-shielded detectors which need to be easily transportable over rough and often soft ground. Some further development of equipment is required here, but technical specifications have been established and conceptual designs illustrated.

2) Clean-up
> ### Forest

For both evergreen and deciduous forests, effective decontamination generally results from removal of leaf litter and humus layers. Cleaning a forest perimeter about 10 m wide at boundaries with living areas gave a significant dose rate reduction (40 to 50%). In the case of dense woods, such as bamboo forests, vacuum suction was efficient for stripping and transport of leaf litter and humus. Special care was needed in the common case of steep slopes, where removal of this cover increases the risk of soil erosion and landslips (the region is regularly exposed to intense rainfall during typhoons). Therefore, in such cases it is necessary to implement measures to reduce soil erosion (such as sandbags or reed mats). The demonstration projects thus provide clear guidelines for wider remediation of forests.

> ### Farmland

More than 80% of radiocaesium in soil is generally found within a depth of about 5cm in the topsoil. Somewhat deeper penetration is found in farmlands that were ploughed just before contamination occurred after the March 11 accident and locally where soil mixing results from biological activity or human actions (*e.g.* tractor wheel ruts). These observations are consistent with Cs being strongly bound to mineral surfaces and hence, to date, "soil washing" approaches have not been attempted – although this might be investigated in the future.

Many different techniques for either soil stripping (possibly with initial solidification or subsequent sorting) or soil profile inversion have been tested in different settings with pros and cons being identified. Deciding between options for a specific site will need to consider not only the local soil Cs profile, but also field size, future agricultural use, desires of the landowner and even weather conditions as different approaches may be appropriate for either wet or cold conditions. Although the demonstration projects provide an extensive knowledge base to support such decisions, incorporation into a user-friendly communication platform will improve future ease of access to this information.

> ### Buildings

A special focus for buildings, especially houses, is cleaning roofs and, especially the hotspots found in gutters, drains and other locations where runoff is captured. Simple manual methods are generally sufficient and, although high pressure water jets can be used, these need to be carefully handled to prevent water penetration of tiled roofs. Porous / weathered cementitious surfaces were trickier, but several methods to remove the thin layer of contaminated material with minimal production of secondary wastes have been tested. Monitoring the effectiveness of decontamination of the surfaces was a challenge that was not completely resolved, but may be handled in the future by improved equipment design (*e.g.* boom-mounted shielded detectors).

> ### Roads

The depth distribution of radiocaesium varied from about 2 to 3 mm for asphalt and dense-graded pavement to about 5 mm in more porous "drainage" paving, which constrains the type of approach used for clean-up. Depending on the required dose rate reduction, potential techniques include high- to ultra-high-pressure jet (up to about 200 MPa), brushing, abrasion and, as a last resort, complete resurfacing. For all such methods, it is important that all fluids and dust are captured and generated wastes are minimised. It was noted that high-efficiency filtration was just as effective as other water decontamination methods including ion-exchangers and flocculants – again compatible with the assumption of Cs being strongly bound to mineral surfaces. Assessing the effectiveness of less disruptive methods may need to be tested on a case-by-case basis, again requiring effective monitoring of progress but with less difficulty of using existing shielded detectors in this case (can be easily mounted on a wheeled vehicle)

3) Waste handling and storage

The standard flexible plastic containers used to collect and transport "raw" wastes represent a pragmatic and cost-effective approach that is familiar to the contractors involved. This is, however, not equivalent to conditioning and packaging required for waste disposal and hence the temporary stores are considered as short-term measures only (about 3 years), providing time to develop and implement plans for longer-term storage (up to about 30 years) and final disposal. Nevertheless, it is important that these temporary stores provide sufficient radiation shielding and prevent any release of radioactivity into air or groundwater.

The designs of stores are tailored to available sites but all include measures to assure mechanical stability (e.g. infilling spaces between containers with sand, graded cover with soil) and prevent releases to groundwater (impermeable base and cap, gravity flow drainage including radiation monitors and catch tanks). Nevertheless, these are simple structures and the contained wastes are labile and vulnerable, in particular, to biodegradation. Gas production may occur and vents are included in the design to avoid pressurisation. Site monitoring also needs to check that structures are not perturbed by external events that can include typhoons, heavy snowfalls, freeze/thaw cycles and earthquakes.

Although stores constructed during the demonstration project have served their initial purpose in terms of concentrating and confining the wastes, their behaviour needs to be continually and comprehensively monitored and, in the event of problems, experience captured is fed back to improve future designs.

4) Informed consent

To obtain the consent from local residents, which are necessary to allow such work to progress, the support from mayors of local municipalities and heads of administrative districts was indispensable. Briefing sessions with communities and use of a clear and simple consent form also helped to facilitate this process. Materials for explaining remediation to stakeholders and providing the basis for establishing dialogue with them have been developed, including plans of remediation and temporary storage and rapid communications on evaluating effectiveness of remedial measures.

Nevertheless, focused efforts are needed to ensure that stakeholders are kept fully informed and encouraged to "buy-into" the work by directly participating in dialogue leading to decision making. During future remediation, an example here is development of a web-based communication platform that allows user-friendly access to the results of the demonstration projects – not just in the form of conventional reports but also using modern media to summarise issues in the form of interactive images, blogs, videos, animations, etc. This platform will be accessible to both computers and smartphones – the latter being very popular in Japan – and will facilitate user feedback and distribution of information via social networks. If successful, this tool could be extended to support the entire regional remediation programme.

CONCLUSIONS AND FUTURE PERSPECTIVE

JAEA demonstration projects have provided a knowledge base including:
➢ experience and tools for planning, coordinating and implementing efficient, safe and cost-effective remediation programmes,
➢ an evaluation of the applicability of existing and newly-developed clean-up technology, with an assessment of the pros and cons of different approaches, and
➢ guidelines for tailoring of projects to the conditions found in different sites.

Practical experience has shown that stakeholder involvement in implementation of clean-up activities is essential. Materials for explaining remediation to stakeholders and providing the basis for establishing dialogue with them have been developed.

During future remediation, experience can be used to constantly update the clean-up knowledge base with the intent of continuous improvement of methodology and toolkits. An advanced communication platform is being implemented to facilitate information exchange between all those involved and, in particular, encourage dialogue with local communities and their involvement in decision-making. Waste management will be a special focus for research and development to support optimisation in terms of both volume reduction and easing/ increasing safety of temporary storage, interim storage and eventual final disposal. The regional work will allow displaced populations to return home to normal lifestyles as quickly as possible and provide the knowledge and experience needed for later decontamination of the Fukushima Dai-ichi site.

ACKNOWLEDGEMENTS

The decontamination pilot projects were funded by the Cabinet Office Government of Japan. The authors wish to thank Dr. Ian McKinley for his technical review comments and Dr. Kenichi Yasue and Mr. Katsuaki Sayama for their technical assistance in preparing figures.

REFERENCES

1. JAEA, *Decontamination pilot projects mainly for evacuated zones contaminated with radioactive materials discharged from the Fukushima Daiichi nuclear plant* (2012) (in Japanese), WWW Document, (http://www.jaea.go.jp/fukushima/kankyoanzen/d-model_report.html).
2. JAEA, *Decontamination pilot projects to establish guidelines of municipal remediation work for residential areas contaminated with radioactive materials discharged from the Fukushima Daiichi nuclear plant* (2012) (in Japanese), WWW Document, (http://www.jaea.go.jp/fukushima/kankyoanzen/guideline_report.html).

Mater. Res. Soc. Symp. Proc. Vol. 1518 © 2013 Materials Research Society
DOI: 10.1557/opl.2013.51

Research and development activities for cleanup of the Fukushima Daiichi Nuclear Power Station

Toshiki Sasaki[1], Shuji Kaminishi[2], Yasuaki Miyamoto[1], and Hideyuki Funasaka[1]
[1]Headquarters of Fukushima Partnership Operations, Japan Atomic Energy Agency,
2-2-2 Uchisaiwai-cho, Chiyoda, Tokyo 100-8577, Japan
[2] Nuclear Fuel Cycle Department, Tokyo Electric Power Company
1-3 Uchisaiwai-cho, Chiyoda, Tokyo 100-8560, Japan

ABSTRACT

The Fukushima Daiichi nuclear power station accident and restoration works have produced significant volume of radioactive waste. The waste has very different characteristics from usual radioactive waste produced in nuclear power stations and it requires extensive research and development for management of the waste. R&D works such as analysis of the waste properties, hydrogen generation by radiolysis and diffusion in a storage vessel and corrosion of storage vessels, etc. have been performed for characterization and safe storage of the waste. The detailed R&D plan for processing and disposal waste will be established by the end of FY2012.

INTRODUCTION

Fukushima Daiichi nuclear power station (NPS) consists of six BWRs. Units 1, 2, and 3 were operating at rated power level while 4, 5, and 6 were in cold shutdown when a magnitude 9.0 earthquake occurred offshore of Japan's east coast in the afternoon on March 11, 2011. The operating units auto-scrammed on reactor protection system trip. The earthquake immediately devastated all off-site AC power supply to the station. On-site emergency diesel generators were automatically kicked in to supply power for emergency systems. However, a series of tsunamis with their greatest height of approximately fifteen meter began hitting the site about forty minutes later the earthquake and killed most of on-site emergency AC/DC power supply by flooding diesel generators, switchgear rooms, DC batteries and so forth. Units 1-3 lost core cooling to remove decay heat that lead to fuel degradation and core melt. Hydrogen produced in damaged cores caused explosions blowing off top of reactor buildings in units 1, 3, and 4.

The accident produced radioactive gaseous, liquid, and solid wastes. TEPCO estimated approximately 500PBq of radioactive iodine and cesium was released into the air and the sea after the accident [1]. Significant gaseous releases have stopped since Reactor Pressure Vessel (RPV) temperatures decreased well below 100 C till the end of 2011 using the temporarily installed core cooling system supplying cooling water to the damaged cores [2].
Approximately 400 tons of cooling water is injected to damaged cores everyday even more than five hundred days after the accident via piping of primary water lines to remove decay heats [3]. Injected cooling water is not lead to any outlet lines after heat removal but spills on Primary Containment Vessel (PCV) floors through accident-made pressure boundary openings, flows through basement floors of reactor/turbine/utility buildings and is pumped out and recycled after decontamination.

Solid wastes consist of secondary wastes from water decontamination treatments such as sludge and filter zeolite, rubble from building explosions, dismantled concrete/metal debris from cleanup work, organic debris from deforestation to make space for waste storages and other cleanup facilities, and radioactive nuclear fuel/debris in damaged units. Total volume of the accumulated rubble and debris have already reached 122,000 m^3 [5] and are expected to increase as cleanup work proceeds. Although Fukushima Daiichi station site is one of the largest nuclear power station sites in Japan, it should not be regarded as inexhaustible as the areas set for settlement of nuclear accidents in other countries [6].

Table 1 shows types, dose rates, and volumes of already retrieved accident wastes.

Table 1. Retrieved accident waste

Generation Process	Waste type	Dose rate [mSv/h][7]	Quantity [5,8]
Cooling water decontamination	Sludge	Approx. 10^3	597 m^3
	Zeolite	Approx. 10^1 at vessel surface	458 vessel
	Concentrated liquid waste	Approx. 10^2 beta ray	Approx. 195,000 m^3
As found / Dismantlement	Concrete / metal debris	Approx. from 10^{-3} to 10^3	Approx. 54,000 m^3
Deforestation	Felled tree, Soil	Approx. from 10^{-3} to 10^{-1}	Approx. 68,000 m^3

This paper concentrates hereafter on concerns for management of secondary wastes, organic debris and rubble stored at and researches for characterization and safe storage, processing, and disposal of the waste.

ON-SITE WASTES AND MAJOR CONCERNS

Organic and inorganic debris and rubble from deforestation, explosion and dismantling are currently accumulated on-site as shown in Table 2. Rubble and other dismantled debris are sorted by dose rate and stored separately. Figure 1 and 2 show temporary storage design for rubble and organic debris. Both designs have radiation shielding reducing dose from waste storage areas to 1mSv/y at the site periphery. Samples are taken periodically from various points of storage area and dismantled locations for future analysis.

Table 2. Amount of debris and felled trees being temporarily stored (as of September 28, 2012)

Amount of debris and felled trees being temporarily stored (as of September 28, 2012)

Storage location	Air dose rate at area border (mSv/h)	Type	Storage method	Storage quantity[*1]	Area occupation rate
Solid waste storage	0.06	Concrete. metal	Container	2,000 m³	36 %
A : north side of site	0.36	Concrete. metal	Temporary storage facility	11,000 m²	98 %
B : north side of site	0.04	Concrete. metal	Container	4,000 m²	98 %
C : north side of site	0.01	Concrete. metal	Stored outside	26,000 m²	83 %
D : north side of site	0.01	Concrete. metal	Stored outside	2,000 m²	86 %
E : north side of site	0.01	Concrete. metal	Stored outside	3,000 m³	90 %
F : north side of site	0.01	Concrete. metal	Container	1,000 m³	99 %
L : north side of site	under 0.01	Concrete. metal	Soil-covered temporarily storage facility	2,000 m³	26 %
Total (Concrete. metal)				54,000 m³	76 %
G : north side of site	0.01	Felled tree	Stored outside	18,000 m³	83 %
H : north side of site	0.01	Felled tree	Stored outside	16,000 m³	93 %
I : north side of site	0.03	Felled tree	Stored outside	11,000 m³	100 %
J : south side of site	0.06	Felled tree	Stored outside	12,000 m³	77 %
K : south side of site	0.04	Felled tree	Stored outside	5,000 m³	100 %
M : west side of site	0.01	Felled tree	Stored outside	6,000 m³	29 %
Total (Felled tree)				68,000 m³	74 %

※1: Totals may not add up since numbers have been rounded for under 10 containers and less than 1,000m3 of volume. The total amount of felled trees is less than 1,000m3 compared with last time so a value of "0" has been entered.

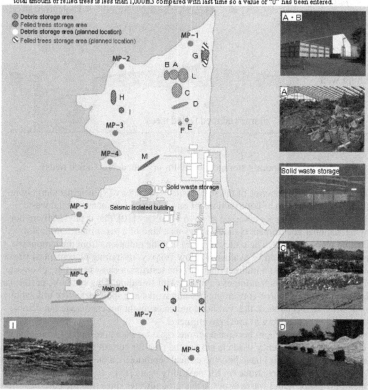

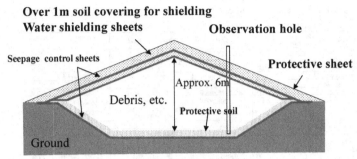

Figure 1. Overview of the temporary storage facility for rubble

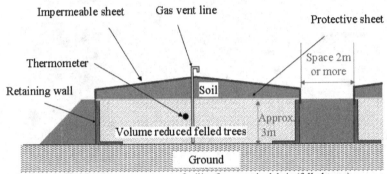

Figure 2. Overview of the temporary storage facility for organic debris (felled trees)

Dose rates at the peripheries of the station site, $\sim 10^2$ mSv/y [9], originate largely in cesium fallout. The summation of contributions of on-site temporally accumulated waste storages to the periphery doses are assessed as up to 10 mSv/y [10], that is physically minor but regarded by the national government cleanup policy as a kind of a performance indicator showing cleanup work progress. The policy requires that the radiations from the temporary storages to the peripheries should be evaluated below 1mSv/y, restricting placement of waste storages near the peripheries in an indirect manner. The restriction causes shortage of on-site waste storage area and/or imposes storages equipped with shielding. This symbolic performance indicator forces time-and-resource-consuming cleanup works and spoils effective utilization of limited on-site area. In this context, this symbolic indicator might not encourage but dispute cleanup works, and evoke anxiety of local governments.

On-site worker dose might become a serious issue for human resource management. Although dose levels of emergency workers are not the case for current cleanup workers, legal limits (50mSv/y and 100mSv/5y) have become critical conditions on the resource management for cleanup works expected to continue for the next forty years [11].

One of the major liquid waste issues is that groundwater sinks into the basement floors and increases the amount of accumulated water. Any agreements for ocean release of surplus water, which is slightly contaminated but could be treated further to below release limit, have not been made due to current environmental policy issues. The surplus is about 500 tons per day and temporally stored on-site [4]. Storage tanks have been multiplying and reducing open on-site areas.

In addition to a series of serious boundary breaks from fuel degradation to building explosion, accident management activities such as seawater/boron injection and subsequent cooling water circulation operation might have influences on volumes, physical/chemical form and mobility of radionuclide by contacting and reacting with a variety of on-site materials. The contamination processes, which determine volumes, characteristics and radioactive inventories of accident wastes, are far from the usual plant contamination process originated in neutron activation of reactor materials. We believe that a brand-new waste management strategy should be considered to handle accumulation, identification, analysis, storage, processing and final disposal for the Fukushima Daiichi case.

Sludge, zeolite and concentrated effluent are produced through cooling water purification treatments. Sampling of these wastes from spent adsorbent vessels or a temporary storage is a never easy task due to their high dose rate and/or hard-to-hollow welded vessels designed, manufactured, and installed by a rush work after the accident.

Hydrogen produced inside spent adsorbent vessels due to radiolysis is a safety issue for long-term storage of the secondary wastes. Chlorides contained in these wastes as a result of tsunami and seawater injection are also a point of safety concern for future geologic disposal.

New decontamination system is under preparation to reduce concentrations of a variety of radionuclides in treated water furthermore for public acceptance of ocean release [12]. The system employs a variety of new and best suited adsorbents and co-precipitation agents. Waste management strategy should also cover these newly born secondary wastes from the system.

WASTE MANAGEMENT STRATEGY AND CLEANUP ROADMAP

Fukushima Daiichi cleanup roadmap is shown in Figure 3. All activities are still in a primary stage of the roadmap that should be further modified and substantiated to clarify the endpoint, boundary conditions and a right way for the entire decommissioning of the station site.

Proper management of wastes from cleanup work is an essential matter for comprehensive decommissioning of the station site. It is, however, facing the above-mentioned social/political issues and highly complicated technical difficulties. United workforce of members from the government, TEPCO, JAEA and other Japanese top authorities are now seriously planning, preparing and performing national level activities for cleanup.

Waste management strategy should contain plans and schedule of waste management, such as characterization, storage, processing, and disposal, for cleanup of the site. The plans approaching toward a suitable endpoint shared among all cleanup authorities and stakeholders should be accomplished. The strategy should also make clear boundary conditions, ensure accessibility to various waste management knowledge, and define resources necessary for acquiring new data and technologies if needed. Waste data and knowledge necessary for reaching the endpoint should be systematized and databased. Radionuclide inventories and other factors with significant impacts on safety of geologic disposal should be evaluated and optimized through system design for waste processing and final disposal using the database. Plans with priorities and schedules for these cleanup steps applied for various areas and materials should be implemented and reviewed periodically till the endpoint is reached.

261

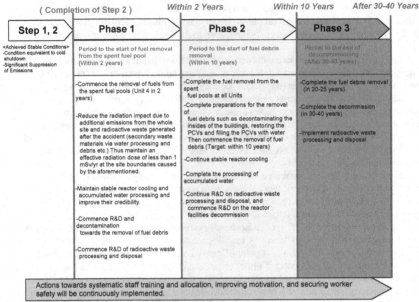

(Completion of Step 2)	Within 2 Years	Within 10 Years	After 30-40 Years
Step 1, 2	**Phase 1**	**Phase 2**	**Phase 3**
<Achieved Stable Conditions> -Condition equivalent to cold shutdown -Significant Suppression of Emissions	Period to the start of fuel removal from the spent fuel pool (Within 2 years)	Period to the start of fuel debris removal (Within 10 years)	Period to the end of decommissioning (After 30-40 years)
	-Commence the removal of fuels from the spent fuel pools (Unit 4 in 2 years) -Reduce the radiation impact due to additional emissions from the whole site and radioactive waste generated after the accident (secondary waste materials via water processing and debris etc.) Thus maintain an effective radiation dose of less than 1 mSv/yr at the site boundaries caused by the aforementioned. -Maintain stable reactor cooling and accumulated water processing and improve their credibility. -Commence R&D and decontamination towards the removal of fuel debris -Commence R&D of radioactive waste processing and disposal	-Complete the fuel removal from the spent fuel pools at all Units -Complete preparations for the removal of fuel debris such as decontaminating the insides of the buildings, restoring the PCVs and filling the PCVs with water Then commence the removal of fuel debris (Target: within 10 years) -Continue stable reactor cooling -Complete the processing of accumulated water -Continue R&D on radioactive waste processing and disposal, and commence R&D on the reactor facilities decommission	-Complete the fuel debris removal (in 20-25 years) -Complete the decommission (in 30-40 years) -Implement radioactive waste processing and disposal

Actions towards systematic staff training and allocation, improving motivation, and securing worker safety will be continuously implemented.

Figure 3. Mid-and-long Term Roadmap Summary

R&D for Resolving Technical Issues

Scope

Japan Atomic Energy Agency (JAEA) has extensively conducted R&D on management of the accident waste based on the Fukushima Daiichi research and development roadmap [9]. The scope of the R&D includes the following major areas.

- Characterization of the waste
- Investigation for safe storage of the waste
- Investigation for processing and disposal of the waste

(1) Characterization of the waste

Although characterization of the waste is essential to plan processing and disposal of the waste, analysis for the characterization is very time-consuming process because of enormous volume and complex composition of the waste caused by the core melt and the subsequent hydrogen explosions. JAEA has analyzed properties of relatively low activity wastes such as contaminated water after treatment, building debris, and felled tree stored at the Fukushima Daiichi NPS since January 2012. Among radioactive nuclides, long-lived alpha and beta nuclides are important for safety assessment of waste disposal. Chemicals in the waste which influence its processing and disposal will be analyzed in FY2013. JAEA is also developing analytical techniques and methods to save time and to improve detection limits.

(2) Investigation for safe storage of the waste

The major wastes stored at the Fukushima Daiichi NPS are, building debris, felled trees, contaminated water, and secondary wastes generated by treatment of the contaminated water. The first three wastes are technically easy to store except a problem of their enormous volumes. The secondary wastes such as spent zeolites and coagulation-sedimentation sludge have some issues for storage. Radioactive cesium concentrated in wastes and saline remained in wastes might cause radiolysis hydrogen fire and corrosion of storage vessels.

JAEA has investigated hydrogen gas generation/diffusion and corrosion of vessels for safe storage of the secondary waste of contaminated water treatment. Figure 4 shows the approach for safe storage of the wastes. The major subjects of the study are characterization of the wastes, hydrogen gas diffusion simulation in storage vessel, corrosion test of storage vessels, and chemical stability test of the waste. JAEA will prepare the measures for safe storage of the spent zeolites and the coagulation-sedimentation sludge in FY 2013.

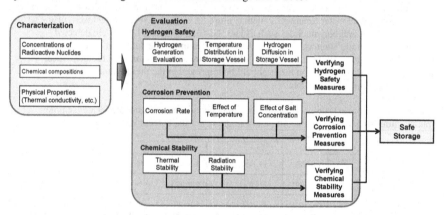

Figure 4. Procedures for safe storage of contaminated water treatment wastes

(3) Investigation for processing and disposal of the waste

There are many issues for processing and disposal of the accident waste because of huge volume, complex radiological and chemical composition, etc. The detailed R&D plan for processing and disposal of the waste is being prepared and will be established by the end of FY2012.The R&D should be conducted by the following steps. First, processing and disposal of the waste using existing technologies will be evaluated. If there are issues for waste processing or disposal using conventional technologies, then new technologies should be investigated to resolve the issues and finally safety of processing and disposal should be verified.

Progress Status

JAEA has conducted R&D on the accident waste management for a year and a half. The investigation is only the beginning of the long way for cleanup of the Fukushima Daiichi NPS and extensive R&D should be needed for the future. This section describes the results of JAEA's study and R&D subjects required for proper management of the accident waste.

(1) Characterization of the waste

Radioactive nuclides contained in the stored waste have been analyzed since January 2012. About 30 nuclides which are important for assessment of waste processing and disposal were selected as objective. At first, nine contaminated water samples from the contaminated water treatment system was analyzed in order to evaluate inventory of the secondary wastes generated from the contaminated water treatment, because it is difficult to sample the secondary wastes due to their extremely high activity. Then, three new contaminated water samples and 12 building debris and five felled trees are being analyzed. The analysis will be conducted to understand overview of distribution of radioactive nuclides in the waste for the coming a few years and then will be conducted to evaluate inventory of the waste.

Table 3 shows concentrations of radioactive nuclides in the contaminated water samples from the contaminated water treatment system (Figure 5). Cesium-137 and Sr-90 were major radioactive nuclides and beta nuclides such as H-3, Ni-63, Se-79, and I-129 were detected. Although alpha nuclides such as U, Np, Pu, Am, Cm were under the detection limit, the detection limits were relatively high because of small sample volume on each radiochemical analysis ($100 \ mm^3$). The detection limits should be improved in the present analysis by increasing the volume of the sample.

Table 3. Activity concentrations of gamma and beta nuclides in contaminated water samples (Alpha nuclides such as U, Np, Pu, Am, and Cm are under the detection limits.)

No.	Sample	Sampling Date	Activity Concentration (Bq/cm^3)						
			^{60}Co	^{137}Cs	^3H	^{63}Ni	^{79}Se	^{90}Sr	^{129}I
1	Concentrated RW basement highly contaminated water	Nov/1/2011	4.9×10^0	7.4×10^5	3.3×10^3	6.3×10^{-1}	8.3×10^0	2.9×10^5	2.5×10^{-1}
2	Water after cesium adsorption device processing (Tandem)	Aug/9/2011	1.7×10^1	1.1×10^4	6.0×10^3	1.5×10^0	2.7×10^0	1.2×10^5	8.3×10^{-2}
3	Water after cesium adsorption device processing (Single)	Nov/8/2011	7.4×10^0	7.7×10^0	4.0×10^3	7.4×10^{-1}	2.5×10^0	2.0×10^5	2.7×10^{-1}
4	Water after decontamination device processing	Aug/9/2011	9.9×10^0	5.3×10^{-1}	6.3×10^3	4.4×10^{-1}	3.1×10^0	1.2×10^4	8.5×10^{-2}
5	Water after 2nd cesium adsorption device processing	Nov/8/2011	4.6×10^{-1}	$<2.7 \times 10^{-1}$	3.3×10^3	$<3.8 \times 10^{-1}$	1.6×10^1	1.0×10^5	1.3×10^{-1}
6	Desalination device outlet water	Nov/1/2011	$<6.0 \times 10^{-2}$	$<1.3 \times 10^{-1}$	3.9×10^3	$<3.1 \times 10^{-1}$	8.1×10^{-1}	4.0×10^1	$<2.1 \times 10^{-2}$
7	Evaporative concentration device inlet water	Nov/1/2011	1.4×10^1	6.6×10^0	6.1×10^3	1.1×10^0	3.0×10^0	2.3×10^4	1.8×10^{-1}
8	Evaporative concentration device outlet water	Nov/1/2011	$<6.1 \times 10^{-2}$	$<1.3 \times 10^{-1}$	5.4×10^3	$<3.2 \times 10^{-1}$	7.8×10^{-1}	3.5×10^{-1}	$<2.1 \times 10^{-2}$
9	Evaporative concentration device concentrated waste water	Nov/3/2011	2.7×10^0	5.3×10^1	6.2×10^3	$<3.1 \times 10^{-1}$	9.4×10^1	3.2×10^3	1.3×10^0

264

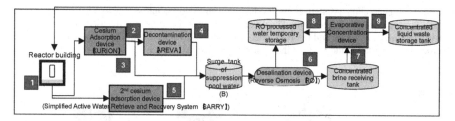

Figure 5. Contaminated water sampling points

In parallel with the radioactive nuclide analysis, new analysis techniques have been developed to deal with high-activity waste samples rapidly and remotely. Figure 6 shows the schematic diagram of laser ablation assisted resonance-enhanced multi photon ionization mass spectrometry. This technique has been developed for analysis of long-lived beta nuclides such as Mo-93. Sample is atomized by laser ablation and aimed nuclide is selectively ionized by resonance-enhanced multi photon ionization and analyzed using the subsequent time of flight mass spectrometer. Concerning alpha nuclides, capillary electrophoresis has been applied for separation of actinides. This technique is suitable for operation in glove box and cell because it can automatically separate actinides and reduce secondary waste significantly.

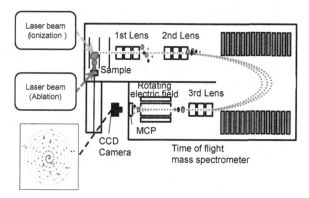

Figure 6. Schematic diagram of laser ablation assisted resonance-enhanced multi photon ionization mass spectrometry

(2) Safe storage of the waste

The secondary waste from the contaminated water treatment system has issues such as containing saline and high concentrations of radioactive nuclides. JAEA has studied hydrogen generation and diffusion in a spent zeolite vessel, corrosion behavior of storage vessels, etc. for safe storage of the waste.

Figure 7 shows analytical results on hydrogen diffusion in the spent zeolite vessel [13]. Analytical conditions were uniform hydrogen production rate of 18.3L/day, 237W of decay heat, and no steam generation. Analytical results show that air is slowly introduced into the vessel bottom through the outlet water pipe. The mixed gas of air and hydrogen is released to outside through the inlet water pipe and the vent tube. This thermal-hydraulic behavior keeps the hydrogen concentration under 4% of the hydrogen explosive limit.

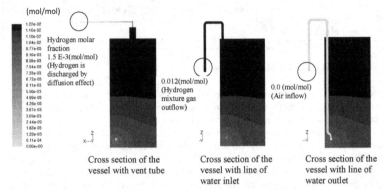

Cross section of the vessel with vent tube   Cross section of the vessel with line of water inlet   Cross section of the vessel with line of water outlet

Figure 7. Analytical results on hydrogen diffusion in a spent zeolite vessel

Salt and radiolysis products such as hydrogen peroxide might accelerate corrosion of storage vessels. The effect of salt concentration on corrosion has been investigated by corrosion test with test pieces and computer simulation. No corrosion was observed in a 500-hour corrosion test using test pieces and artificial seawater (2,000 ppm of Cl⁻). Corrosion test for evaluating effect of gamma ray radiation is being prepared. The basic results on storage vessel corrosion will be summarized by the end of FY 2013.

(3) Investigation for processing and disposal of the waste
The detailed R&D plan for processing and disposal of the waste is being discussed and it will be established by the end of FY2012 (March 2013). The plan will aim the following goals.

- 2017: Evaluating applicability of existing methods in accordance with waste characteristics
- 2021: Confirming safety predictions of waste processing and disposal

The investigation plan should contain wide range of topics such as characterization, inventory estimation, storage, processing, transportation, disposal concept, disposal safety assessment, etc.

CONCLUSIONS

➢ The Fukushima Daiichi nuclear power station accident produced significant volume of radioactive waste.
➢ The waste has very different characteristics from usual radioactive waste produced in nuclear power stations due to core melt, hydrogen explosions, seawater/boron injection, etc.

- At present, the biggest issue is shortage of storage area for the waste by government policy.
- The unusual waste provides a lot of issues on storage, processing, final disposal, etc. and it should require a brand new management strategy and extensive R&D.
- R&D works such as radionuclide analysis of the waste, hydrogen generation/diffusion analysis in the storage vessel, etc. have been performed in JAEA for characterization and safe storage of the waste.
- The detailed R&D plan for processing and disposal of the waste will be established by the end of FY2012.

REFERENCES

1. Tokyo Electric Power Company, Inc., "Fukushima Nuclear Accident Analysis Report", http://www.tepco.co.jp/en/press/corp-com/release/betu12_e/images/120620e0104.pdf, June 20,2012
2. Nuclear Emergency Response Headquarters Government-TEPCO Mid-and-long Term Response Council, "Progress Status of Mid-and-long Term Roadmap towards the Decommissioning of Units 1-4 of TEPCO Fukushima Daiichi Nuclear Power Station (Outline)", http://www.tepco.co.jp/en/nu/fukushima-np/roadmap/images/m121022-e.pdf, October 22, 2012
3. Tokyo Electric Power Company Inc., "Effort for treating radioactive accumulated water ~Outline Version~", http://www.tepco.co.jp/en/nu/fukushima-np/images/handouts_111022_02-e.pdf, October 22, 2011
4. Tokyo Electric Power Company, Inc., "Situation of storing and treatment of accumulated water including highly concentrated radioactive materials at Fukushima Daiichi Nuclear Power Station (70th Release)", http://www.tepco.co.jp/en/press/corp-com/release/betu12_e/images/121024e0301.pdf, October 24, 2012
5. Nuclear Emergency Response Headquarters Government-TEPCO Mid-and-long Term Response Council, "Progress Status of Mid-and-long Term Roadmap towards the Decommissioning of Units 1-4 of TEPCO Fukushima Daiichi Nuclear Power Station (Outline)", http://www.tepco.co.jp/nu/fukushima-np/roadmap/images/m121022_05-j.pdf, October 22, 2012 (in Japanese)
6. Ministry of Ukraine of Emergencies, "<<Twenty-five Years after Chernobyl Accident: Safety for the Future>> National Report of Ukraine", ISBN 978-966-1547-64-2,2011
7. Tokyo Electric Power Company, Inc., "Development of technologies for the processing and disposal of radioactive waste", http://www.tepco.co.jp/en/nu/fukushima-np/roadmap/images/m120314_03-e.pdf, March 14, 2012
8. Tokyo Electric Power Company, Inc., "Situation of storing and treatment of accumulated water including highly concentrated radioactive materials at Fukushima Daiichi Nuclear Power Station (70th Release)", http://www.tepco.co.jp/en/press/corp-com/release/betu12_e/images/121024e0301.pdf, October 24, 2012
9. Ministry of Education, Culture, Sports, Science and Technology, "Results of Airborne Monitoring in Restricted Areas and Deliberate Evacuation Areas", http://radioactivity.mext.go.jp/en/contents/5000/4709/24/203_0224_e.pdf, February 24, 2012

10. Tokyo Electric Power Company Inc., "Instructions on managing radiation exposure through radioactive materials and objects contaminated by radioactive materials and disposing radioactive waste", http://www.tepco.co.jp/cc/press/betu12_j/images/121019j0304.pdf, October 19, 2012 (in Japanese)

11. Nuclear Emergency Response Headquarters Government and TEPCO's Mid-to-long Term Countermeasure Meeting, "Mid-and-long-Term Roadmap towards the Decommissioning of Fukushima Daiichi Nuclear Power Station Units 1-4, TEPCO", http://www.tepco.co.jp/nu/fukushima-np/roadmap/images/t120730_02-j.pdf, July 30, 2012 (in Japanese)

12. Tokyo Electric Power Company Inc., "Radioactive liquid waste treatment facility and related facilities/equipments", http://www.tepco.co.jp/cc/press/betu12_j/images/121019j0305.pdf, October 19, 2012 (in Japanese)

13. H. Nakamura, "Research for Treatment and Disposal of Secondary Waste Produced by the Processing of Contaminated Water", Kankyo Gijutsu, **41**, 365 (2012). (in Japanese)

Mater. Res. Soc. Symp. Proc. Vol. 1518 © 2013 Materials Research Society
DOI: 10.1557/opl.2013.75

Decontamination of School Facilities in Fukushima-city

Hideki Yoshikawa[1,5], Kazuki Iijima[1,5], Hiroshi Sasamoto[1,5], Kenso Fujiwara[1,5], Seiichiro Mitsui[1,5], Akira Kitamura[1], Hiroshi Kurikami[1,5], Takayuki Tokizawa[2,5], Mikazu Yui[1,4], and Shinichi Nakayama[3,5]

[1] Geological Isolation Research and Development Directorate, Japan Atomic Energy Agency

[2] Ningyo-toge Enviromnental Engineering Center, Japan Atomic Energy Agency

[3] Policy Planning and Administration Department, Japan Atomic Energy Agency

[4] Nuclear Science Research Institute, Japan Atomic Energy Agency

[5] (Present affiliation) Fukushima Environmental Safety Center, Headquarters of Fukushima Partnership Operations, Japan Atomic Energy Agency, Okitama-cho 1-29, Fukushima

ABSTRACT

Following the release of radionuclides into the environment as a result of the accident at Fukushima Daiichi nuclear power plant, Japan Atomic Energy Agency (JAEA) had to develop an immediate and effective method of reducing the dose rate received by students in school facilities. A demonstration of a reducing method was carried out by JAEA at a junior high school ground and kindergarten yard in the center of Fukushima-city. Dose rates of the released radionuclides are largely controlled by the ground level contamination and accumulation of mainly Cesium137 (Cs-137) and Cesium 134 (Cs-134) in populated areas. An effective means of reducing dose rate was to remove the surface soil and to bury it on-site under fresh uncontaminated soil or soil collected under deep depth at the site for shielding. The dose rate at 1 m above ground level was reduced from 2.5 μSv/h to 0.15 μSv/h.

INTRODUCTION

Two explosions at the Fukushima Daiichi nuclear power plant on 12th and 14th March 2011 caused an uncontrolled release of radionuclides into the environment. The majority of the released radionuclides were deposited on the ground and accumulated at various surfaces such as plants, roofs, roads, and soils near the nuclear plant. In populated areas, dose rates related to the released radionuclides are largely controlled by the surface contamination and accumulation of Cesium137 (Cs137). While elevated dose rates recorded after the accident generally continued to decrease daily, elevated dose rates were recorded in May 2011 in Fukushima city, located some 60 km NW of the Fukushima Daiichi plant.

The Fukushima University Junior High School and the Kindergarten attached to Fukushima University are located in the central part of Fukushima city. About 450 students and about 70 pupils go to the school and the kindergarten. The junior high school covers a ground surface of about 15000m^2 and the kindergarten an area of about 1200m^2. They are surrounded by big trees, such as cherry and zelkova of over the 10m height, and make up a good educational environment. Both buildings of the junior high school and the kindergarten are made of concrete, therefore, air dose rate in these buildings are with less than 0.1μSv / h lower than outside. After the accident, an educator and parents had declined to exercise in the outside in order to reduce the radiation dose for students. The Ministry of Education, Culture, Sports, Science and Technology-Japan approached the Japan Atomic Energy Agency (JAEA) regarding development of an immediate and effective method of reducing the dose rate received by students in Fukushima-city's school facilities.We report here on details of the dose rate reduction project carried out at the abovementioned school.

EXPERIMENT 1

Radioactivity distribution at the site

1 Determination method of radionuclides

In order to identify the radionuclides contaminating the unpaved schoolyard and sandbox surface, sand samples were analyzed by using γ-ray spectroscopy (Seiko EG & G) of a monitoring car. Five samples were collected on the surface of sand in the sandbox court of the school facilities (Figure 1).

2. Air dose rate

Air dose rates were measured at 165 locations in the kindergarten yard and the junior high school grounds by NaI scintillation survey meter TCS-161 manufactured by Aloka. On the junior high school grounds, the measurement points were set up in a 10 m-mesh to cover the complete area. The measurements were carried out at three elevations above ground at each location. One location was at 1 cm above ground to determine the surface dose rate, the other measurement positions were at 50 cm and 1 m above ground to determine exposure dose rates of students, considering different body heights.

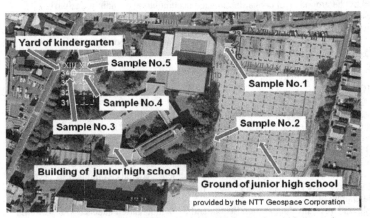

Figure 1. Picture of sampling locations in junior highschool and kindergarden

3. Depth profiles of the radionuclides

We performed a survey on the depth profiles of the radionuclides in order to narrow down the location of the contaminated soil to be removed from the school grounds and confirmed the effectiveness of the proposal measures through a demonstration project at the kindergarten yard. Radioactive I and Cs were determined by γ-ray spectrometry in the same manner as described above, and the air dose rates were also assessed out using a NaI scintillation survey meter (TCS-161). To measure the local surface value as accurately as possible we used a detector laterally surrounded with thin lead plates to shield γ-radiation from afar (from outside the school yard). A square hole with a depth of about 20 cm was excavated in the ground to measure the depth profiles of the radionuclides. The procedure is detailed below (Figure 2):

1) Outer of a square area of 50cm edge length, plastic sheeting was used to prevent contamination of soil particles from the surroundings. A screen panel was also set up as a windbreaker.
2) A surface soil sample was collected for gamma-ray spectral analysis in a plastic container.
3) We excavated the ground to a depth of 5cm below the surface. The excavated soil was placed in a plastic bag, stirred uniformly, and taken as a sample in a plastic container for analysis by γ-ray spectrometry.
4) Samples were also collected at 10cm and 20cm depth.

Figure 2. Soil sampling and dose rate measurement for the depth plofifes of the radionuclide

EXPERIMENT 2

Demonstration of dose rate reduction method

It is clear that cesium is the major radionuclide released due to the accident and is remaining in the environment. The dose rate is not easily reduced over short timeframes because Cs-137 has a half-life of 30 years and Cs-134 has a halflife of 2.1 y. Therefore, it was required to remove the radioactive cesium from the surface of the school yard. However, appropriate techniques were not immediately available, and in addition, it was difficult to prepare storage or disposal sites for large amounts of contaminated soil. Still, immediate measures were required to reduce the dose rate simply and quickly. The "turn over replacement method" that has been used in the remediation of heavy metal contaminated soil also proved to be an effective means of reducing the dose rate. This method involves removing the top soil and burying it on-site under fresh uncontaminated soil (including soil collected at greater depth at the site) for radiation shielding. A demonstration of this method was carried out at the kindergarten yard.

RESULTS AND DISCCUSION

Radioactivity measurement distribution at the site

It was found that mainly Cs-137 and Cs-134 occurred at high concentrations, caused by the Fukushima Daiichi nuclear power plant accident (Table I). Because the two Cs isotopes which have different half-lifes occurred at a ratio of unity, it can be concluded that the measured Cs did not result from fall-out-Cs of a past Chinese nuclear test but were precipitated from the atmosphere in each sandbox by the accident at the power plant.

Table I. Radionuclide concentration in sand (Bq/kg)(Sampling date: 7th ,May, 2011)

nuclide		Sample name				
	halflife	No.1	No,2	No.3	No.4	No.5
I-131	8.02 d	1100 ± 29.5	849 ± 27.2	277 ± 14.6	517 ± 20.9	296 ± 18.3
Cs-134	2.06 y	12400 ± 71.6	10400 ± 64.5	3640 ± 36.4	8310 ± 55.1	6470 ± 47.1
Cs-136	13.16 d	183 ± 17.7	144 ± 18.2	60 ± 10.1	112 ± 14.1	98.1 ± 12.2
Cs-137	30.17 y	14900 ± 85.2	13000 ± 78.5	4650 ± 44.9	10100 ± 66.1	7740 ± 56.6
Te-129m	33.6 d	6620 ± 521	6010 ± 486	1230 ± 261	3870 ± 422	3810 ± 351
K-40		419 ± 62.4	451 ± 68.9	527 ± 63.4	567 ± 67.5	263 ± 50.9
La-140	40.27h	26.3 ± 6.19	L.T.D	L.T.D	32.9 ± 7.26	21.7 ± 5.78
134/137Cs	-	0.83	0.80	0.78	0.82	0.83
136/137Cs	-	0.012	0.011	0.013	0.011	0.013
131I/137Cs	-	0.074	0.065	0.060	0.051	0.038
129m Te/137Cs	-	0.44	0.46	0.26	0.38	0.49
129mTe/ 131I	-	6.0	7.1	4.4	7.5	12.8

(L.T.D : <3σ, N.D : not detected)

272

The abundance ratios of I-131 and Cs-137, as well as of Te-129m and Cs-137, for each sample are also shown in the Table I. These ratios will be equal if each radionuclide has the same origin, but will be changed when samples correspond to different rainfall conditions or have experienced artificial disturbance, based on the differences in chemical properties of each radioelement. The data of I-131/Cs-137 and Te-129m/Cs-137 ratios show that No.3 and No.5 sandboxes had different conditions regarding rainfall and artificial disturbance during the timeframe from the accident to the sampling date.

At any location, air dose rates were decreasing as a function of distance from the surface. The values at ground level were about 1.7 ~ 4.5µSv/h (white column), whereas at the positions at 50 and 100 cm approximately 2.0 ~ 3.6µSv/h (hatched column) and about 2.0 ~ 3.0µSv/h (solid column) were detected, respectively (Figure 3). Average values were 3.1µSv/h on the surface, and about 2.9 and 2.5µSv/h at 50 cm and 1m elevation, respectively. Since almost the similar surface dose rate is observed regardless of location, it is considered that radionuclides are deposited uniformly in the entire area with similar soil.

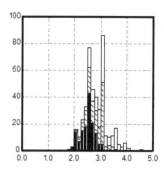

Figure 3. Distribution of dose rates at junior high school ground. The vertical axis is the frequency and the horizontal axis is the air dose rate.

Table II. Radionuclide concentrations in the school grounds (Sampling date: 7th ,May, 2011)

Nuclide (Bq/kg)	halflife	Sampling location				L.T.D
		Surface	0-5 cm	5-10 cm	10-20 cm	
I-131	8.02 d	2080 ± 67.5	94.8 ± 12.2	N.D	N.D	15.5±1.3
Cs-134	2.06 y	38500 ± 169	637 ± 23.7	94.3 ± 9.5	L.T.D	19.6± 3.2
Cs-136	13.16 d	403 ± 42.7	N.D	N.D	N.D	17.5 ± 0.7
Cs-137	30.17 y	46400 ± 203	785 ± 28.5	120 ± 11.2	L.T.D	19.5 ± 4.3
Dose rate		2.0 µSv/h	0.4µSv/h	0.3 µSv/h	0.1 µSv/h	

The radionuclide depth profile was shown in Table II. The dose rate at the surface before replacement of topsoil was 2.0 μSv/h, and 0.4 μSv/h, 0.3 μSv/h, and 0.1 μSv/h at 5cm, 10cm, and 21cm, respectively. It was found that the dose rate was reduced to approximately 1/10 at about 10 cm depth gelow the surface.

Demonstration of the effectiveness of the dose rate reduction method

Two square holes with 80 cm edge length, A and B (135cm distant), were selected as demonstration location at the kindergarten yard. Firstly, hole A was excavated to a depth of 50 cm, and then the surface soil of hole B was placed at the bottom of hole A. The hole A was backfilled with the soil from hole B finally. We judged to reach target depth when the hole A was excavated to 50cm and a clay-like layer could be observed, since the dose rate had decreased to 0.10 μSv/h at this position.

A dose rate of 2.10 μSv/h was obtained at the surface of hole B before excavation. The dose rate was also measured for four stages of backfilling (i.e., as a function of depth) at the bottom of hole A. These measurements gave 0.69 μSv/h, 0.36 μSv/h, and 0.22 μSv/h at about 23.0 cm, 13cm and 9cm, respectively. After backfilling was complete, the dose rate was 0.20 μSv/h at the surface of hole A, which corresponds to a reduction by a factor of about 10. Therefore, it is found that the dose rate will be reduced by about factor about 10 by replacement of the top soil and the radiation shielding effect of the fackfilled soil.

CONCLUSIONS

A demonstration of a dose rate reduction method was carried out by JAEA at a junior high school and kindergarten yard in the center of Fukushima-city. From this demonstration, the following can be concluded:
1. Distributions of radionuclides and dose rate were almost uniform throughout the junior high school grounds and kindergarten yard.
2. The radionuclides mainly exist at only surface of the grounds.
3. The dose rate at ground level was reduced significantly from 2.10 to 0.20 μSv/h (reduction to 1/10 of original value) by removal of the top 5-10 cm of soil and replacement by uncontgaminated soil collected from a depth of 50 cm.

Excavated soil was placed into underground trenchs at a selected location of the yard, taking into consideration the future management of the contaminated soil. Through this measure, the dose rate was reduced to 1/10-1/20 of the values measured before the decontamination. The dose rate at a level of 1 m above ground was reduced from 2.5 μSv/h to 0.15 μSv/h.

ACKNOWLEDGEMENTS

The authors would like to deeply appreciate the Fukushima University attached Junior High School and Kindergarten, and JAEA staff.

Mater. Res. Soc. Symp. Proc. Vol. 1518 © 2013 Materials Research Society
DOI: 10.1557/opl.2013.393

Investigation and Research on Depth Distribution in Soil of Radionuclides Released by the TEPCO Fukushima Dai-ichi Nuclear Power Plant Accident

Haruo SATO[1], Tadafumi NIIZATO[1], Kenji AMANO[2], Shingo TANAKA[2] and Kazuhiro AOKI[2]
[1] Headquarters of Fukushima Partnership Operations, Japan Atomic Energy Agency (JAEA), 1-29 Okitama, Fukushima, Fukushima 960-8034, Japan
[2] Geological Isolation Research and Development Directorate, Japan Atomic Energy Agency (JAEA), 432-2 Hokushin, Horonobe-cho, Hokkaido 098-3224, Japan

ABSTRACT

The accident of the TEPCO Fukushima Dai-ichi Nuclear Power Plant occurred by the 2011 off the Pacific coast of Tohoku Earthquake on 11 Mar. 2011. It is estimated that totally $1.2\text{-}1.5\times10^{16}$ Bq for ^{137}Cs and $1.5\text{-}1.6\times10^{17}$ Bq for ^{131}I were released until the beginning of Apr. and those radionuclides (RN) were deposited on soil surface and forest etc. widely around Fukushima Pref. This work was carried out as one of the investigations for making the distribution maps of radiation dose rate and soil contaminated by RNs which the MEXT promotes. The Geoslicer investigation on the depth distribution of RNs in soil was performed after 3 months from the accident. The investigation was conducted at 11 locations in Nihonmatsu City, Kawamata Town and Namie Town, and soil samples of depth 50 cm to 1 m were taken. Both of ^{134}Cs and ^{137}Cs were detected in all investigated locations, and ^{129m}Te and ^{110m}Ag were detected only in locations where radiation dose rates are high. At many locations investigated, radiocaesium more than 99 % distributed within a depth of 10 cm in soil in the surface layer. On the other hand, RNs tended to distribute to deeper part in soil at locations that are supposed to have been used as farmland than in soil in the surface layer, and radiocaesium more than 99 % in soil at locations that are supposed to have been used as farmland also distributed within a depth of around 14 cm. The apparent diffusion coefficients (D_a) of RNs derived from penetration profiles near the surface layer showed a tendency to be higher in soil at locations that are supposed to have been used as farmland ($D_a=0.1\text{-}1.5\times10^{-10}$ m^2/s) than in soil in the surface layer ($D_a=0.65\text{-}4.4\times10^{-11}$ m^2/s), and most D_a-values were nearly 10^{-11} m^2/s. The distribution coefficients (K_d) by a batch method were in the range of $K_d=2,000\text{-}61,000$ ml/g for Cs and $K_d=0.5\text{-}140$ ml/g for I. Although the K_d-values are different between cation (Cs$^+$) and anion (I$^-$), the D_a-values (^{134}Cs, ^{137}Cs, ^{129m}Te and ^{110m}Ag) were similar levels. This is considered to be due to that the D_a-values were controlled by dispersion by flow of rain water.

INTRODUCTION

The accident of the Tokyo Electric Power Company (TEPCO) Fukushima Dai-ichi Nuclear Power Plant (FD1-NPP) occurred by the 2011 off the Pacific coast of Tohoku Earthquake on 11 Mar. 2011. Part of the radionuclides (RN) in atomic reactors was released and deposited on soil surface and forest, etc. widely around Fukushima Pref. by rain or snow after being transported by wind. The total amount of RNs released until 5 Apr. by the FD1-NPP accident is estimated to reach 1.2×10^{16} Bq for ^{137}Cs and 1.5×10^{17} Bq for ^{131}I according to the publication of the Nuclear Safety Commission on 12 Apr. [1]. The JAEA also re-evaluated the total amount of RNs released for the same period and reported to be 1.3×10^{16} Bq for ^{137}Cs and 1.5×10^{17} Bq for ^{131}I on 12 May [1]. The Nuclear and Industrial Safety Agency estimated the total amount of RNs released until 15 Mar. to be 1.5×10^{16} Bq for ^{137}Cs and 1.6×10^{17} Bq for ^{131}I [1]. Although these values are a little scattering, it is approx. the same and is considered to be reliable. These radioactivity amounts are all evaluated lower than 1/10 of the 5.2×10^{18} Bq released by the Chernobyl NPP accident occurred in Apr. 1986 [2]. However, the FD1-NPP accident is evaluated as the worst level 7 on the International Nuclear Event Scale (INES).

Receiving such serious situation, the Cabinet Office's Council for Science and Technology Policy (CSTP) published the enforcement policy of project by promotion budget for strategy on science and technology (Science and Technology Policy) on 19 May [3]. The

Ministry of Education, Culture, Sports, Science and Technology (MEXT) initiated the monitoring of radiation dose rate and project for making the distribution maps of radiation dose rate and soil contaminated by RNs based on the CSTP's enforcement policy. This work was carried out as one of the investigations for making the distribution maps of radiation dose rate and soil contaminated by RNs which the MEXT promotes. In this work, an investigation on the depth distribution of RNs in soil as of about 3 months after the accident was carried out.

With respect to depth distribution of RNs in soil, e.g., Kato et al. carried out an investigation using a scraper plate in Yamakiya District, Kawamata Town where is located 40 km north-west from FD1-NPP in around the end of Apr. 2011, and it is reported that approximately 80 % of the deposited radiocaesium was absorbed by the surface of soil within the upper 2 cm [2]. Also as of 1 year after the FD1-NPP accident, it is said that most % of the radiocaesium which is considered to be predominating the radiation dose rate is trapped in soil within the upper 5 cm. On the other hand, migration of RNs in soil strongly depends on the sorption property of RNs onto soil such as whether it is reversible or irreversible, and RNs are possible to gradually migrate in soil depending on reversible and/or irreversible property in sorption reaction. However, quantitative data relating to migration in soil for RNs released by the FD1-NPP accident are quite limited.

In this paper, the authors describe investigation results on the depth distribution of RNs in soil as of about 3 months after the FD1-NPP accident and data relating to retardation and migration of RNs such as distribution coefficient (K_d) and apparent diffusion coefficient (D_a).

INVESTIGATION

Outline of the investigation

The Geoslicer investigation was carried out in mid-June 2011 after about 3 months from the FD1-NPP accident. Table I shows the investigation items. In the Geoslicer investigation, totally 29 slicers were drilled at 11 locations in Nihonmatsu City, Kawamata Town and Namie Town (the 20-60 km-range from FD1-NPP), and plate-shaped soil samples of depth approximately 50 cm to 1 m were obtained. After obtaining the soil samples, the surface was carefully scraped using a stickle and a brush to clean and remove any contamination resulting from sampling process, and then soil observation and soil description (including photo) were carried out. Then, radioactivity distribution (gamma activity) using the Imaging Plate was conducted to visualize the radioactivity distribution. The measurement was carried out in Fukushima City of low background. After the Imaging Plate measurement, soil samples for the measurements of physical property, sorption experiments and RN analysis (gamma decay nuclides) were taken from the plate-shaped samples. The sampling for the measurements of physical property and sorption experiments was conducted from 2 different depths based on the

Table I. The Investigation Items

Item	Method / condition
Investigation period	7-19 June 2011 (negotiation, sampling, pull out)
Sampling period	10-16 June 2011
Drilling method	Wide-sized Geoslicer (1.1m wide x 1m deep x 10cm thick): 1 slicer
	Handheld-sized Geoslicer (10cm wide x 1m deep x 2cm thick): 28 slicers
Investigation location	11 locations: Nihonmatsu City (1 location), Kawamata Town (3 locations), Namie Town (7 locations)
Soil observation / description	29 slicers, corrosion (existence of organic substances and root), humidity, clay content,
	grain size of soil & sediment (i.e., soil, mud, sand), color of soil (yellow soil, red soil, etc.)
Visualization of radioactivity distribution by the Imaging Plate	5 slicers, exposure time: 12 hours
Physical property	12 slicers, 2 samples/slicer
	Wet density, dry density, porosity, solid density, water content
Distribution coefficient (K_d)	Batch method (AESJ (2002.10) standard method), element: Cs (Cs-137), I (I-131)
Depth distribution of RNs	16 slicers (11 locations)
	Sampling pitch: 2cm pitch up to 10cm deep, 4cm pitch between depths 10cm and 30cm,
	10cm pitch in the part deeper than 30cm

results of soil observation and soil description. The soil samples for RN analysis were taken with 2 cm pitch up to a depth of 10 cm, 4 cm pitch between depths 10 cm and 30 cm and 10 cm pitch in the part deeper than 30 cm, and gamma decay nuclides were analysed for 60 min with a Ge semiconductor detector. The gamma analysis was carried out by the Japan Chemical Analysis Center.

Selection of investigation sites

Figure 1 shows the investigation sites plotted on the map of monitoring results of airborne measurement by the MEXT and USDOE [4]. Areas where a high deposition of ^{134}Cs and ^{137}Cs were confirmed and ^{131}I was expected to be remaining even taking into account its physical attenuation were firstly selected as candidate investigation areas (Kawamata Town, Namie Town, Iitate Village, etc.). Topographical viewpoint was also considered in the selection of investigation areas, because the pathway of the plume of RNs was considered to be affected by the topography. Moreover, the following items were considered in the detailed site selection.

(1) Land utilization → public land, no structures around the site and no traces of the recent human disturbance, non-cultivated site covered with plants such as grasses, etc.
(2) Hydrologic topography → no streams near the site, topographically plain, etc.
(3) Kind of soil → relatively soft to be able to drill by Geoslicer

Finally drilling points were determined by examining the hardness of soil using a soil auger (Kendojo) in respective candidate sites. Thus, 11 locations in Nihonmatsu City, Kawamata Town and Namie Town were finally selected as shown in Fig. 1 and Table I.

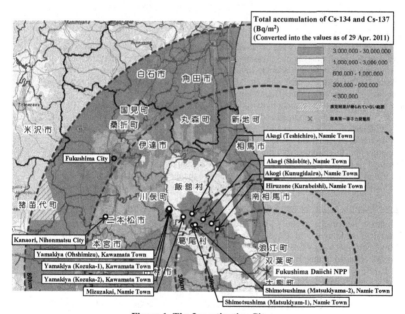

Figure 1. The Investigation Sites
(Plotted on the map of monitoring results of airborne measurement by the MEXT and USDOE (6 May 2011) [4])

Laboratory experiments

The measurements of physical property and sorption experiments were carried out in the laboratory. In the former, wet density, dry density, porosity, solid density and water content were measured and further the degree of saturation was calculated from those results. The measurements of dry density and porosity were carried out by a water saturation method [5], and the solid density was measured by a pycnometer method (JIS A 1202) [6]. These parameters were used for determining the concentration per unit soil volume (Bq/cm^3).

Sorption experiments onto soil were carried out for Cs and I by a batch method (AESJ (2002) standard method [7]). As a tracer in the sorption experiments, ^{137}Cs ($^{137}CsCl$ solution) and ^{131}I ($Na^{131}I$ solution) were used. Before experiments, soil sample was classified into particles under 2 mm after drying at 49 °C in an oven. An experiment solution for the sorption experiments was prepared by contacting the screened soil particles with deionized water with a solid-liquid ratio of 1:10 (g/ml) for 24 hours. The experiment solution was finally obtained by separating solid phase from suspension by centrifugation and 0.45 μm syringe filter.

The batch sorption experiments were carried out by contacting soil particles (under 2 mm) with the experiment solution with a solid-liquid ratio of 1:10 (g/ml) for 7 days at 25 °C in triplicate. Initial concentrations of both nuclides in the sorption experiments are 100 Bq/ml for ^{137}Cs and 200 Bq/ml for ^{131}I. During the reaction, all reaction bottles (polypropylene) were shaken with 100 rpm and pH and Eh were also monitored. After 7 days, solid-liquid separation was carried out by 0.45 μm syringe filter and the filtered solution was counted by NaI(TI) counter (ALOKA, AccuFLEX-ARC-7000). The K_d-values of Cs and I onto soil were obtained from the differences between initial and final concentrations of respective elements in solution.

RESULTS AND DISCUSSIONS

Depth distributions of RNs in soil

^{134}Cs, ^{137}Cs, ^{129m}Te and ^{110m}Ag were detected from radioactivity analysis. Both of ^{134}Cs and ^{137}Cs were detected in all investigated locations, and ^{129m}Te and ^{110m}Ag were detected only in locations where radiation dose rate and the inventories of RNs have been evaluated to be high. Particularly ^{110m}Ag was detected only in Akōgi District (Kunugidaira) and Hiruzone District (Kurabeishi), Namie Town. The penetration distributions of ^{134}Cs and ^{137}Cs in soil were approximately overlapped, and the radioactivity ratio of ^{137}Cs to ^{134}Cs was approximately 1:1 at all investigated locations. This indicates that the behavior of Cs is entirely the same independent on nuclide.

At many locations investigated, radiocaesium more than 99 % distributed within a depth of 10 cm excepting soil at locations that are supposed to have been used as farmland. RNs tended to distribute to deeper part in soil at locations that are supposed to have been used as farmland than in soil in the surface layer. Radiocaesium more than 99 % in soil at locations that are supposed to have been used as farmland also distributed within a depth of around 14 cm.

Distribution coefficients (K_d) of Cs and I onto soil

Table II shows the results of the K_d-values of Cs and I onto soil obtained by a batch method. Although a wide range of the K_d-values for Cs were obtained in the range of $K_d=2,000-61,000$ ml/g, many K_d-values were in the range of $K_d=2,000-20,000$ ml/g, and it was confirmed that soil obtained in this work wholly has a fairly high sorption property.

Whilst, a wide range of the K_d-values for I were obtained in the range of $K_d=0.5-140$ ml/g, and many K_d-values were in the range of $K_d=20-30$ ml/g. Therefore, it was confirmed that even anion such as I is sorbed onto soil. The cause of a wide range of the K_d-values is considered to be variations in the kind and content of clay mineral contained in soil and the content of organic matters, etc. The effect of these factors on sorption is future work.

Table II. The Results of the K$_d$-values of Cs and I onto Soil

Location	Soil	Sampled depth interval (cm)	Representative depth (cm)	Distribution coefficient (K$_d$)							
				Cs-137				I-131			
				pH	K$_d$ (ml/g)	Error (±)	N	pH	K$_d$ (ml/g)	Error (±)	N
Yamakiya (Kozuka), Kawamata Town	Sandy	15-20	17.5	5.8	1.87E+04	6.68E+03	3	5.8	4.44E+00	1.66E-01	3
		35-40	37.5	6.4	8.67E+03	1.63E+03	3	6.2	2.84E+01	2.92E-01	3
Yamakiya (Ohshimizu), Kawamata Town	Sandy	5-15	10.0	6.8	1.51E+04	6.52E+03	2	6.6	1.25E+00	8.73E-02	3
		33-41	37.0	7.3	4.31E+04	4.14E+04	3	7.1	5.52E-01	1.25E-01	3
Kamori, Nihonmatsu City	Sand	4-12	8.0	6.2	3.30E+03	2.38E+02	3	6.1	1.36E+00	1.34E-01	3
	Clayish	21-38	29.5	6.5	6.10E+04	9.75E+04	3	6.5	9.24E-01	1.31E-01	3
Shimotsushima (Matsukiyama-1), Namie Town	Sand	8-19	13.5	6.7	4.01E+04	3.56E+02	3	6.6	1.20E+01	1.90E-01	3
	Sandy	55-65	60.0	6.2	2.17E+04	1.60E+04	3	6.2	1.31E+01	1.54E-01	3
Hiruzone (Kurabeishi), Namie Town	Sandy	10-25	17.5	6.5	2.16E+03	1.08E+02	3	6.6	2.08E+01	3.16E-01	3
	Clayish	33-46	39.5	6.5	3.53E+04	4.79E+04	2	6.5	2.58E+00	1.47E-01	3
Akogi (Kunigidaira), Namie Town	Sandy black	10-15	12.5	6.0	2.08E+03	9.85E+01	3	6.1	1.19E+01	1.00E-01	3
		30-35	32.5	6.0	2.86E+03	1.85E+02	3	6.1	1.03E+01	1.86E-01	3
Akogi (Teshichiro), Namie Town	Sandy	7-15	11.0	6.5	2.98E+03	1.92E+02	3	6.4	9.81E+01	6.53E-01	3
		29-36	32.5	6.1	7.43E+03	1.25E+03	3	6.2	5.99E+01	5.18E-01	3
Mizuzakai, Namie Town	Sandy black	8-18	13.0	5.8	2.24E+03	1.19E+02	3	5.8	3.08E+01	5.13E-01	3
	Black	40-50	45.0	5.5	2.37E+03	1.26E+02	3	5.5	2.33E+01	4.12E-01	3
Akogi (Shiobite), Namie Town	Sandy	10-23	16.5	5.8	2.16E+03	1.10E+02	3	5.7	1.42E+02	1.47E+00	3
		31-44	37.5	5.7	2.18E+03	1.07E+02	3	5.6	1.67E+01	3.08E-01	3
Shimotsushima (Matsukiyama-2), Namie Town	Black (Kuroboku soil)	5-15	10.0	5.6	2.84E+03	1.86E+02	3	5.6	8.78E+00	1.35E-01	3
		20-30	25.0	5.6	2.97E+03	2.03E+02	3	5.6	1.30E+01	1.65E-01	3
Yamakiya (Kozuka-2), Kawamata Town	Sandy	5-15	10.0	6.1	2.88E+03	1.73E+02	3	6.2	1.84E+01	2.53E-01	3
		21-30	25.5	7.5	3.37E+03	2.40E+02	3	6.4	2.76E+01	2.53E-01	3
Yamakiya (Kozuka-1), Kawamata Town	Sandy	10-20	15.0	6.5	1.30E+04	5.15E+03	3	7.6	1.76E+01	1.85E-01	3
	Sand	32-43	37.5	6.2	4.24E+04	8.75E+04	3	6.1	2.37E+01	2.25E-01	3

Apparent diffusion coefficients (D$_a$) of RNs near the surface layer

The D$_a$-values of RNs were derived from penetration profiles near the surface layer assuming that all radioactive materials averagely moved in soil. Table III shows the estimated results of the D$_a$-values. The D$_a$-values showed a tendency to be higher in soil at locations that are supposed to have been used as farmland (D$_a$=0.1-1.5x10^{-10} m^2/s) than in soil in the surface

Table III. The Estimated Results of the D$_a$-values of RNs near the Surface Layer

Location	Radionuclides	Apparent diffusion (dispersion) coefficient D$_a$ (m^2/s)	
Yamakiya (Kozuka-2), Kawamata Town	Cs-134	1.704E-11	4.433E-11
	Cs-137	1.705E-11	4.305E-11
	Te-129m (Te-129)		9.848E-11
Yamakiya (Ohshimizu), Kawamata Town	Cs-134	1.037E-11	
	Cs-137	1.045E-11	
Kamori, Nihonmatsu City	Cs-134	2.995E-11	
	Cs-137	3.048E-11	
Shimotsushima (Matsukiyama-1), Namie Town	Cs-134	7.358E-12	1.054E-11
	Cs-137	7.421E-12	1.033E-11
	Te-129m (Te-129)	9.228E-12	1.090E-11
Hiruzone (Kurabeishi), Namie Town	Cs-134	6.523E-12	1.896E-11
	Cs-137	6.523E-12	1.912E-11
	Te-129m (Te-129)	6.691E-12	1.168E-11
	Ag-110m	6.696E-12	1.249E-11
Akogi (Kunugidaira), Namie Town	Cs-134	1.454E-10	1.639E-11
	Cs-137	1.467E-10	1.616E-11
	Te-129m (Te-129)	3.052E-11	1.311E-11
	Ag-110m	1.109E-10	5.412E-11
Akogi (Teshichiro), Namie Town	Cs-134	8.389E-12	
	Cs-137	8.360E-12	
	Te-129m (Te-129)	8.155E-12	
Mizuzakai, Namie Town	Cs-134	1.313E-11	2.172E-11
	Cs-137	1.287E-11	2.211E-11
Shimotsushima (Matsukiyama-2), Namie Town	Cs-134	1.031E-11	
	Cs-137	1.046E-11	
	Te-129m (Te-129)	1.347E-11	
Yamakiya (Kozuka-1), Kawamata Town	Cs-134	2.458E-11	
	Cs-137	2.464E-11	

layer (D_a=0.65-4.4x10^{-11} m^2/s), and most D_a-values were nearly 10^{-11} m^2/s. Although the K_d-values are significantly different between cation (Cs$^+$) and anion (I$^-$), the D_a-values were similar levels for all detected RNs (^{134}Cs, ^{137}Cs (Cs$^+$), ^{129m}Te (HTeO$_3^-$ and/or TeO$_3^{2-}$ [8]) and ^{110m}Ag (Ag$^+$)). This is considered to be due to that the D_a-values were controlled by dispersion caused by flow of rain water.

Since the movement of RNs in soil is strictly concerned with K_d discussed above, minerals constituting soil, those contents, and further the content of organic matters are important factors. Particularly mineralogy for clay minerals directly affects sorption property. Also the effect of those factors on the movement of RNs is future work together with K_d.

CONCLUSIONS

The Geoslicer investigation on the depth distribution of RNs in soil was performed at 11 locations in Nihonmatsu City, Kawamata Town and Namie Town after about 3 months from the FD1-NPP accident, and quantitative data relating to retardation and migration of RNs (physical property, K_d, D_a, etc.) were obtained by field investigation and laboratory experiments. The conclusion is summarized as follows.

(1) Both of ^{134}Cs and ^{137}Cs were detected in all investigated locations, and ^{129m}Te and ^{110m}Ag were detected only in locations where radiation dose rates are high. At many locations investigated, radiocaesium more than 99 % distributed within a depth of around 10 cm in soil in the surface layer. Whilst, in soil at locations where are supposed to have been used as farmland, radiocaesium more than 99 % distributed within a depth of around 14 cm.

(2) The K_d-values onto soil obtained by a batch method were in the range of K_d=2,000-61,000 ml/g for Cs and K_d=0.5-140 ml/g for I, and no trend for the kind of soil was found.

(3) The D_a-values of RNs derived from penetration profiles near the surface layer showed a tendency to be higher in soil at locations that are supposed to have been used as farmland (D_a=0.1-1.5x10^{-10} m^2/s) than in soil in the surface layer (D_a=0.65-4.4x10^{-11} m^2/s), and most D_a-values were nearly 10^{-11} m^2/s. Although the K_d-values are significantly different between cation (Cs$^+$) and anion (I$^-$), the D_a-values were similar levels for all detected RNs (^{134}Cs, ^{137}Cs, ^{129m}Te and ^{110m}Ag). This is considered to be due to that the D_a-values were controlled by dispersion caused by flow of rain water.

The movement of RNs in soil is strictly concerned with sorption property, and it strongly depends on minerals constituting soil (especially clay minerals), those contents and the content of organic matters. The effect of these factors on the movement of RNs is future work.

ACKNOWLEDGEMENTS

The authors would like to thank Drs. Tsuyoshi NOHARA and Teruki IWATSUKI, Messrs Hiroaki MURAKAMI and Noboru NAKATSUKA for assisting the field work. Dr. Masashi NAKAYAMA and Messrs Yutaka SUGITA and Hironobu ABE are also thanked for assistance in editing document. In addition, this study was carried out as a contract work of the MEXT.

REFERENCES

1. T. Ohara, Y. Morino and A. Tanaka, J. Natl. Inst. Public Health, 60 (4) (2011) [in Japanese].
2. H. Kato, Y. Onda and M. Teramage, J. Environmental Radioactivity, in press.
3. Council for Science and Technology Policy (CSTP), Science and Technology Policy, 19 May 2011 (http://www8.cao.go.jp/cstp/budget/h23kidou_housya.pdf).
4. MEXT, Results of Airborne Monitoring by the MEXT and U.S. Department of Energy (May 6, 2011), 6 May 2011.
5. For example, H. Sato, JNC Technical report, JNC TN1400 2002-022 (2003).
6. Japanese Standards Association, JIS A 1202:1999 (1999) [in Japanese].
7. Atomic Energy Society of Japan (AESJ), AESJ-SC-F003: 2002 (2003).
8. D. G. Brookins, Eh-pH Diagrams for Geochemistry, Springer-Verlag, Berlin Heidelberg (1988).

AUTHOR INDEX

SUBJECT INDEX

absorbent, 217
actinide, 197
additives, 217
adsorption, 225, 277

barrier layer, 217
biomaterial, 191

ceramic, 67, 91, 97, 111
chemical composition, 157
chemical synthesis, 97
Cl, 91
composite, 3
corrosion, 3, 139, 145, 151, 173
Cs, 245, 269, 277

diffusion, 191, 277

electron microprobe, 47
environmentally protective, 245, 269
extended x-ray absorption fine
structure(EXAFS), 59

Fe, 59

geologic, 123
government policy and funding, 257
grain boundaries, 151

hot isostatic pressing (HIP), 67, 79

I, 15, 79, 85
infrared (IR) spectroscopy, 53, 79
ion-exchange material, 67
ion-solid interactions, 231

kinetics, 133

microstructure, 117, 167
Mo, 21

nanostructure, 185
Ni, 231

nuclear magnetic resonance (NMR),
167
nuclear materials, 9, 21, 41, 103,
111, 117, 123, 145, 151, 157,
179, 185, 197, 203, 237, 257

oxidation, 103

packaging, 85
Pd, 231
Pu, 73
purification, 103

Ra, 179
radiation effects, 9, 41, 73, 179
Raman spectroscopy, 53, 179, 225

scanning electron microscopy
(SEM), 47
scanning transmission electron microscopy (STEM), 15
second phases, 21, 139
simulation, 133, 211, 237
steel, 211
storage, 203, 245

transmission electron microscopy
(TEM), 117

U, 133, 139, 145

waste management, 3, 9, 41, 53, 59,
73, 79, 85, 91, 97, 111, 123,
157, 167, 173, 185, 191, 197,
203, 211, 225, 237, 257, 269
water, 179

x-ray diffraction (XRD), 15, 47

Zr, 173